Radar Signal Processing and Adaptive Systems

For a complete listing of the *Artech House Radar Library*,
turn to the back of this book.

Radar Signal Processing and Adaptive Systems

Ramon Nitzberg

Artech House
Boston • London

Library of Congress Cataloging-in-Publication Data
Nitzberg, Ramon.
Radar signal processing and adaptive systems / Ramon Nitzberg.
p. cm. — (Artech House radar library)
Includes bibliographical references and index.
ISBN 1-58053-034-6 (alk. paper)
1. Radar. 2. Adaptive signal processing. I. Title. II. Series.
TK6575.N59 1999
621.3848—dc21 99-22870
CIP

British Library Cataloguing in Publication Data
Nitzberg, Ramon
Radar signal processing and adaptive systems. – (The Artech House radar library)
1. Radar 2. Signal processing
I. Title
621.3'848

ISBN 1-58053-034-6

Cover design by Elaine Donnelly

International Standard Book Number: 1-58053-034-6
Library of Congress Catalog Card Number: 99-22870

10 9 8 7 6 5 4 3 2 1

Contents

	Preface	**xi**
1	**Radar System Fundamentals**	**1**
1.1	Radar Introduction	1
1.1.1	Radar Function	2
1.1.2	Target Detection, Resolution, and Clutter	3
1.2	Conceptual Implementation of Basic Radar	6
1.2.1	Surveillance Radar	6
1.3	Maximum Detection Range	8
1.3.1	Horizon and Multipath Effects on Detection Range	10
1.4	False Alarm Probability Effects	12
1.5	System Parameter Inter-relations	14
1.5.1	Transmit/Receive Antennas	14
1.5.2	Phased Array Antennas	18
1.6	System Performance Equations	19
1.6.1	Maximum Detection Range in Thermal Noise	20
1.6.2	Maximum Detection Range in Jamming Noise	22
1.6.3	Detection Probability	23
1.6.4	Probability of Detection Equations	25
1.6.5	False Alarm Sensitivity and Constant False Alarm Rate Processors	27

1.7	Detection of Targets in Clutter	31
1.7.1	Maximum Range in Clutter	33
1.7.2	Delay Line Cancelers	34
1.7.3	Doppler Filter Bank	38
1.7.4	Some Doppler Filtering Details	41
1.7.5	Noncoherent Summing	42
1.7.6	Some Implementation Details	43
1.8	Resolution and Measurement Accuracy	45
1.8.1	Ambiguity	47
1.8.2	Doppler Ambiguity	52
1.9	Transmit Waveforms	53
1.9.1	Pulse Coding Concept	54
1.9.2	Candidate Waveforms	54
1.9.3	Multiple Subcarrier (Frequency Diversity) Waveforms	56
1.10	Tracking Radar Techniques	58
1.10.1	Range Tracking	59
1.10.2	Tracking Radar Angle Tracking Techniques	63
1.10.3	Track-While-Scan Data Processing	66
1.10.4	Track-While-Scan Track Initiation	68
1.10.5	Track-While-Scan Track Maintenance	72
1.11	Summary and System Trade Areas	74
1.11.1	System Trade Areas	75
1.12	Bibliography	77
2	**Waveforms and Matched Filtering**	**79**
2.1	Fourier Transforms	80
2.1.1	The Autocorrelation Function	84
2.1.2	The Delta Function	84
2.2	Matched Filter Processing	86
2.3	Matched Filtering Examples	88
2.3.1	Simple Pulse Example	88
2.3.2	Right Triangle Example	89
2.3.3	sin x/x Type Spectrum Example	90
2.4	The LFM Pulse	91
2.4.1	LFM Pulse Doppler Tolerance	93
2.4.2	LFM Generation Techniques	96
2.4.3	Biphase-Coded Waveforms	99
2.5	Very Wideband LFM Waveforms	103
2.5.1	Linear Frequency Sweep	103

2.5.2 Bandspan Waveforms 103
2.6 Waveform Summary 106

3 **Antennas** **109**

3.1 Line Antenna 109
3.1.1 Uniformly Illuminated Line Antenna 110
3.1.2 Properties of sin x/x Functions 111
3.2 Planar Antenna 111
3.3 Uniformly Illuminated Planar Antenna 112
3.3.1 Parabolic Reflectors 113
3.4 Linear Phased Array 114
3.4.1 Grating Lobes 117
3.4.2 Beam Steering 119
3.4.3 Matched Filter Comparison 120
3.4.4 Introduction to Eigenvectors 123
3.5 The Maximum Detection Range of Multiple Beam Radar 125
3.6 Sidelobe Reduction Weighting 127
3.7 Effect of Steering Weight Errors 133
3.8 Earth Effects on Antenna Patterns 135
3.8.1 Multipath Effect on Antenna Pattern 136
3.8.2 Spherical Earth and Low-Altitude Targets 139
3.9 Waveform Sidelobe Reduction 140

4 **Target Detection in Noise** **143**

4.1 Target Scattering Models 144
4.2 Target Detection Theory 145
4.3 Fluctuating Cross-section Targets 148
4.4 Additional Swerling Cross-section Fluctuation Models 151

5 **Target Detection in Clutter** **155**

5.1 Introduction 155
5.2 Clutter Discriminants 156
5.2.1 Target Elevation 156
5.2.2 Target Velocity 156
5.2.3 Transmission Polarization-Sensitivity 156
5.2.4 Carrier Frequency Sensitivity 157
5.2.5 Azimuth Angle Sensitivity 157
5.2.6 Signal Bandwidth 157

5.3	Parameter Definition for Ground and Sea Returns	158
5.3.1	Sea Return	160
5.3.2	Ground Clutter	162
5.3.3	Weather Return	162
5.4	Doppler Filtering Techniques	164
5.4.1	MTI Delay Line Cancelers	165
5.4.2	Clutter Spectral Density	167
5.4.3	Clutter Attenuation and Improvement Factor	169
5.4.4	Three-Pulse Cancelers	173
5.4.5	Generalized Canceler Weights	176
5.4.6	Delay Line Canceler Shortcomings	177
5.4.7	Stagger-Coded MTI Waveforms	178
5.4.8	Implementation Errors Effect on Improvement Factor	183
5.4.9	A/D Converters	190
5.4.10	A/D Quantization Effects	194
5.4.11	Special Doppler Processing for Airborne Radar	198
5.4.12	Airborne Radar Doppler Compensation Techniques	199
5.5	Doppler Filter Banks	202
5.5.1	Statistical Independence of Detection	206
6	**Sidelobe Canceler (SLC)**	**207**
6.1	SLC Introduction	207
6.2	SLC Concept	208
6.3	Residue Power Minimization	210
6.4	Cancellation Degradation Effects	212
6.4.1	Propagation Delay Effect upon Cancellation	213
6.4.2	Amplifier Mismatch Effect upon Jammer Cancellation	216
6.5	Optimum Weight Estimation	220
6.5.1	Multiplication by Mixing and Filtering	220
6.5.2	SLC Weight Estimation Using LMS Algorithm	222
6.5.3	LMS Convergence Time Analysis	223
6.5.4	LMS Algorithm with Hard Limiter	225
6.5.5	Normalized LMS Algorithm (NLMS)	227
6.5.6	Digital LMS Algorithm	227

6.5.7	Digital LMS Algorithm with Limiter	231
6.5.8	Digital NLMS Algorithm Analysis	232
6.6	Weight Noise	232
6.6.1	LMS Algorithm Weight Noise	233
6.6.2	LMS with Limiter Weight Noise	234
6.6.3	NLMS Algorithm Weight Noise	235
6.7	Multiple Sidelobe Canceler	236
6.8	Sidelobe Blanking	237
	Appendix 6A: Differentiation Theorem	239
7	**Constant False Alarm Rate Processors**	**241**
7.1	Introduction	241
7.2	CFAR Concept	242
7.2.1	Cell-Averaging CFAR	243
7.2.2	Homogeneous Environment CACFAR CFAR Loss	247
7.3	Nonhomogeneous Environments	248
7.3.1	CACFAR Performance in Mismatched Environments	249
7.3.2	Mismatched Environment Example	249
7.4	Geometric Mean CFAR	254
7.5	CFAR Problems and "Solutions"	256
7.5.1	Greatest-of CFAR	257
7.5.2	Multiple Target Effect upon CACFAR	258
7.6	Clutter Map CFAR	260
7.6.1	Clutter Map Time Constant	262
7.7	Composite CFAR Techniques	264
8	**Multiple Sidelobe Cancelers**	**267**
8.1	Optimum MSLC Weight Derivation	268
8.2	Covariance Matrix and Correlation Vector Evaluation	270
8.3	Optimum Weight Performance Evaluation	271
8.3.1	MSLC Evaluation Examples	272
8.3.2	Channel-to-Channel Mismatch Effect	274
8.3.3	Channel-to-Channel Match Specification	278
8.3.4	Channel-to-Channel Match Evaluation Examples	279
8.3.5	Bandwidth Effect Analysis	281

8.4 Algorithms 284
8.4.1 LMS Algorithm 285
8.4.2 SMI Algorithm 286
8.4.3 SMI Performance Analysis 288

9 Signal-to-Interference Plus Noise Ratio Maximization 291

9.1 SINR Weight Maximization Derivation 292
9.2 Array Covariance Matrix Computation 296
9.3 Optimum Weighting of a Doppler Processor 297
9.3.1 Doppler Covariance Matrix Computation 298
9.3.2 Optimum Array/Doppler Examples 299
9.3.3 Multiple-Jammers/Clutter Frequencies 301
9.4 Relative Performance of Several Doppler Filters 303
9.5 Algorithms 308

10 Super-Resolution 313

10.1 Array Angle Resolution 313
10.2 Frequency Resolution 316
10.3 Application to Adaptive Doppler Processing 317
10.3.1 Modern Spectral Estimation and Adaptive Doppler Processing 318
10.3.2 Predicted Performance 320
10.3.3 Mismatch Loss Computation Equations 321
10.3.4 Estimation Error Loss 324
10.4 The Burg Algorithm 325

11 Polarization Rejection of Clutter 327

11.1 Reflected Wave Polarization 329
11.1.1 Receive Antenna Polarization Response 330
11.1.2 Reflectivity Examples 330
11.2 Circular Polarization Clutter Cancellation 332
11.2.1 Error Effects on Cancellation 332

About the Author 335

Index 337

Preface

This book discusses various aspects of radar signal processing. The material is presented in several levels of detail. The first chapter takes the form of an executive summary. This initial material is introductory in nature and does not require an extensive engineering background. The material includes the development of equations for the maximum radar range in free space and as affected by multipath and the horizon as well as various aspects of signal processing such as resolution, ambiguities, antennas, waveforms, the theory of detecting targets in noise and/or clutter, and tracking using data processing. This chapter could serve as a general introduction to the concepts of radar. The remaining chapters present a more detailed development of the material in two levels of detail. Most of this material requires only a standard engineering mathematics background. However, some of the detail requires a more advanced background. For these cases the derived equations and graphs are discussed and then the math background is introduced and the pertinent equations derived. Thus, the results for this material do not require the esoteric math background but the derivations are presented to any who wish it.

The second chapter discusses radar waveforms, including LFM and biphase coded waveforms, range resolution, and matched filter design and performance.

Chapter 3 develops the theory and performance of reflector antennas and phased arrays. Antenna sidelobes are considered in detail. Grating lobes of phased arrays are discussed. Techniques for reducing sidelobes by weighting are analyzed and the relative performances are determined. The effects of weighting errors are analyzed. The application of sidelobe reduction techniques to waveforms is also considered.

Chapter 4 develops equations for the detection of targets in noise for several different cross-section fluctuation models. Chapter 5 develops the concepts of signal Doppler processors to detect targets in a clutter environment. The theory of the performance of delay line cancelers and Doppler filter banks is developed, and the relative performance is determined. The effect of equipment implementation errors is analyzed, and equations for the performance degradation are developed. Special techniques that are used for airborne radar are described.

Chapter 6 discusses the application of the sidelobe cancelers to jamming environments. The optimum weighting equations for the residue power minimization criterion are derived, and the performance in the jamming environment is evaluated. It is shown that constant false alarm receiver techniques are required even for optimum weighting. Because optimum weights are not realizable, several iterative algorithms that approach the optimum performance are discussed and evaluated. The convergence properties of the algorithms, including the effect of weight noise, are analyzed.

Chapter 7 evaluates the optimum CFAR technique for a noise jamming environment. Since there is no optimum technique for clutter environments, several useful techniques are discussed and analyzed.

Chapter 8 introduces the multiple sidelobe canceler. It is usually implemented since the SLC is inadequate for complicated noise jamming environments. The matrix equations for the optimum weighting vector are derived. The effect of radar waveform bandwidth on jammer cancellation is derived and evaluated for typical cases. Equations for the degradation of cancellation caused by channel mismatch are derived. The generalization of the iterative SLC algorithms to approximate the optimum weight vector is analyzed and typical cases are evaluated. The sample matrix inversion algorithm is discussed.

Chapter 9 analyzes optimum processors that maximize the output signal-to-interference plus noise ratio. This criterion is appropriate for either arrays or Doppler filters. The optimum weights equation is derived. Examples of the performance for arrays and Doppler processors are evaluated. The performance improvement for adaptive Doppler processors as compared to fixed weighting techniques such as delay line cancelers or Tchebychev weighting is evaluated. Algorithms that can approximate the optimum performance are discussed.

Chapter 10 discusses super-resolution techniques as applied to the array angle domain and to the frequency domain. The theoretical application to Doppler processing is discussed. The performance of an adaptive Doppler processor when implemented using the Burg Modern Spectral Estimation algorithm is evaluated.

The last chapter gives a cursory introduction to the theory of electromagnetic polarization so as to discuss the applicability of polarization techniques to the rejection of clutter. The effect of errors upon the achievable cancellation is analyzed.

1

Radar System Fundamentals

This chapter is a general introduction and overview of radar. The material is usually presented without derivation and is intended as an "executive summary." The subjects are treated in greater detail in the succeeding chapters. Though the material emphasizes radar based on the propagation of electromagnetic energy, much of it is also applicable to other sensors such as active sonar based on the propagation of acoustic energy in water or geological sensors based on the propagation of acoustic energy in the ground.

1.1 Radar Introduction

A radar operates by transmitting energy into the environment and obtaining information concerning the location and identification (classification) of objects by detecting reflections of the transmitted energy. This concept is identical to the familiar one of shouting and hearing an acoustic echo from a nearby object. The range to the reflector is determined by measuring the time difference, t_D, between transmission and reception of the echo. For a velocity of propagation of v, the target range is

$$R = (v/2)t_D \tag{1.1}$$

where the factor is 2 because of the two-way propagation to the reflector. For electromagnetic radiation, the velocity of propagation in a vacuum, denoted as c, is approximately $3\ (10)^8$ m/s. For acoustic energy, the velocity of propagation in air (the shouting sensor) is approximately 330 m/s; the velocity of propagation in water (an active sonar) is approximately 1450 m/s.

Object location can also be performed using passive sensors. A passive sensor (sonar or IR) does not transmit energy but detects energy generated by the objects. Range measurement for passive sensors can be performed using triangulation of angle of arrival measurements from multiple passive sensor locations.

The target can be detected only if the returned energy is large compared to interfering energy. There are several sources of interfering energy. One is random energy fluctuations, termed *noise*. This noise can be both natural and man-made. Another is reflections from other targets that are not of interest. These are called *clutter returns* for radar and *reverberation* for sonar. The energy reflected to the radar decreases as the distance to the target increases; and in general, it becomes more difficult to detect long-range targets. The radar designer maximizes the detectability of long-range targets by procedures such as *signal processing* and *data processing*. The role of signal processing is to maximize the ratio of the reflected energy of the desired target to that of the interference. The data processor performs target tracking and future location prediction. The dichotomy between these procedures is fuzzy and often a function termed signal processing in one radar system is performed by the data processor in another system.

1.1.1 Radar Function

The primary function of a radar system is target position prediction. One example is a military radar system whose function it is to intercept a vehicle with another vehicle. The two vehicles can be aircraft, missiles, satellites, or any such combination. Another example is an air traffic control radar system. Its function includes preventing aircraft collisions and landing aircraft under poor weather conditions. The methodology of target tracking is to combine present position measurements with previous position measurements so as to perform track initiation, track updates, and future position prediction.

Several types of radar are used to accomplish this goal. One is the search radar. As its name implies, its function is to perform general surveillance of a volume and report the location of all detected targets. Depending upon the detailed design of the radar, the position report can be in terms of two, three, or more dimensions. Thus, most surveillance radar measure the range and azimuth angles of detected targets. Depending upon the antenna design, some radar are also capable of measuring elevation angle. Depending upon details of the transmission waveform and signal processing procedures, measurements of vehicle radial velocity and/or vehicle dimensions can sometimes be performed. Similarly, depending upon details of the transmit and receive antennas' polarization characteristics, information concerning target shape can be derived.

Another class of radar is the tracking radar. Its function is to obtain more accurate location information than can be obtained by the search radar. It is often a dedicated tracker, meaning that it focuses only on a single vehicle that has been previously identified by a search radar as being of interest. Because it can devote all its energy to gather information about a single target and because its design—such as antenna beamwidths, choice of carrier frequency and transmission waveform—can be optimized for this goal, the tracker-derived location information is more accurate. Since the track radar antenna usually has narrow beamwidths, it needs to receive moderately accurate target location designation from the search radar in order to acquire the target. The difficulty of the acquisition phase of the tracking radar depends strongly upon whether the search radar can supply two- or three-dimensional location information. Thus, the design of the radar combined system must be optimized as opposed to optimizing each radar's design separately.

The track-while-scan (TWS) radar is a radar system that combines the operations of search and track. Modern versions accomplish this using very high speed digital circuitry for data processing, which contrasts with the conventional tracking radar where the tracking circuits are often implemented using analog techniques. In addition, often the TWS radar's antenna design uses phased array techniques. The TWS radar tends to be an expensive implementation compared to separate search and track radar, but the increased expense is coupled with increased performance.

A major constraint for either the search radar or the TWS radar is that they both need to initiate and maintain tracks. This is a complicated procedure that is very computationally stressing. This tracking function, often the bottleneck of any radar system, is a finite resource that needs protection, and must not be wasted by attempting to set up tracks on false detections. Thus, as will be discussed in detail later, it is mandatory that false detections due to thermal noise be minimized. Similarly, it is mandatory that techniques to prevent jammers from overloading the tracking channels be implemented.

1.1.2 Target Detection, Resolution, and Clutter

Radar energy is usually transmitted as a pulse or group of pulses at a carrier frequency ranging between tens and thousands of megahertz. The bandwidth of each transmitted pulse is of the order of megahertz. The detection decision is made by amplifing the received waveform and removing the carrier frequency using envelope detection. For a simple radar, the operator display looks like that shown in Figure 1.1. The target present or absent decision is made by comparing the output voltage of the envelope detector to a threshold. When the envelope detector output voltage is large, a target present decision is made; otherwise a target absent decision is made. A target present or absent decision is made for each resolution cell. The concept of a resolution cell is approximate. If two targets

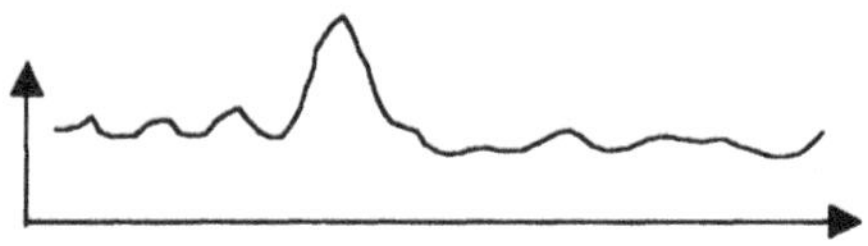

Figure 1.1 Operator display of envelope detector output.

are very closely spaced in range, the detector output voltage is so similar to that resulting from a single target that the two targets cannot easily be distinguished. Resolution refers to the minimum displacement such that two targets' waveforms are visually recognized as due to separate reflectors. The resolution dimension depends upon the domain. For range resolution, the cell's time duration equals approximately the inverse of the transmitted bandwidth. Thus, the decision rate of target absent or present is of the order of megahertz. For angle resolution, the resolution dimension is approximately equal to the antenna's beamwidth.

A common class of waveform is a simple pulse. For it, a sinusoidal oscillator is gated on for a time duration of τ, and τ is of the order of microseconds. The bandwidth, B, of the simple pulse is approximately $1/\tau$. Two targets displaced in range-time by τ are easily resolved. From (1.1) the range resolution δ_R, is

$$\delta_R = (c/2)\tau \tag{1.2}$$

More generally, a sinusoidal oscillator can be amplitude modulated (gated on) and phase modulated. The phase modulation causes the bandwidth to be greater than the reciprocal of the pulse duration. The signal processor implements a technique called pulse compression or matched filtering. The duration of the compressed (also termed collapsed) pulse, the cell's range-time resolution, equals the reciprocal of the transmitted bandwidth. Thus, the generalization of (1.2) for a matched filter processor is

$$\delta_R = (c/2)/B \tag{1.3}$$

Because of the presence of thermal noise, the detection decision can have errors. If a target present decision is made when there is no target, it is called a false alarm. The reverse error is that a target absent decision is made when a target is actually present. The radar performance is summarized by the probability of a false alarm and the probability of a target detection. The object of radar design is to maximize the probability of a target detection while maintaining a small false alarm probability. The detectability of an object depends upon the reflected energy. When all other radar parameters are held constant, the received target energy is increased by transmission and reception of multiple pulses. Usually the multiple pulses are generated with a constant time interval between each pulse. This interval, T, is known as the pulse repetition interval (PRI) or pulse repetition period. The frequency of the pulses, the reciprocal of the PRI, is called the pulse repetition frequency (PRF). The PRI determines the design maximum range, which is also referred to as the instrumented range. This is the range such that a reflection from a transmitted pulse is received at the instant of the next transmitted pulse. From (1.1), this range, R_{MI}, is given by

$$R_{\text{MI}} = cT/2 \tag{1.4}$$

where the MI subscript denotes maximum or instrumented. Reflections from objects at ranges greater than the instrumented range will appear as if they are at a much closer range. This is known as a range ambiguity or the "second-time-around" effect.

There are many reflecting objects in the environment. Usually there are reflections from objects that are not of primary interest to the radar. These objects are called clutter. The desired reflecting objects are often referred to as targets because of the military background of radar usage. Note that the definition of desired and undesired is arbitrary and can differ from user to user. In order to be able to discriminate between "targets" and "clutter," there must be some physical discriminant that allows for sorting. Often a useful discriminant is the fact that clutter has low velocity and targets have high velocity. The velocity causes a Doppler shift to the reflected waveform, and the differential Doppler shift can be used to sort between targets and clutter. Another possible physical discriminant is the polarization of the electromagnetic radiation. The reflected polarization from aircraft targets tends to be that of the incident polarization, but there is usually a significant cross-polarized component. The relative cross-polarized component from rain tends to be smaller. Therefore, transmitting and receiving circular polarization attenuates the reflections from rain more than those from aircraft targets. Other useful physical discriminants will be discussed later.

The detectability of targets of interest is degraded by the presence of jamming in addition to thermal noise and clutter. The jamming can be intentional or nonintentional. The latter is usually denoted as radio frequency interference (RFI). Intentional jamming is either noise or false targets. Noise jamming attempts to duplicate the action of thermal noise to degrade the detectability of the target. False target jamming generates many duplicates of the target. If the many duplicates overload the tracking capability of the radar and a track of a true target is not maintained because of the overload, the radar is not effective.

1.2 Conceptual Implementation of Basic Radar

This section discusses conceptual implementations and performance of search (surveillance) and tracking radar. A block diagram of a surveillance radar is shown and discussed. An example of a tracking radar using the early-late gate method of range tracking is given in Section 1.10.1. Conical scan and monopulse methods of angle tracking are introduced. Fuller discussions of angle tracking are presented in Section 1.10.2.

1.2.1 Surveillance Radar

The function of a radar system is to perform surveillance of a volume, track all detected objects of interest, and hand off this information to other systems. The other systems could be radar, such as track radar, or weapon systems, for example. The search radar performs surveillance of the volume, generates detections, and reports on targets.

A search radar operates by forming steered beams with its transmit and receive antennas and transmitting and receiving energy from the steered direction. These beams are sequentially steered to different portions of the surveillance volume. Radar that use reflector antennas form fixed beams that are approximately perpendicular to the face of the antenna. The beam steering, or scanning, is done by mechanically rotating the antenna. The scan period is the time period required to complete a mechanical rotation. Because of the constraint due to mechanical scanning, the interrogation of the surveillance volume is performed by sequentially steering in azimuth. Radar that use phased array antennas do not have the constraint of adjacent azimuth angle cells and must be scanned sequentially. However, on average, they also are configured such that the surveillance volume is scanned during a fixed scan period.

A conceptual implementation of a basic surveillance radar is shown in Figure 1.2. A single antenna is shown since most radars use the same antenna to both transmit and receive. A switch, the duplexer, is used to switch between the transmit and receive modes. The preferred transmit technique, as shown, generates a waveform at low power level (approximately milliwatts) by modulating a sinusoidal oscillator, and the generated waveform is amplified to the megawatt level for transmission. Different transmit waveforms can be used as a function of differing performance requirements and environment conditions. The simplest transmit waveform is the repetitively gated sinusoid. The detection of targets is enhanced by transmission and reception of multiple pulses.

The received waveform is at a very low power level and needs amplification before processing can be completed. The waveform transmission is often done

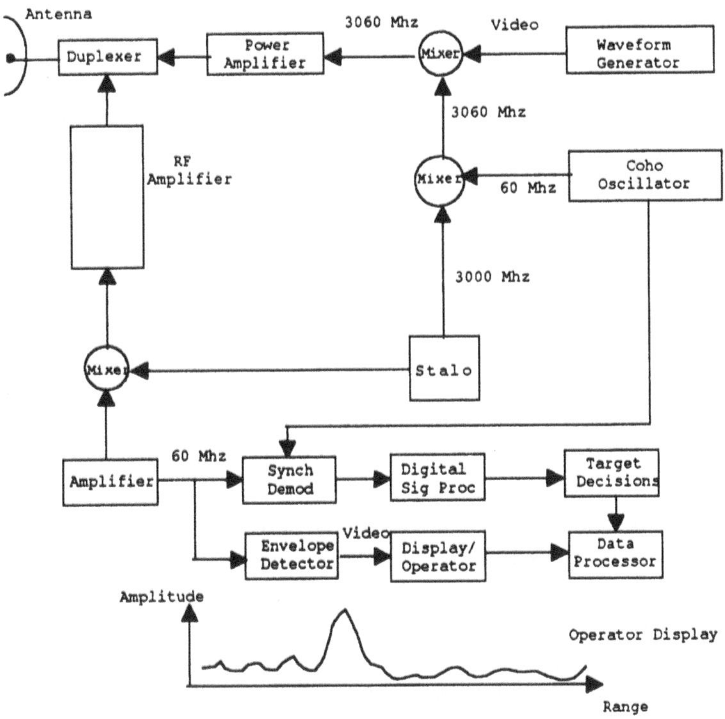

Figure 1.2 Conceptual implementation of a basic search radar.

at a relatively high carrier frequency, and it is usually necessary to translate to a lower frequency to simplify the implementation of the amplifiers. The frequency translation is performed by the mixer. A common method of performing the frequency translation for both transmission and reception is indicated in the figure with typical center frequency values indicated at several points.

The waveform generator creates a video waveform. This is mixed to the 3060-MHz transmit carrier frequency, which is created by mixing the stable local oscillator (STALO) frequency of 3000 MHz with the coherent local oscillator (COHO) frequency of 60 MHz. The STALO is also used to translate the received waveform to an intermediate frequency (IF) that is convenient for amplifier implementation. After amplification, the signal is presented to an operator and/or signal processor. The operator/signal processor inputs information to the data processor concerning the presence and absence of targets. A typical operator display of amplitude of the signal envelope versus range is shown at the bottom of the figure. Large amplitudes, relative to the random noise fluctuation, indicate the presence of targets.

1.3 Maximum Detection Range

Detection of an object requires that the received reflected energy is much larger than the background energy. For any active or passive sensor system, the energy received from a reflector decreases as the object range increases. Thus, the difficulty of reflector detection increases with range.

There are many contributors to the background energy. The most fundamental is the random motion of electrons and the resulting induced random energy fluctuations. Another contributor is the presence of many other reflectors (e.g., ground, weather, birds, and insects) whose reflected energy tends to obscure that of the reflector of interest. The reflector of interest is denoted as a target, whereas reflectors not of interest are denoted as clutter. The energy due to random electron effects is known as thermal noise.

To illustrate the form of dependence upon sensor parameters, an equation for the maximum detection range of a conceptual radar will be determined. The initial derivation shown in this section is further manipulated in Section 1.6.1. The two forms of the equation are considered because the inter-relation of radar parameters can cause confusion as to the dependence of maximum range on particular parameter values.[1]

Usually a radar transmits a train of pulses in order to perform target detection. For initial simplicity the maximum range equation is derived by assuming the radar generates a single pulse of energy, transmits it via a directional transmit antenna, receives reflected energy via a directional receive antenna, and detects a target if the observed energy at any time delay is large compared to the background noise energy. A pulse transmission is used to resolve reflections from targets at different ranges. The directional characteristic of the transmit antenna causes the energy transmitted in the direction of the interrogated angle to be greater than the value of the energy generated by the power amplifier. The ratio of the two energies is the transmit antenna gain, denoted G_T. Similarly, the directional characteristics of the receive antenna cause a receive antenna gain, denoted G_R.

Since the surface of a sphere of radius R equals $4\pi R^2$, an antenna transmitting equally in all directions (omnidirectional) generates a peak power density at a range of R equal to $P/(4\pi R^2)$, where P is the peak pulse transmit power. Because

1. The intent of this section is to introduce the radar parameters that affect the maximum detection range. Since there are many losses in addition to these parameters, the derived equation is not adequate for an accurate maximum range prediction. The preferred computational method uses the Blake chart as described by D. K. Barton in *Modern Radar System Analysis*, Norwood, MA: Artech House, 1988, p. 20.

of the transmit antenna gain, the peak transmit power density in the interrogated angle direction is $PG_T/(4\pi R^2)$. The product PG_T is called the effective radiated peak power. The transmitted pulse is a gated portion of a sinusoidal source. The sinusoid (or carrier) frequency, f_0, ranges between a few megahertz and tens of thousands of megahertz depending upon the particular application.

When the transmitted waveform impinges upon a target, energy is reflected back toward the radar. The reflection can be viewed as a result of an equivalent pulse transmitter and transmit antenna in the target. The equivalent effective radiated power of the reflector equals the product of the power density impinging on the target and the target scattering cross section, σ. Thus, the target's effective radiated power, P_E, is

$$P_E = PG_T\sigma/(4\pi R^2) \tag{1.5}$$

The reflected power density at range R equals $P_E/(4\pi R^2)$. The power induced in the receive antenna, P_R, equals the product of the power density impinging on the antenna and the antenna effective capture area, A_R. The antenna effective capture area is approximately equal to the physical area of the antenna when the antenna is steered to the target direction. Thus,

$$P_R = PG_T\sigma A_R/(4\pi R^2)^2 \tag{1.6}$$

As an example of the magnitude of various parameters, typical radar system values are a peak transmit power of 1 MW, transmit antenna gain equal to 1000, target cross section of 1 m^2, effective receive antenna area of 10 m^2, and a target range of 200 km. From the last equation, the power induced in the receive antenna evaluates to 3.96×10^{-14} W.

The target is detectable if the received peak power is large compared to the background power, which is due to the sum of thermal noise, jamming noise, and clutter. The thermal noise is always present and is the fundamental limitation on the radar's maximum range. The other sources of interference will be initially neglected. The average power of the background thermal noise depends upon the bandwidth of the amplifiers used in the receiver subsystem. The amplifier bandwidth is approximately equal to the transmitted waveform's bandwidth, B. The average thermal noise power in watts equals[2]

$$P_n = kT_S B \tag{1.7}$$

2. There are also many other noise contributors (e.g., receiver noise, galactic noise, etc.), so the total noise is often denoted as F_n times that computed using (1.7); F_n is denoted as the noise factor.

where B is in Hertz and kT_S is a constant approximately equal to $4 \times (10)^{-21}$ and is the product of Boltzmann's constant, k, and the effective system noise temperature in degrees kelvin, T_S. For the assumed transmit waveform of 1 μs, the bandwidth is approximately 1 MHz. Using (1.7), the average noise power evaluates to $4 \times (10)^{-15}$. Thus, for example, the peak power of the target reflection is approximately a factor of 10 larger than the average noise power and the probability of target detection is high. If the target range increases by a factor of 2, the reflected power decreases by a factor of 16 and is less than the average noise power and the probability of target detection is very low. The ratio of reflected target peak power to the thermal noise power is termed the signal-to-noise ratio (SNR).

A common method of increasing the maximum detection range is to transmit and process a train of N pulses rather than the single pulse transmission considered previously. For optimum signal processing, the performance improvement is equivalent to increasing the peak transmitter power by a factor of N. The maximum range in thermal noise increases by a factor of $N^{1/4}$.

The previous equations can be combined to relate the maximum radar detection range to the radar parameters. Thus, denote the minimum SNR required for reliable target detection as $\gamma_{\min}$. Combining (1.6) and (1.7) yields the maximum radar detection range as

$$R^4 = (PG_T\sigma A_R N)/[\gamma_{\min}(4\pi)^2 kT_S B] \tag{1.8}$$

This equation is a common form of the maximum range equation. However, it is deceptive and must be interpreted with care since various parameters are interrelated. As an example, from the equation it appears that the radar's maximum range depends upon the bandwidth of the transmitted waveform and increased range can be obtained using low-bandwidth transmission waveforms. As will be shown later, this is incorrect. It is shown that the radar's maximum range is independent of the waveform bandwidth.

1.3.1 Horizon and Multipath Effects on Detection Range

The previous derivation of the maximum range equation neglects the effect of reflections from the surface of the Earth; it is called the free space equation. As depicted by Figure 1.3, the presence of the Earth affects the maximum detection range for two reasons. The first is due to the combination of the curvature of the Earth and the target's low altitude. Microwave energy propagates along straight lines in a homogeneous medium. Thus, the far-range target indicated to the left of the figure is below the optical horizon and, to a first-order approximation, cannot be detected. This occurs even though the computed SNR for this range target is large enough to predict easy detectability.

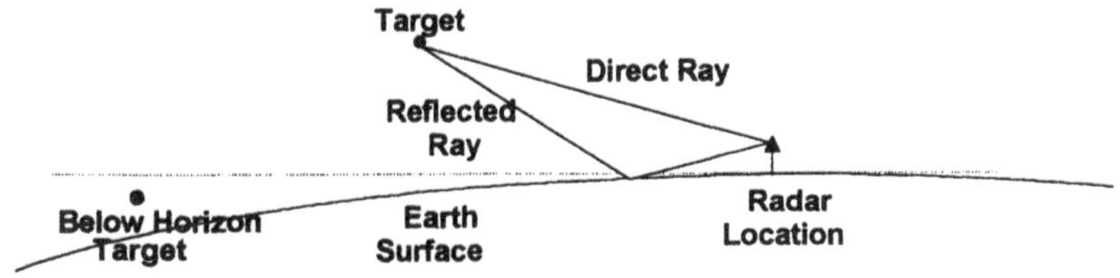

Figure 1.3 Geometry of horizon and multipath effects.

Since the atmosphere is slightly nonhomogeneous with altitude, microwave energy is bent (refracted) toward the Earth and targets that are only slightly below the optical horizon can be detected. The degree of refraction is atmosphere state dependent. An approximation of the difference between the radar horizon and the optical horizon can be obtained assuming the Earth's radius is larger than its physical measurement. For a "standard atmosphere" assumption, the approximation is that the radar horizon is obtained assuming the Earth's radius equals 4/3 of its physical value. However, this assumption often substantially underestimates the detectability of far-range targets. The standard atmosphere assumption is valid primarily for the temperate zones of the Earth. The tropics or the polar regions often have nonstandard atmospheres, and much greater ray bending can occur. This is known as anomalous propagation. It can cause the detection of very long range targets and, due to the "second" time around effect discussed later (range ambiguity), the targets range is mismeasured as a value much closer to the radar. The effect is also significant because the existence of long-range "second" time around clutter has an impact on the allowable choice of transmission waveforms.

The other aspect of the figure shows the effect of multipath. The free space detection range analysis assumes there is only a single propagation path between the radar antennas, transmit and receive, and the reflector. Because of reflections from the ground, there can be many paths. The diagram shows a single reflection wave, though in general, there are several reflection paths. The field intensity at the reflector is due to the additive combination of the direct and indirect waves. The relative magnitude of the direct and indirect waves depends upon the antenna gains in the respective directions and the coefficient of reflection for the surface. Usually, as the antenna is steered in the approximate direction of the target, the antenna gain for the direct wave is approximately equal to the maximum antenna gain, G_T or G_R, respectively. The indirect waves are usually launched at substantially different angles, so the antenna gain for this path is usually much less. For each reflected wave, the phase difference between the direct and indirect waves depends upon the additional path length of the indirect wave and the phase change induced due to the reflection coefficient. The reflection coefficient depends upon the angle of incidence of the wave with respect to the surface and the polarization. For the usual multipath situation, the angle of incidence is small and the reflection coefficient is approximately equal to -1 for all polarizations. When there is only a single indirect wave, as shown in Figure 1.3, the direct and indirect waves will sometimes combine constructively and sometimes destructively. Constructive combination means that the field strength at the reflector is larger than the free space value and the detection range increases. The field strength addition of the direct and reflected waves is the same on transmission and reception. Thus, neglecting the differing antenna gains, for constructive combination the reflected wave has a field strength equal to four times its free space value. The reflected power increases by a factor of 16, so the maximum detection range increases by a factor of 2.

Destructive combination implies a decreased field strength and decreased range. As will be shown later, to a first-order approximation, this effect causes all the antenna patterns of all radar systems to have a null at the horizon, so targets cannot be detected at this angle. Also, any target whose trajectory is close to the horizon is difficult to detect. As will be discussed in detail later, the transitions between twice free space range and zero range cause a modified antenna pattern. The modified pattern has a maximum at all elevation angles that corresponds to constructive combination of the direct and reflected waves and nulls at all elevation angles corresponding to destructive interference. However, it is emphasized that these results are based on the assumption of a single reflected wave. An actual environment is considerably more complex and requires detailed analysis.

1.4 False Alarm Probability Effects

Target detection is accomplished by comparing the detector output voltage to a threshold value. When the voltage is larger than the threshold value, a target present decision is made. When it is smaller, a target absent decision is made. Small threshold values increase the probability of correctly deciding a target is present. However, small threshold values increase the probability of incorrectly deciding a target is present because a large noise voltage exceeds the threshold, a false alarm. The bad aspect of false alarms is that the false alarms saturate the data processor and prevent the tracking of true targets. The acceptable false alarm probability depends upon details of the data processor hardware, such as the size and speed of the implemented computer and of the software algorithms used to establish and update a track. An approximation to the allowable value can be obtained as follows.

A target present or absent decision is made in each resolvable volume cell within the surveillance volume. The number of resolvable cells equals the product of the number of resolved range cells and the number of resolved azimuth cells and the number of resolved elevation cells. The number of resolved range cells equals the ratio of the maximum instrumented range, given by (1.4), and the range resolution, given by (1.2) or (1.3). For the simple pulse radar transmit waveform, the number of range cells, N_R, is

$$N_R = T/\tau \tag{1.9}$$

From (1.3), the generalization of (1.9), which is valid for an arbitrary radar waveform of bandwidth B, is

$$N_R = BT \tag{1.10}$$

When the radar's maximum instrumented range equals 200 km, the pulse repetition interval equals 1333 μs. For the example of Section 1.3, the transmit bandwidth equals 1 MHz, so the number of resolvable range cells equals 1333. Assuming the surveillance angle is the upper hemisphere, the number of resolvable angle cells, N_{ANGLE}, approximately equals

$$N_{ANGLE} = 2\pi/\theta_A\theta_E \tag{1.11}$$

where θ_A and θ_E are the azimuth and elevation angle beamwidths, respectively. These beamwidths are approximately related to the receive antenna physical dimensions as

$$\theta_A = \lambda/D_A \tag{1.12}$$

and

$$\theta_E = \lambda/D_E \tag{1.13}$$

where λ is the carrier frequency wavelength related to the carrier frequency, f_0, and velocity of light, c, by

$$\lambda = c/f_0 \tag{1.14}$$

and D_A and D_E are the antenna width and height, respectively. The receive antenna capture area is approximately equal to the physical area, the product of D_A and D_E. Continuing the example of Section 1.3, the receive antenna area equals 10 m^2. For a square antenna and a carrier frequency of 3 GHz, each beamwidth equals 0.0316 radians (1.81 degrees); so the number of angle cells equals 6283. The total number of resolution cells equals the product of these two values,

$8.4 \times (10)^6$. If the false alarm probability equals $1. \times (10)^{-6}$, 8.4 false targets are reported to the data processor per radar scan. If the false alarm probability equals $1. \times (10)^{-5}$, 84 false targets are reported to the data processor per radar scan. How many false alarms per scan is tolerable depends upon the computer used in the data processor. However, since the latter figure is quite large, the implication is that the false alarm probability should be of the order $1. \times (10)^{-6}$ or less.

1.5 System Parameter Inter-relations

As mentioned at the end of the derivation of the radar range equation, (1.8), the equation is subject to misinterpretation concerning the dependence of the maximum range upon system parameters due to inter-relations between the parameters. This section discusses the inter-relations and derives a more meaningful form of the maximum range equation.

An important system parameter of the radar's data processor is the data update period—the period between target position measurements. For a mechanically rotating radar, this is the scan period. For some digitally controlled phased arrays, the update period can be variable since the beam can be directed to arbitrary angles at arbitrary times. Short update periods are desired to track maneuvering targets. However, as will become clear, short update periods tend to decrease the per-scan detectability of targets. This is an example of a trade-off parameter. There is no "optimum" value to maximize system performance. The chosen value depends upon the relative importance of long-range detection versus the ability of tracking maneuvering targets. As will be seen, there are many other cases where it is necessary to perform a parameter trade-off.

1.5.1 Transmit/Receive Antennas

There are two types of antennas. One, sketched in Figure 1.4, is a parabolic antenna. It is one of the earliest designs. When used as a transmit antenna, the transmitter is connected to the antenna feed. Power is reflected from the surface of the parabolic surface. A single feed produces a single beam.

The location of the feed in the vicinity of the parabola's focal point causes the power to be radiated as a directive beam. As will be shown later, the power is radiated in a direction that is approximately perpendicular to the plane of the antenna. The beam is steered in azimuth by the mechanical scanning of the antenna.

The beams exact pointing direction depends upon the placement of the feed. When the feed is along the axis of the focal point of the parabola, the beam is perpendicular to the parabolic surface. If the feed point is above (below)

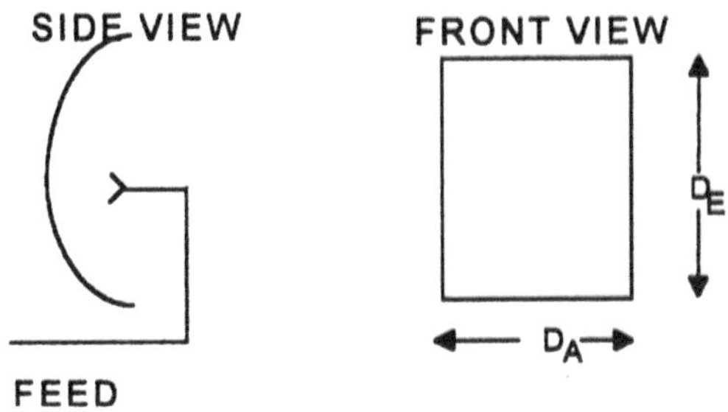

Figure 1.4 Parabolic reflector antenna.

the axis, the beam is directed to a lower (higher) elevation angle. The action on receive is similar. The received energy reflects from the surface of the parabola and is focused at the feed point. The focusing action produces a directive receive antenna pattern. Multiple beams, stacked in elevation, are produced using multiple feeds bracketing the parabolic focal point.

The typical beam shape is depicted by the sketches of Figure 1.5. The left-hand portion shows contour lines of the gain normalized to the maximum gain with respect to azimuth and elevation angle axes. The contours shown are for −3 dB, −10 dB, and −20 dB. The beamwidths are conventionally defined as the two-sided −3-dB beamwidths. The azimuth beamwidth is shown as the angular displacement between the ellipse intersections with the azimuthal axis. Similarly, the elevation beamwidth is the angular displacement between the ellipse intersections with the elevation axis.

From (1.12) and (1.13), the azimuth beamwidth depends upon the antenna width and the elevation beamwidth depends upon the antenna height. If the antenna is square, the elevation and azimuth beamwidths are equal and the contours are circles. In the figure, the contours are ellipses, indicating that the azimuth and elevation beamwidths differ.

The right-hand side of the figure shows the normalized antenna gain along the azimuth cut that passes through the maximum gain point. This pattern is termed the azimuth antenna gain pattern. The elevation gain pattern is taken along the cut perpendicular to the azimuthal cut. The patterns are similar in shape and differ only because of the differing antenna beamwidths.

The equation for the antenna pattern depends upon the detailed design of the feed structure. For a particular feed design called uniform illumination, the power gain pattern is given by

$$\begin{aligned} G(\Phi_A, \Phi_E) = G_0\{&[\sin(\pi D_A \Phi_A/\lambda)/(\pi D_A \Phi_A/\lambda)] \\ &\times [\sin(\pi D_E \Phi_E/l)/(\pi D_E \Phi_E/\lambda)]\}^2 \end{aligned} \tag{1.15}$$

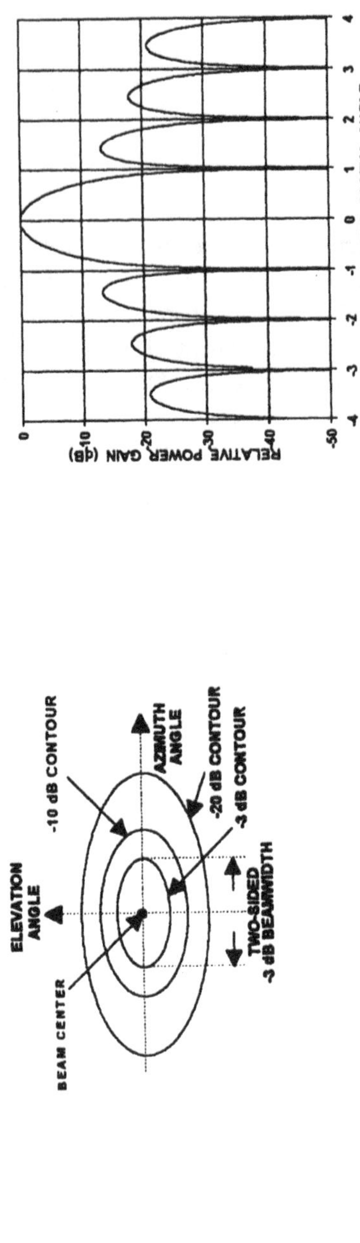

Figure 1.5 Antenna pattern.

where Φ_A and Φ_E are the spatial angles measured with respect to the steered direction and G_0 is the maximum gain. It can be seen that the pattern is the product of a term that depends only upon Φ_A and another term that depends only upon Φ_E and that the two terms are similar in form. Either term equals unity when the spatial angle equals zero. Thus, the maximum gain occurs when both angles equal zero, and this is the beam steered angle. The gain pattern shown by Figure 1.5 is based on (1.15). The antenna pattern shown is valid for either the azimuth cut or the elevation cut. The spatial angle horizontal axis of Figure 1.5 is normalized with respect to the beamwidth, as discussed in the next paragraph.

The pattern decomposes into two regions. The region of maximum gain is defined as the mainlobe gain region. The angular displacement between the maximum gain and the first null is called the natural beamwidth and is approximately equal to the two-sided −3-dB beamwidth. The angular displacement between the two nulls on either side of the mainlobe region equals twice the natural beamwidth. The angular region outside the mainlobe is known as the sidelobe region. Note that the angular displacement between adjacent nulls in the sidelobe region equals the natural beamwidth. An "ideal," though physically impossible, antenna pattern would have nonzero gain within the mainlobe region and zero gain elsewhere. The nonzero gain of the sidelobe region can cause many problems for a radar. One example is an angle ambiguity. Specifically, when a target is detected by the signal/data processor, by assumption the target angle is reported as the steered angle. However, if the target's radar cross section is very large, it can be detected even when it is in the sidelobe region and the data processor track will be based upon erroneous data. Another example of a sidelobe region-induced problem is that it represents a vulnerability to jamming. For the "ideal" receive antenna pattern, a jammer would have to be within the mainlobe region in order to generate jamming within the radar. This occurs only when the antenna is pointing at the jammer and would restrict the time period of jammer effectiveness. However, because of the nonzero sidelobe gain, if the jammer power is sufficiently large, it can always generate jamming within the radar and the jammer's effectiveness is substantially increased. The antenna pattern shown in Figure 1.5 has a relatively high sidelobe gain. The peak sidelobe gain equals −13.2 dB, caused by the uniform illumination pattern induced by the antenna feed. Alternate feed designs are usually used to reduce the sidelobe gain. However, reducing the sidelobe gain causes other problems. One is that the peak gain decreases, which reduces the detectability of targets. A second is that the mainlobe width increases and, as will be discussed later, this increases the received clutter energy. Increased clutter energy also degrades the detectability of targets. Thus, antenna design is an additional trade-off area.

From (1.15) the gain equals its maximum value when Φ_A and Φ_E equal zero. The gain equals zero when either

$$\begin{aligned} D_A\Phi_A/\lambda &= \text{positive or negative integer} \\ D_E\Phi_E/\lambda &= \text{positive or negative integer} \end{aligned} \tag{1.16}$$

As previously discussed, the angular displacement between the peak gain angle and the first null, the "natural" beamwidth, is approximately equal to the two-sided −3-dB beamwidth. The natural beamwidth is usually simpler to evaluate than the two-sided −3-dB beamwidth and, therefore, is a useful analytic concept. The beamwidths defined by (1.12) and (1.13) are natural beamwidths.

If the antenna pattern was actually equal to the ideal pattern, the maximum gain would be related to natural beamwidths as

$$G_0 \approx 4\pi/\theta_A\theta_E \tag{1.17}$$

The physical area of the antenna equals the product $D_A D_E$, so the area is related to the maximum gain as

$$A = D_A D_E \approx G_0\lambda^2/4\pi \tag{1.18}$$

Thus, the receive antenna capture area is

$$A_R \approx G_R\lambda^2/4\pi \tag{1.19}$$

1.5.2 Phased Array Antennas

The phased array technique is a much newer technology than that of reflectors. The antenna consists of many elements such as dipoles. Each element of a phased array is phase shifted to steer the beam to the desired direction. Depending upon the array design, the number of array elements could be a few or a few thousand. Because phase shifters and associated equipment are relatively expensive, the technique is expensive to implement. Its advantage is that the technique has significantly better performance than the parabolic reflector. Thus, the use of this technique requires a performance-implementation cost trade-off.

In general, the phased array is implemented either as a planar array or as a line array. A line array can have a narrow beamwidth in only one angle dimension since the beamwidth, given by an equation similar to (1.12), is a function of the array's physical size. Since the array is one dimensional, its size in the other dimension is essentially zero, so it has a wide beam in the other angle dimension. A planar array can have a narrow beamwidth in both angle dimensions but the number of array elements, and thus implementation cost, is

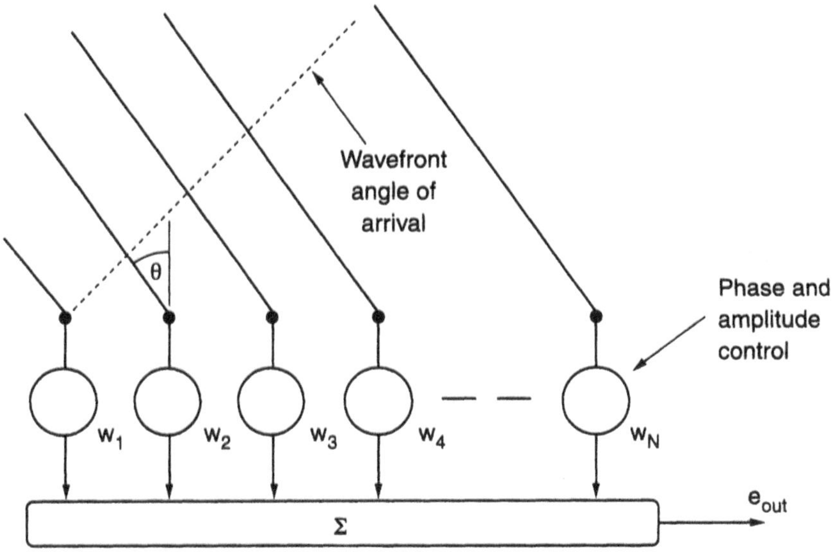

Figure 1.6 Conceptual block diagram of line array antenna.

of the order of the square of that of the line array. One technique of using a line array is to form a narrow beam in azimuth and use the phase shifters to steer the beam in elevation. Mechanical rotation is used to scan the beam in azimuth over the desired surveillance angle. When the antenna is phase steered in both angle axes, this is termed phase-phase steering.

Figure 1.6 shows a line array used as a receiving antenna. Each element is multiplied by a gain that phase shifts and amplifies the elemental received signal. The array output is the sum of the weighted elemental signals.

1.6 System Performance Equations

Combining (1.8) and (1.19) gives a common form of the maximum range equation

$$R^4 = (P G_T G_R \lambda^2 \sigma N)/[\gamma(4\pi)kT_S B] \tag{1.20}$$

As indicated after the derivation of (1.8), the equation is misleading because it implies, among other things, that the maximum detection range depends upon the transmitter's peak power and radar's bandwidth. This is incorrect because various parameters of (1.20) cannot be independently specified. This section manipulates this maximum detection range equation into a form that shows performance as a function of fundamental, independently specified, system parameters. The final result, though based upon several approximations, indicates that the detection range in thermal noise is independent of the transmitted waveform bandwidth, pulse repetition frequency, carrier frequency, and other waveform details. The only waveform parameter that determines the maximum detection range is the average power of the transmitter.

Three different cases are considered. The first case is for interference due only to thermal noise. The second is for interference due only to jamming noise. This is a simplification of the actual case when the total interference is the sum of thermal noise plus jamming noise and the received jamming power is much greater than the thermal noise power. The third is for total interference due only to clutter. This is a simplification of the actual case when interference is the sum of thermal noise plus clutter and the received clutter power is much greater than the thermal noise power. This case is more complicated to consider because there are signal processing techniques that can be used to reduce the interfering effect of the clutter. Thus, the consideration of this case will be deferred.

1.6.1 Maximum Detection Range in Thermal Noise

The operational system concept is that the antenna is pointed to a specified azimuth-elevation resolution cell and the waveform of N pulses is transmitted. After allowing sufficient time for energy to be received from the maximum instrumented range (the dwell time), the antenna is resteered to a different azimuth-elevation region. This action is continued until all angles have been interrogated and then the cycle is reinitiated. This action is called step scan because the antenna discretely steps from angle cell to angle cell. Step scan is exactly correct for the phased array antenna using phase-phase steering. It is only approximately correct for a mechanically scanned antenna.

Note that target detection is only a secondary goal of the radar system. The primary goal is initiating and maintaining target tracks. For this purpose an important system parameter is the data processor update period T_U. The update period is the time between detection/position measurement data reports to the data processor from each angular resolution cell. This time strongly affects the target tracking error and is especially important in establishing the maximum target maneuver rate that can be tracked. Short update periods improve target maneuver tracking but degrade target detectability since the update period is related to other terms in the maximum range equation. Specifically, the update period depends upon the antenna scanning mechanism.

When the radar antenna is mechanically scanned and rotates at a constant rate, the update period is equal to the antenna scan period. For a phased array radar, there is not necessarily a fixed scan period because the antenna can be used to interrogate a specified angle cell at any desired time instant. However, usually the operation is such that an approximately fixed scan period is used so that, in general, the update period equals the antenna scan period.

The scan period depends upon the pulse repetition interval, the number of interrogation angle cells, and the dwell period at each angle cell. Usually, the time between pulses is constant at the pulse repetition interval of T, the value of which depends upon the maximum instrumented target range and is given by (1.4). For N pulses transmitted and received from each angle cell, the dwell time, T_D, equals NT. The update period equals the number of resolvable angle dwell positions multiplied by the dwell period. The number of resolvable angle cells, N_{ANG}, is

$$N_{ANG} = \theta_{SA}\theta_{SE}/\theta_A\theta_E \tag{1.21}$$

and θ_{SA} and θ_{SE} are the azimuth and elevation angular surveillance regions, respectively. Thus, the update period, T_U, is

$$T_U = T_D N_{ANG} = NT(\theta_{SA}\theta_{SE}/\theta_A\theta_E) \tag{1.22}$$

In addition, for a pulse transmission of duration τ, the average transmit power, P_{AVE}, is

$$P_{AVE} = P\tau/T \tag{1.23}$$

For a simple pulse waveform, the waveform bandwidth, B, is approximately equal to the reciprocal of the pulse duration τ. Combining this approximation with (1.17), (1.19), (1.22), and (1.23) and substituting into (1.20) gives

$$R^4 = P_{AVE} A\sigma\, T_U/(\gamma 4\pi\, k\, T_S\theta_{SA}\theta_{SE}) \tag{1.24}$$

The antenna area is denoted as A, without subscript, to emphasize that the equation is applicable for a monostatic radar using the same antenna for transmit and receive. An interesting interpretation of (1.24) is that the system's maximum detection range (in a thermal noise environment for a simple pulse transmission waveform) depends directly upon these radar parameters:

- Transmitter average power;
- Antenna area;
- Data processor update period;
- Angular surveillance region;
- SNR required for detection/false alarm control.

In addition, though (1.24) was derived specifically for a simple pulse waveform, as will be shown later, it is true in general. Thus, the maximum range is completely independent of the transmitted waveform. Also note that the maximum range equation is expressed in terms of an energy ratio. Thus, the product $P_{AVE} T_U$ is the total energy transmitted during the surveillance of the volume. The term kT_S is the noise power density. It has the dimension of watts/hertz, or joules, and is the noise energy. This observation is used to derive an equation for the detection range in a noise-jamming environment.

1.6.2 Maximum Detection Range in Jamming Noise

A noise jammer transmits a noiselike waveform to the radar, so the total noise level of the radar is increased. The increased noise level will degrade the detectability of targets and thus improve the capability of an enemy successfully completing its mission.

The pertinent maximum detection range equation evolves from (1.24) by substituting the jammer energy value (jammer noise power density measured in watts/Hertz) for the thermal noise energy term.

Thus, the jammer radiates a total power P_J toward the radar via an antenna with gain G_J. The jammer power is spread over a bandwidth B_J. There are two types of noise jammers. One, the spot jammer, attempts to increase the radiated power density by restricting its bandwidth approximately to the radar's bandwidth. In order to accomplish this, the jammer needs a receiver that can accurately measure the radar's center frequency. This is not sufficient to exactly match the jammer's bandwidth to the radar's bandwidth, but it does ensure a reasonable match. Measuring the radar's center frequency is not a simple task when, as is often true, the jammer receiver is within line-of-sight of hundreds of radar. Also, often a radar will attempt to deny spot jammer capability using a transmit waveform that transmits on several carrier frequencies simultaneously or constantly changes carrier frequency. Thus, the jammer's action is termed electronic countermeasures (ECM) and the radar's response is termed electronic countercountermeasures (ECCM).

Instead of operating in a spot jammer mode, an easier strategy for a noise jammer is to act as a barrage jammer. Then the jammer bandwidth is much greater than the radar bandwidth. The compensating feature is that the jammer no longer needs to measure the radar's frequency and is not vulnerable to the ECCM action of changing carrier frequency.

For a jammer range of R_J, the jammer's transmit energy density, E_D, received at the radar is

$$E_D = AG_{rs}(P_J G_J/B_J)/(4\pi R_J^2) \tag{1.25}$$

where G_{rs} is the ratio of the radar antenna's gain (usually a sidelobe gain) in the direction of the jammer to the peak mainlobe gain. Substituting this energy for the thermal noise energy, kT_S, in (1.24) gives the range equation for the noise jamming environment. Thus,

$$R^4 = P_{AVE}\sigma T_U / \left[\gamma \theta_{SA} \theta_{SE} G_{rs} (P_J G_J / B_J) / R_J^2\right] \tag{1.26}$$

There are two possible jammer geometries. For one, the self-screening jammer, the jammer is located on the target, so R_J equals R. Then, (1.26) reduces to

$$R^2 = P_{AVE}\sigma T_U / [\gamma \theta_{SA} \theta_{SE} G_{rs} (P_J G_J / B_J)] \tag{1.27}$$

Note that the maximum range varies as the square root of parameters rather than as the fourth root of parameters as for the thermal noise environment. As an example of the impact of this functional form change, assume that the transmit power decreases by 1 dB due to aging. In a thermal noise environment, the maximum range decreases by 5.5%; in a noise jamming environment, the maximum range decreases by 11%.

1.6.3 Detection Probability

The previously developed maximum range equations are based on the inherent assumption that detection range is a dichotomy. That is, the equations assume that targets whose range is less than a particular value are detected, whereas targets whose range is greater than a particular range are not detected. This is a gross oversimplification. Detection is a statistical phenomenon. Thus, at any particular range, there is a particular probability of detecting the target.

There are two major factors that contribute to the statistical variation. One is due to the statistical nature of the additive interference (the sum of thermal noise, jamming noise, and clutter returns). The randomness of the noise is what causes the detector output voltage to vary randomly and causes false alarms when noise voltage is sufficiently large to cause a threshold crossing. The second factor is the statistical nature of the target's cross section. The energy reflected from a target emanates from many points on the body. For most radar targets, the radar's range resolution does not resolve the individual reflection points so that the cross section is caused by the additive combination of these reflectors and the relative phase of the reflector returns is important. Small changes of phase, due to small vehicle aspect angle changes, cause large changes of radar cross section. The reflectivity pattern is similar to an antenna pattern. Specifically, from (1.16), the pattern nulls or peaks are spaced at angle differences of λ/L for a target of length L. As an example, consider a radar whose carrier frequency is 3 GHz so that its wavelength equals 0.1m. For a 10-m-long target, the null/peak separation of the cross-section pattern equals 0.01 radians (0.6 degrees). This means that the cross section can vary from about 100 m^2 to almost zero in the span of 0.6 degrees. Aircraft aspect variations of 0.6 degrees are easily caused by atmospheric-induced effects such as wind gusts. Thus, the target's radar cross section is essentially nonpredictable and well modeled as a random variable.

The probability of detection depends upon the target and interference statistics. One form of statistical variation for a target is nonfluctuating. This occurs despite the summation of many reflector arguments given previously if the target seems identical when viewed from any aspect angle. A perfect sphere has this property. The original investigation of detection probability of radar targets was done by Marcum for nonfluctating targets. However, the radar cross section of most targets varies in a statistical manner and different targets have different statistical variations. A common model used to analyze target probability of detection is a Swerling model, initially described by Peter Swerling. Curves of the probability of detection versus the SNR, γ in the equations for maximum detection range, are given in Figure 1.7 for Swerling 1 and nonfluctuating targets for false alarm probabilities of 1×10^{-6} and 1×10^{-8}. It can be seen that as the false alarm probability decreases, the SNR required for any specified detection probability increases. Also, for detection probabilities greater than a few tenths, the SNR required to detect a fluctuating target is larger than that required for a nonfluctuating target. The SNR difference is called the fluctuation loss and increases as the probability of detection increases.

The Swerling target curves are obtained from the nonfluctuating target curves by averaging the probability of detection over the Swerling model. As a simple example, assume that a target's range and cross-section fluctuation are

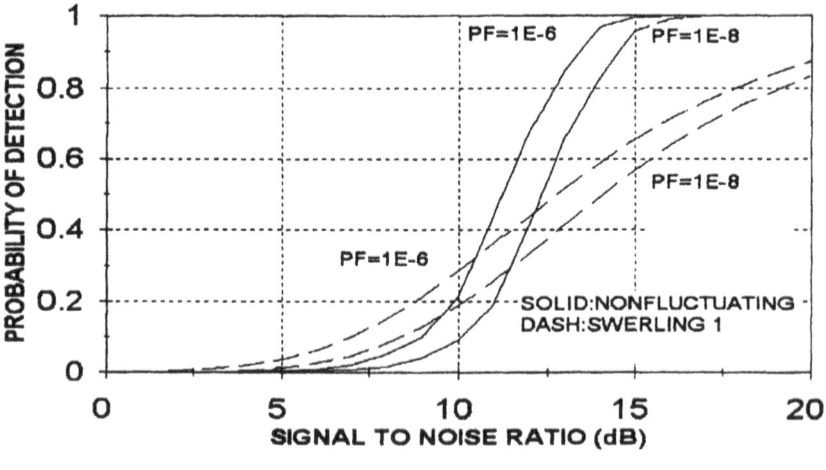

Figure 1.7 Probability of detection for fluctuating and nonfluctuating targets.

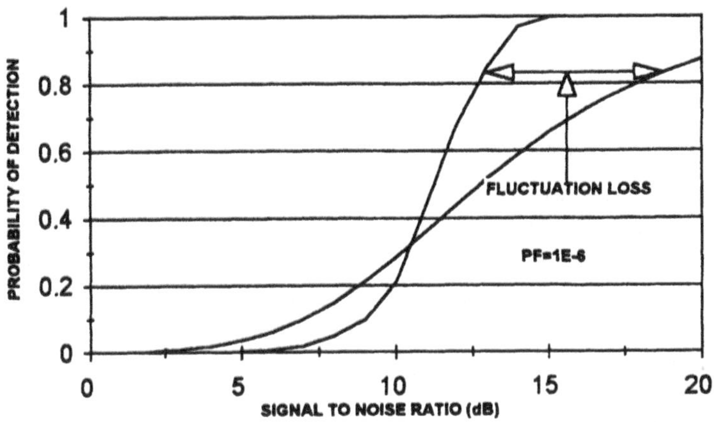

Figure 1.8 Fluctuation loss.

Table 1.1
Fluctuation Loss Table

Detection Probability	Loss (dB)
0.3	0.0
0.5	1.0
0.7	3.8
0.9	8.2
0.95	11.0

such that the probability of the SNR being equal to 10 dB, 12 dB, and 14 dB is 0.3, 0.5, and 0.2, respectively. For a false alarm probability of 1×10^{-6}, the corresponding values of detection probability for these three SNR values is 0.21, 0.68, and 0.97, respectively. Then the probability of detection for this fluctuating target equals the summed product of $0.3^*0.21 + 0.5^*0.68 + 0.2^*0.97$, which equals 0.60.

Figure 1.8 is obtained from Figure 1.7 by extracting the curves for the nonfluctuating and Swerling 1 targets for a false alarm probability of 1×10^{-6} in order to show the fluctuation loss more clearly. Table 1.1 presents values of the fluctuation loss.

As will be discussed later, there are special transmit waveforms that are often used to decrease the fluctuation loss.

1.6.4 Probability of Detection Equations

The probability of detection equations depends upon the statistics of the interference and the fluctuation characteristics of the target. For thermal noise, the interference statistics are Gaussian. For other interferences, such as clutter, non-Gaussian statistics are sometimes appropriate.

The target fluctuation characteristics, as previously mentioned, are nonfluctuating or fluctuating. There are several fluctuation models. One aspect of the fluctuation model deals with the rate of change of the cross section. The slow fluctuation model is defined as the case where the target cross section varies randomly from scan to scan but is constant over the dwell period. Thus, the SNR value is the same for each pulse of the dwell, but the SNR value varies randomly from scan to scan. For the fast fluctuation model, the SNR of each pulse of the dwell is not constant but varies randomly. In addition, there are several possible assumptions as to the statistics of the fluctuation model. As one example, the Swerling 1 model is for a slow fluctuation model and the probability density function of the random SNR, γ, is given by the exponential probability density. Thus, the density function $p(\gamma)$ is given by

$$p(\gamma) = \exp(-\gamma/\gamma_{\text{AVE}})/\gamma_{\text{AVE}} \tag{1.28}$$

and γ_{AVE} is the average SNR value. The Swerling 3 model is a fast fluctuation model, and the probability density function is also given by (1.28).

The average value of the SNR depends upon the average radar cross section of the target. It is computed by manipulating one of the maximum range equations previously derived. From (1.24), for a thermal noise environment,

$$\gamma_{\text{AVE}} = P_{\text{AVE}} A \sigma_{\text{AVE}} T_{\text{U}} / (4\pi k T_{\text{S}} \theta_{\text{SA}} \theta_{\text{SE}} R^4) \tag{1.29}$$

where now the average value of the target cross section is used. Note that (1.24) was derived for an N-pulse dwell. The inherent assumption in (1.24) and (1.29) is that the N returned pulses are processed by a signal processor such that the SNR for the N pulses is N times the SNR of a single pulse. Since the target is usually in motion, its range changes on a pulse-to-pulse basis. The range change is sufficiently small such that the per-pulse SNR is constant. However, the range change causes a significant difference in the relative phase of each received pulse. The SNR of the combined pulses equals N times the SNR of each pulse only when the signal processor compensates for this phase variation. This requires a special type of signal processor termed coherent integration or Doppler processing since moving targets cause a Doppler shift of the reflected waveform. These processors are partially discussed in Section 1.7 and in greater detail in the following chapters.

The various target fluctuation models are chosen because they are relatively close approximations to the measurements obtained for real targets and because they are analytically tractable. Thus, it is relatively simple to derive equations for the probability of detection when using (1.28) as a model. Specifically, the voltage at the output of the envelope detector of Figure 1.2 is observed. For conceptual convenience, it is useful to assume that the radar signal processor is implemented using a digital processor so that the output voltage is discretely sampled in time. Then each sample can be denoted as the voltage observed in a test cell. If the test cell voltage exceeds a threshold value T, a target present is declared. The probability of declaring a target present when a target is absent is the false alarm probability. The probability of detection, p_D, for the Swerling 1 model of (1.28) is given by

$$p_D = \exp[-T/P_n(1 + \gamma_{AVE})] \tag{1.30}$$

where P_n is the thermal noise power and γ_{AVE} denotes the average SNR of the N coherently integrated pulses. If the statistics of the combination of thermal noise, jamming noise, and clutter also are Gaussian, (1.30) is valid when P_n is replaced by the total interference power, P_I.

When there is no target in the test cell the average SNR equals zero. From (1.30) the false alarm probability, p_F, equals

$$p_F = \exp[-T/P_n] \tag{1.31}$$

The threshold value is chosen to produce the desired false alarm probability as

$$T = P_n \ln[1/p_F] \tag{1.31a}$$

Combining (1.30) and (1.31) yields

$$p_D = p_F^{[1/(1+\gamma_{AVE})]} \tag{1.32}$$

The Swerling 1 curves of probability of detection in Figures 1.7 and 1.8 are computed using (1.32).

1.6.5 False Alarm Sensitivity and Constant False Alarm Rate Processors

The threshold value T is designed and implemented under the assumption that the interference power level (the total of thermal noise, jamming noise, and clutter) is known. Often, this assumption is erroneous, so the actual interference power differs from the design power level. Denote the design interference power as P_{ID} and the actual interference power as P_{IA}. Under those circumstances, the attained false alarm probability, p_{FA}, differs from the design false alarm probability, p_{FD}. From (1.31), the design and actual false alarm probabilities are given by

$$\begin{aligned} p_{FD} &= \exp[-T/P_{ID}] \\ p_{FA} &= \exp[-T/P_{IA}] \end{aligned} \tag{1.33}$$

Combining these two equations gives

$$p_{FA} = p_{FD}^{(P_{ID}/P_{IA})} \tag{1.34}$$

As an example, if the actual interference power is twice the design value, the actual false alarm probability equals the square root of the design false alarm probability. Thus, a design false alarm value of 1×10^{-6} becomes an actual value of 1×10^{-3}. More generally, the relation between design and actual is given by the curves of Figure 1.9 for design false alarm probabilities of 1×10^{-6} and 1×10^{-8}.

It can be seen that if the actual interference power level is only 1 dB greater than the design power level, a design false alarm probability of 1×10^{-6} becomes an actual false alarm probability of $1 \times 10^{-4.77}$. As previously discussed, increasing the false alarm probability by as little as an order of magnitude can cause catastrophic overloading of the data processor and is unacceptable. Thus, even "small" additional jamming power or clutter power can cause data processor saturation and destroy the radar's function. As will be seen, signal processors that "cancel" (a misnomer; "significantly attenuate" is more accurate) jamming and clutter are often implemented. However, even the best possible jammer and/or clutter cancellation techniques cause the residue power to be larger than

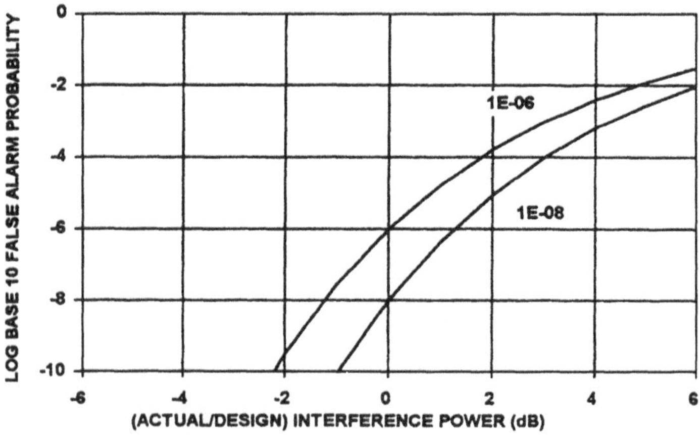

Figure 1.9 Actual false alarm probability.

the thermal noise power in a nonpredictable manner. These residue powers would thus cause data processor saturation. Methods of maintaining the false alarm probability even when the interference or residue power level changes is a mandatory feature of a radar signal processor. These signal processors are termed constant false alarm rate (CFAR) processors.

For a thermal noise environment, the threshold, T, as given by (1.31a), equals a gain constant times the thermal noise power. The gain constant depends upon the desired false alarm probability. When the environment also includes clutter and/or jamming, the total interference power is unknown and possibly varying with range. If a precomputed threshold is used, the threshold value will usually be based upon the wrong interference power level. If the threshold value is based upon the thermal noise power level when the actual level is higher, the resultant false alarm probability will be high, and so the data processor saturates. If it is based upon an interference power level that is too high, the threshold value will be too high such that it becomes more difficult to detect desired targets. An adaptive threshold based upon an estimate of the actual residue interference level is required. This is implemented by estimating the total interference power and multiplying the estimate by a new constant. The estimate is performed by a CFAR processor. For a noise jamming environment, there is a unique best processor called the cell averaging CFAR (CACFAR). As will be discussed in detail, because of the great variability of clutter environments, there is no best clutter environment CFAR processor and there are many variants.

A conceptual block diagram of the CACFAR is shown in Figure 1.10. The voltage denoted as v_0 is the detector output for the test cell. The other voltages are from the cells bracketing the test cell and are termed the reference cells or reference window. The processor forms an estimate of the interference power of the test cell by averaging the voltages of the M reference cells. A target present decision is made if the test cell voltage, v_0, exceeds the sum of the M reference cells multiplied by a gain factor, g_a. Since, in general, the goodness of an estimator tends to improve as the number of samples used for the estimate increases, it should be expected that the performance of the technique improves as M increases.

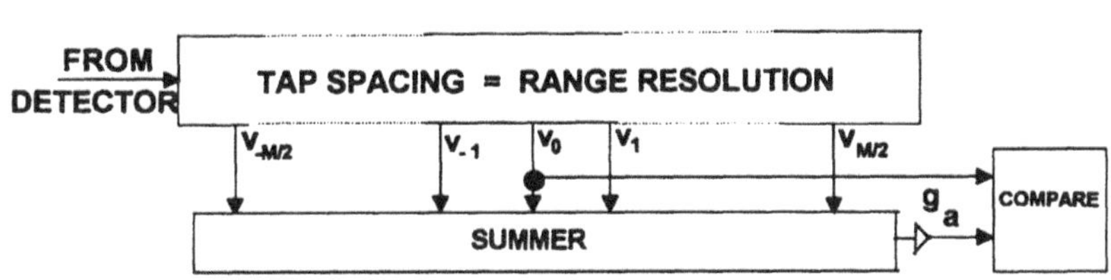

Figure 1.10 Conceptual block diagram of CACFAR processor.

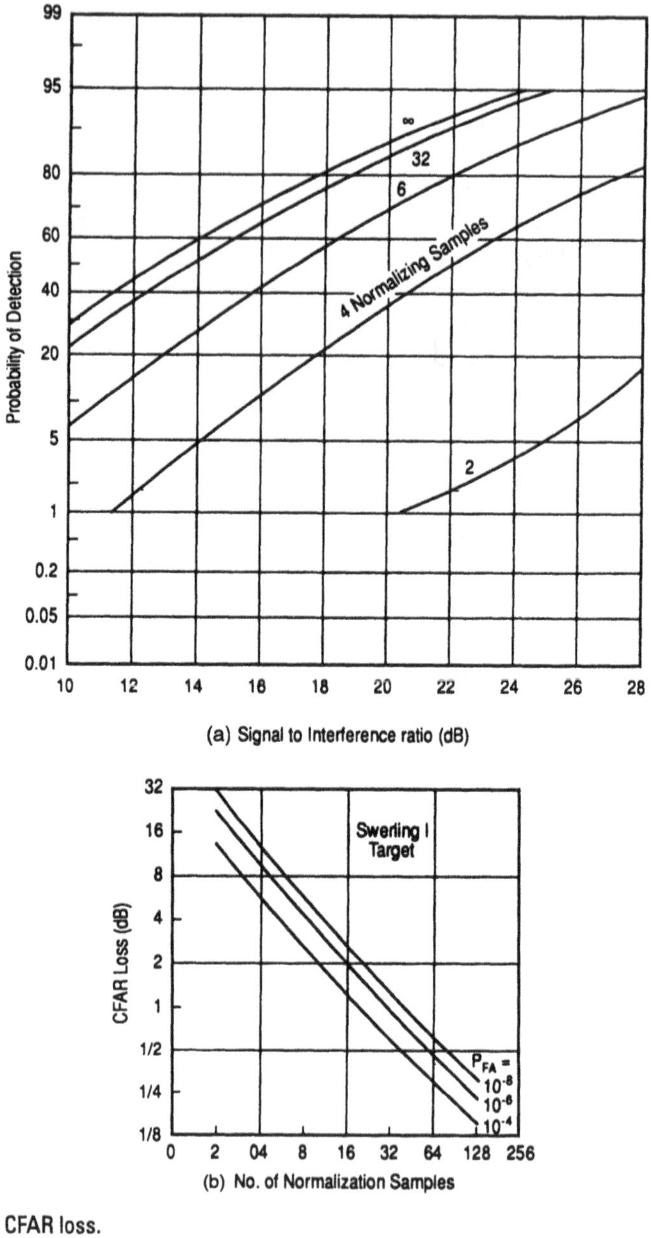

Figure 1.11 CFAR loss.

The performance curves of the CACFAR technique are shown in Figure 1.11 for a Swerling 1 target. The upper portion of the figure shows the detection probability versus the signal-to-interference ratio (SIR), measured in decibels for a false alarm probability of 1×10^{-6}. The interference is the summed power of the thermal and jamming noises. The curves are parametric with the number of reference cells. The curves are for M equal to 2, 4, 8, 16, 32, and infinity. It can be seen that for any specified SIR the probability of detection increases as M increases. The highest curve is for an infinite number of reference cells. This curve is identical to the curve for the known interference power level environment, given by (1.32). The difference between the curves for finite M and infinite M is termed the CFAR loss. Note that for M greater than about 4, the curves are approximately parallel. This indicates that the CFAR loss is approximately independent of the probability of detection. Curves of the CFAR loss are shown by the lower portion of the figure for several false alarm probabilities. The curves are approximately parallel straight lines for the axes chosen. Both axes are logarithmic. The vertical axis is the loss expressed in decibels. The horizontal axis is the number of reference cells. The reference cells also are often termed normalization cells for reasons that will be explained later. It can be seen that the CFAR loss, expressed in decibels, is halved when the number of reference cells is doubled. As a specific example, the loss for 16 reference cells and a false alarm probability of 1×10^{-6} equals 2 dB. The loss increases to 4 dB if the number of reference cells decreases to 8; the loss decreases to 1 dB if the number of reference cells increases to 32. Always for a noise environment, the probability of target detection increases as the number of reference cells increases. This is not usually true for a clutter environment because as the number of reference cells increases, it becomes more probable that the environment becomes nonhomogeneous over the window. The interference power estimation technique of the CACFAR technique is only valid for homogeneous environments. Other CFAR processing techniques are usually implemented for clutter environments.

1.7 Detection of Targets in Clutter

In any usual radar environment, there are many reflectors. Some of these are reflectors that are of interest to the radar system called "targets." Some of these are of no interest to the radar system and are called "clutter." Reflectors that are of no interest to the radar system cause difficulty in two manners: (1) erroneously reporting clutter as target tends to overload the data processor's capacity and (2) clutter can prevent the detection of desired targets. The degree of detectability of targets in a noise environment is controlled by the ratio of the reflected signal energy and the thermal noise power. Often, however, even when the target's SNR value is very large, it is difficult to detect it because the reflected energy is less than that reflected from the clutter. That is, if the radar cross section of the clutter in the resolution volume containing the target is greater than the radar cross section of the target, it may be impossible to detect the presence of the target since it is obscured by the clutter. This is a problem similar to a noise jammer obscuring the return from a target.

A target can be detected even when it is obscured by clutter if the radar reflections can be distinguished because there is some physical discriminant between the reflectors. However, not all physical discriminants are measureable by the radar. As an example, two reflectors that are identical except that one is colored red and the other is colored green cannot be distinguished by a radar because the color difference does not affect the radar return. Examples of possible discriminants are Doppler shift and polarization of the electromagnetic wave.

Thus, there are two related problems. The first is to prevent clutter from being reported as a target. The second is to detect targets despite the presence of clutter in the same resolution cell. Note that if a signal processor, based on a discriminant, attenuates the clutter return while not affecting the target return, both problems are alleviated. There are two primary potential radar discriminants. One is reflector velocity because of its effect upon the carrier frequency of the reflected waveform. When a target has a radial velocity relative to the radar, the carrier frequency of the return differs from the transmitted frequency. The difference is termed the Doppler shift, f_D, which is given by

$$f_D = f_0 v/(c/2) = 2v/\lambda \tag{1.35}$$

where f_0 is the carrier frequency of the transmission, v is the radial velocity of the reflector, and c is the velocity of light. Often, targets of interest have a nonzero radial velocity whereas ground clutter has approximately zero radial velocity.

The usefulness of the Doppler discriminant increases as the dwell time increases. However, as discussed while deriving the equations for maximum radar range, increasing the dwell time affects other inter-related system parameters. Thus, increasing dwell time increases the energy on target, which increases the target's detectability. However, it also increases the scan, or update, period and this affects the ability of the data processor to track maneuvering targets. This is another example of a trade-off area.

An alternate method of differentiating between types of reflectors is via the polarization characteristics of reflectors. As mentioned earlier, if a raindrop is approximately spherical, polarization can be used to discriminate between weather clutter and aircraft targets. An advantage of polarization techniques as compared to Doppler techniques is that it, in theory, can differentiate targets with a single pulse so that it does not require the long dwell time of the Doppler technique. In addition, the two techniques are compatible and can be cascaded. The major disadvantage of the polarization technique is that it has high implementation cost because of the increased antenna complexity.

1.7.1 Maximum Range in Clutter

Consider the case when the test cell contains both target and clutter. An equation for the maximum detection range is derived for the special case where noise is neglected and a single pulse is transmitted and detection occurs if the received target energy exceeds the received clutter energy.

Since the target and clutter are at the same range, the returns are

$$P_{\text{TARGET}} = PG_{\text{T}}\sigma A/(4\pi R^2)^2 \tag{1.36}$$

$$P_{\text{CLUTTER}} = PG_{\text{T}}\sigma_{\text{C}} A/(4\pi R^2)^2 \tag{1.37}$$

where σ denotes the target's radar cross section and σ_{C} denotes the clutter's radar cross section. Comparing (1.36) and (1.37) they imply that target detection occurs when $\sigma \gg \sigma_{\text{C}}$.

Two classifications of clutter are point clutter and distributed clutter. Point clutter is the reflection from isolated point objects such as a water tower or a mountain. Point targets have physical dimensions that are much less than the dimensions of the resolution volume. Distributed clutter is due to the reflection from homogeneous regions such as rain or sea or some types of ground. For distributed clutter, the radar cross section of the clutter depends upon the size of the resolution region. The radar cross section of volume clutter, such as weather or chaff, is given by the product of a reflection coefficient η and the resolution volume. Define the range resolution as δ_{R}, the azimuth resolution dimension as $\theta_{\text{A}} R$, and the elevation resolution dimension as $\theta_{\text{E}} R$. Thus, the radar cross section, assuming that the reflector fills the resolution volume, is

$$\sigma_{\text{C}} = \eta(\theta_{\text{A}} R)(\theta_{\text{E}} R)\delta_{\text{R}} \tag{1.38}$$

Similarly, the radar cross section of ground clutter is given by

$$\sigma_{\text{C}} = \sigma^0(\theta_{\text{A}} R)\delta_{\text{R}} \tag{1.39}$$

where σ^0 is the reflectivity constant of the ground and $(\theta_{\text{A}} R)\delta_{\text{R}}$ is the resolution area of the ground. Note that the dimensions of η are m^2/m^3 and those of σ^0 are m^2/m^2. The clutter cross sections that are calculated using (1.38) and (1.39) are usually much greater than the cross section of many targets of interest. Thus, the target is not detectable when the target is in a resolution cell containing the clutter unless some improvement in the signal-to-clutter ratio (SCR) can be achieved by a signal processing technique that has differing gains to the target and clutter.

As numeric examples, assume the cross section of a typical small aircraft is 1 m^2. The parameters of a typical radar, θ_E and θ_A, both equal 2 degrees and the transmit bandwidth equals 1 MHz, so δ_R equals 150m. A reasonable assumption for the ground clutter reflectivity is that σ^0 equals -24 dB (4×10^{-3}) so that the ground clutter cross section equals 836 m^2. Since the clutter power is much greater than the aircraft cross section, some type of clutter cancellation technique is required. In order to detect the aircraft target with high probability, the SCR after cancellation should equal approximately 10 dB. Since the precancellation ratio equals -29 dB, the required cancellation equals at least 39 dB.

For rain clutter, the reflectivity coefficient equals approximately 1×10^{-8}, so the clutter cross section equals 18.3 m^2 and the required cancellation equals approximately 22 dB. This is a significantly smaller cancellation requirement than that for ground clutter, but, as will be seen, it is much more difficult to attain.

1.7.2 Delay Line Cancelers

The fundamental concept of Doppler processing is that the range to a stationary reflector is the same for two consecutive pulse transmissions while it changes for a nonstationary reflector. The left-hand portion of the sketch of Figure 1.12 depicts a transmitter that can be used to exploit this differential characteristic of targets and clutter. The right-hand portion of the figure shows two pulses of the transmitted waveform. The left-hand portion shows the transmitted waveform. The top waveform shows two pulses of the repetitive pulse train of the pulsed modulating waveform. It gates the output of the sinusoidal oscillator. The lower waveform is two sinusoidal pulses of the transmitted repetitive pulsed sinusoidal waveform.

The received waveform, neglecting thermal noise, is given by Figure 1.13. The upper waveform indicates the reflection due to the transmission of the first

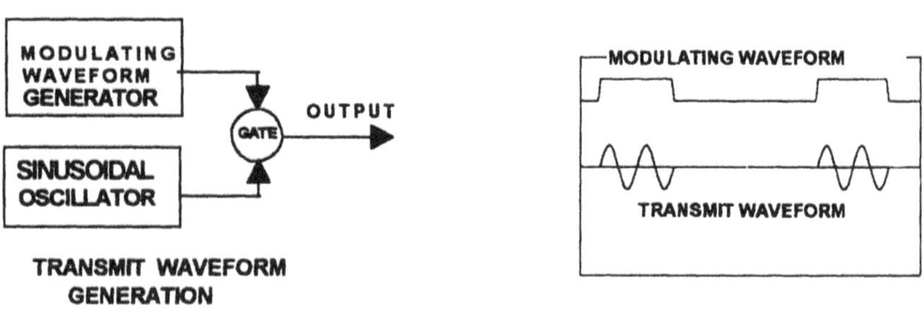

Figure 1.12 Generated waveform.

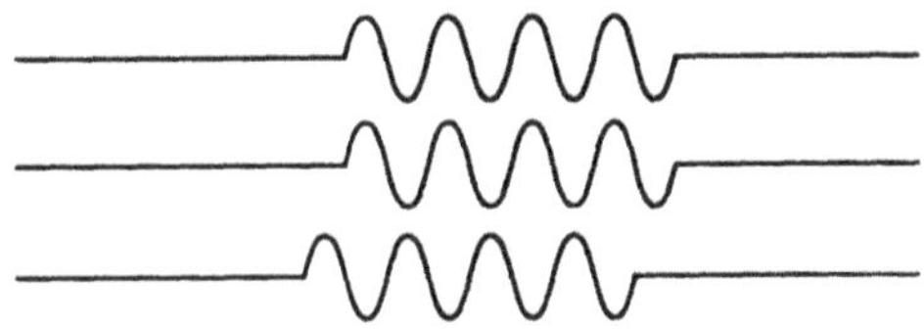

Figure 1.13 Reflections from stationary and moving targets.

pulse. The middle waveform is due to a reflection of pulse 2. Its time origin has been shifted to be identical to that of the first pulse in order to emphasize the effect of target range. Note that for the middle waveform the reflected pulse occurs at the same time as that due to the first transmitted pulse, which indicates that the reflection is from a stationary target. The lower waveform depicts the reflection of the second pulse for a decreasing range target. Note that the pulse occurs at an earlier time. As will be discussed later, the earlier time of arrival is identical to a differential pulse-to-pulse phase shift, or Doppler frequency translation, and forms the basis for discriminating between stationary and moving targets.

There are two types of signal processors used to exploit the Doppler characteristics of targets and clutter. One is the delay line canceler. The second is the Doppler filter bank. The canceler is relatively simple to implement, but its performance is inferior to that of the Doppler filter bank. A conceptual block diagram of a delay line canceler is shown in Figure 1.14. It consists of a delay line whose delay equals the pulse repetition interval, T, a subtracter, and an amplifier of gain $1/\sqrt{2}$. This gain value is chosen so that the output noise power equals the input noise power. The amplifier is not actually used in an implementation but is included here to simplify the interpretation of the SNR performance equations. In addition, the circuit is discussed as if the circuit input is at the carrier frequency, f_0. In practice, the implementation is not done at the carrier frequency but is implemented either at an IF frequency, such as 60 MHz, or at baseband using synchronous demodulation. Thus, this design and all others that are shown in this and later chapters are intended as conceptual, and not actual, implementations. Since, as shown in Figure 1.13, the returns for the stationary target for the two pulses are identical, the output of the subtracter is zero for stationary targets. However, the output of the subtracter is not zero for moving targets since subtracting the upper waveform from the lowest

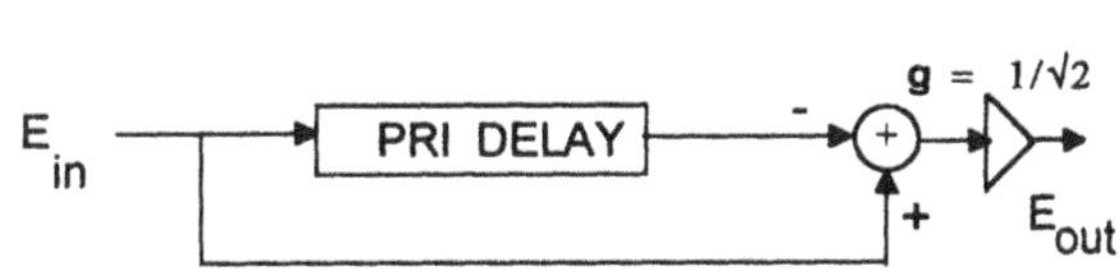

Figure 1.14 Conceptual block diagram of delay line canceler.

waveform produces a nonzero result. Thus, a signal processor that exploits the Doppler effect can differentiate between moving and stationary targets.

Though not emphasized in the preceding discussion, it is mandatory that the transmit waveform have a special property in order for the signal processor to obtain a zero output for stationary targets. Thus, note that the waveform shown in Figure 1.13 has each transmit pulse starting at the same phase. If this were not true, the output magnitude after subtraction of adjacent pulse returns for a stationary target would be nonzero and dependent upon the amplitude of the returned waveform. Since it is dependent upon the clutter's cross section, it is not possible to compensate by simply adding or subtracting a fixed voltage to generate a zero output for clutter. Thus, constant pulse-to-pulse phase is mandatory. One way to satisfy this requirement is to derive the carrier frequency and the pulse repetition frequency from a fundamental clock frequency. Then the carrier frequency is an integer multiple of the pulse repetition frequency and the required phase lock is obtained. Modern radar with klystron amplifier transmitters often use this technique. Radar using magnetron transmitters cannot use this procedure because a magnetron is a pulsed oscillator that does not have pulse-to-pulse phase coherence. The required coherent starting phase is obtained by locking the phase of a local oscillator to the starting phase of each pulse. Exact phase locking is difficult to attain, which is one of the reasons that magnetron implementations do not work as well as the simple theory predicts.

A difficulty with the processor in Figure 1.14 is that the output magnitude equals zero for some nonzero target velocities in addition to equaling zero when the reflector's radial velocity equals zero. As depicted in Figure 1.13, when the radial velocity is zero, the returns are at the same time delay relative to the transmitted pulses and the subtraction yields zero. However, note that if the reflector's velocity is such that the return for the second pulse is advanced (or delayed) by a full cycle, the subtraction yields zero everywhere except for the leading and trailing cycles of the output waveform. Usually, the number of cycles in a pulsed waveform is very large and the two nonzero leading and trailing cycles are negligible. Specifically, the number of cycles in a pulse is the product of the carrier frequency and the pulse duration. For the typical radar parameters considered previously, a carrier frequency of 3 GHz and a pulse width of 1 μs, the number of cycles in the waveform equals 3000. Thus, the output waveform for this target radial velocity is essentially indistinguishable from that of a stationary target. This velocity is called a blind speed to emphasize that the radar cannot detect a target with this velocity when using Doppler processing. As discussed later, this shortcoming is not a function of the form of the Doppler processor. It is inherent in any Doppler processor.

Similarly, if the target velocity equals twice the blind speed velocity, the output waveform is zero except for two leading and two trailing residual cycles

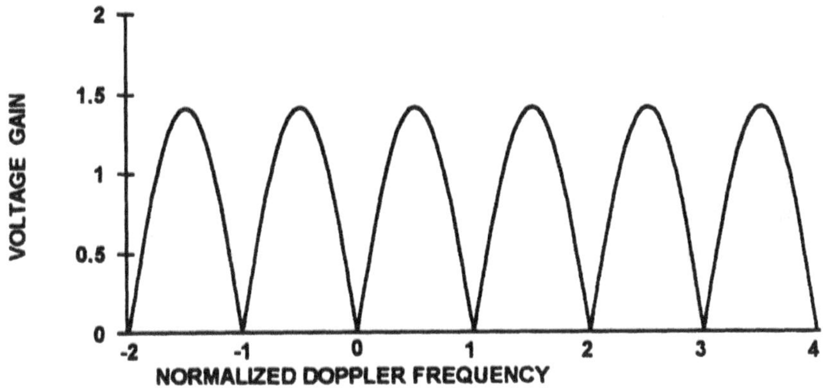

Figure 1.15 Delay line canceler Doppler gain.

and this also is a blind speed. Neglecting the effect of the leading and trailing residual cycles, in general the magnitude of the output of the delay line canceler varies sinusoidally as a function of Doppler frequency. A graph of the output magnitude is shown in Figure 1.15. The equation for the magnitude of the voltage gain, e_M, is

$$\begin{aligned} e_M &= |E_{OUT}/E_{IN}| = \sqrt{2}|\sin(\pi\nu)| \\ \nu &= f_D/\text{PRF} = f_D T \end{aligned} \tag{1.40}$$

and the horizontal axis is the normalized Doppler frequency, ν. There is a zero response at zero Doppler frequency. Also note that the zero output response occurs at nonzero Doppler frequencies and, therefore, nonzero target velocities. These velocities are the blind speeds. The blind speeds occur when the normalized Doppler frequency equals an integer so that the blind Doppler frequencies are at any positive or negative multiple of the PRF. The blind speeds occur because, as will be discussed in more detail later, the repetitive pulsed waveform has a Doppler ambiguity at multiples of $1/T$ just as it has a range ambiguity at range-time delays of multiples of T. The values of the velocities for the blind speeds depend upon the carrier frequency via (1.35) and the PRF. As a numerical example, for a radar operating at a carrier frequency of 3 GHz and a pulse repetition frequency of 1000 Hz, the blind speeds are at multiples of 50 m/s (97.2 knots). Increasing the PRF to 10 kHz (or decreasing the carrier frequency to 300 MHz) causes the blind speeds to be at multiples of 500 m/s (972 knots).

The maximum value of e_M equals $\sqrt{2}$. Thus, the energy increases by a factor of 2. As the circuit gain is adjusted such that the output noise power equals the input noise power, the maximum output SNR value is twice the input SNR value. As two pulses are being processed, this increase is consistent with the target energy increase associated with multiple pulse transmissions assumed when deriving the maximum range equations. However, note that for many other Doppler frequencies, the value of e_M is less than unity and, thus, the SNR value of the output is less than that of the input. Thus, the previously derived maximum range equations are valid for the delay line canceler processor only after incorporating a velocity-dependent loss factor.

1.7.3 Doppler Filter Bank

The SNR decrease incurred when using a delay line canceler can be avoided by using a Doppler filter bank. This circuit processes N pulses with N generally ranging between 6 and 32. Just as for the delay line canceler, the circuit has a low response to ground clutter but a high response to targets at most other Doppler frequencies. The ratio of output-to-input SNR is approximately N at almost all Doppler frequencies. The filter bank is implemented as a number of parallel filters. Multiple parallel filters are required because the gain of a single filter is high only over a relatively narrow Doppler band. Thus, it is necessary to implement a bank of parallel filters in order to obtain high gain over the complete Doppler band.

The conceptual implementation of a single filter is shown in Figure 1.16 for an N-pulse processor. The processor delays the returns due to the first transmitted pulse by a time of $(N-1)T$ and weights it by a gain of g_1, delays the returns due to the second transmitted pulse by a time of $(N-2)T$ and weights it by a gain of g_2, and so on. The delay line is used to obtain time coincidence of the returns for the N transmitted pulses for all range cells. The gain values are used for two different purposes. The magnitudes control the sidelobe response of the gain characteristic versus Doppler frequency. The phases determine where the peak response occurs. Figure 1.17 shows the filter response curve of an 8-pulse filter for the case when all of the filter weights equal the reciprocal of the

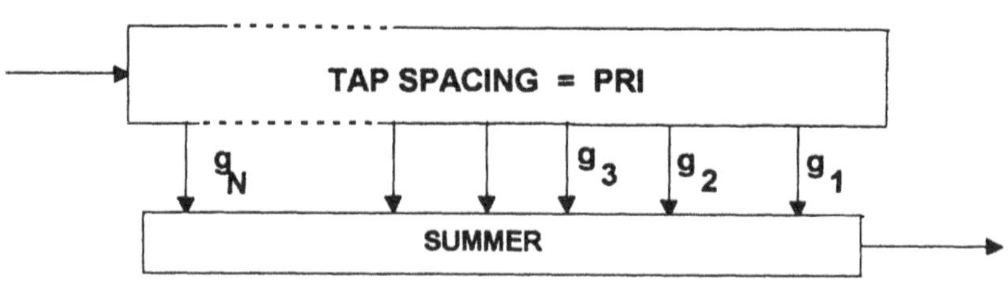

Figure 1.16 Conceptual Doppler filter block diagram.

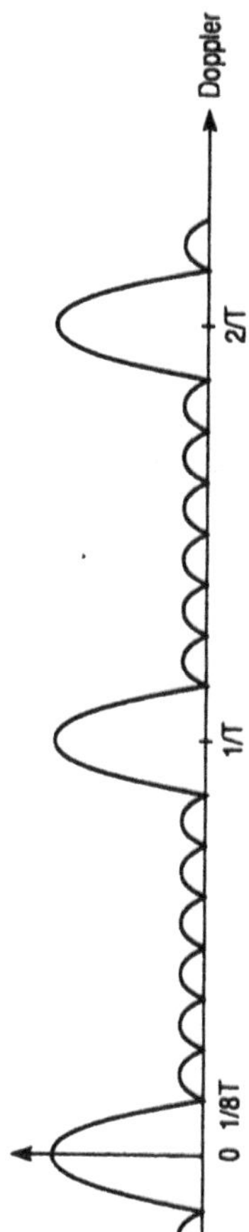

Figure 1.17 Doppler response of single 8 pulse filter.

square root of N. This magnitude choice causes the output thermal noise power to equal the input thermal noise power. Just as for the delay line canceler discussion, this simplifies the understanding of the filter's operation as the gain at any Doppler frequency then equals the SNR gain of the filter. In general, the Doppler frequency gain equals the SNR gain when the sum of the squared magnitude of the weights equals unity. It can be shown that equal magnitude weights maximize the target's detectability when its Doppler frequency equals the filter's center frequency. However, this choice causes large sidelobes and is another trade-off area.

When the filter weights have the same phase, the output sum is a maximum for a zero Doppler center frequency. This can be seen by referring back to Figure 1.9. Zero Doppler frequency means that all of the returned pulses are at the same range, and adding them without phase shift causes a maximum possible output. The peak voltage gain equals the square root of N, so the peak SNR gain equals the number of pulses, N. Examining Figure 1.17, it can be seen that just as for the delay line canceler, the response is periodic with the periodicity equal to the pulse repetition frequency. Each period of the filter response consists of a mainlobe and six sidelobe regions. The response for this filter is maximum at zero Doppler frequency and the maximum voltage gain equals $\sqrt{N}$. The maximum energy gain equals N, so the maximum SNR gain equals N. The sketch is drawn for an 8-pulse processor and the first null of the response is at a frequency of $1/8T$. In general, the first null is at $1/NT$. Thus, the filter energy gain is approximately equal to N over a frequency band of only approximately $1/NT$. To achieve large gain over the total unambiguous Doppler band, it is necessary to use a bank of filters with each filter tuned to a different Doppler frequency. Usually a bank of N filters is used to process N pulses. Tuning a filter to a different Doppler frequency is done using delay line tap gains that include phase shifts. Each of the filters is configured using the same magnitude weights but different sets of phase shift. The filters are rarely implemented using equal magnitude weights at each of the delay line taps because this causes relatively high, and therefore undesirable, sidelobe response. The preferred weight sets for low sidelobes are either Tchebychev or Taylor. The disadvantage of these weights is that they are not the optimum weights to maximize target detectability and thus cause a loss. There is a trade-off between low sidelobes and low loss.

Note that the delay line canceler uses a single filter to cover the Doppler band whereas the bank uses N filters. The latter implementation is significantly more complex but has an SNR gain of close to N whereas the SNR gain of the canceler is usually much less. Thus, this is another trade-off between equipment simplicity and improved detectability of targets.

1.7.4 Some Doppler Filtering Details

The voltage output of a Doppler filter for a sinusoidal input can be computed by the N component sum

$$e_n = \sum_m g_m \exp[j(\theta + m2\pi f_D)] \tag{1.41}$$

where $\exp[j(\theta + m2\pi f_D)]$ depicts samples at the PRF rate of the sinusoidal input and g_m is the weight at the mth tap of the delay line. The weight is given as the product of its magnitude and phase as

$$g_m = |g_m| \exp(-jm2\pi f_S) \tag{1.42}$$

where f_S denotes the steered center frequency of the filter. Typically the N values of f_S are 0, $1/NT$, $2/NT, \ldots, n/NT, \ldots, (N-1)/T$. A block diagram of a filter bank configured with N filters is shown in Figure 1.18. Each of the parallel filters is cascaded with a CFAR circuit. The outputs of the CFAR circuits are target present or absent decisions and are passed to the data processor along with angle of location measurements. Specific examples of the filter responses are shown in Figure 1.19 for a 4-pulse processor implemented with equal magnitude weights. The crossovers of adjacent filters are at relative gains of −3.9 dB. A method of decreasing this loss is the expensive option of using more than N filters.

The filter responses of Figure 1.19 also demonstrates that Doppler processing improves as the dwell time or number of pulses transmitted and processed increases. The cross section of ground clutter is usually much larger than that

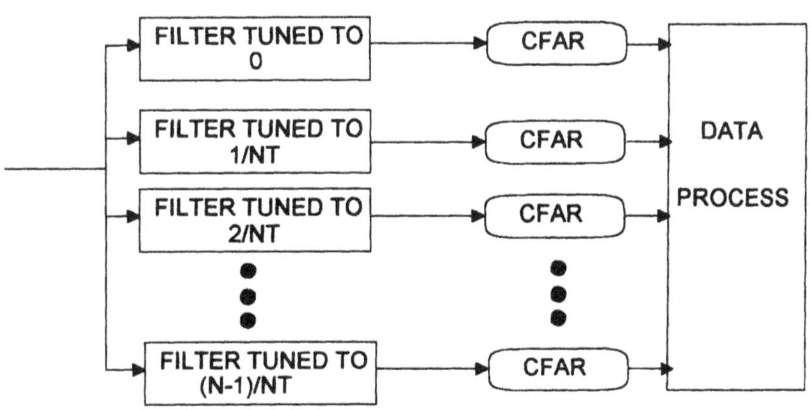

Figure 1.18 Conceptual block diagram of Doppler filter bank.

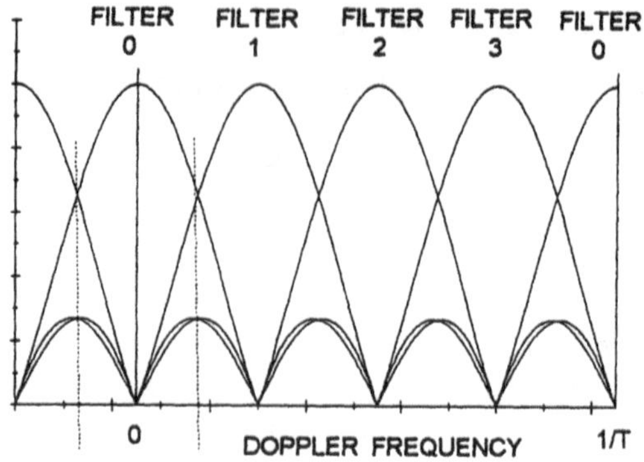

Figure 1.19 Multiple Doppler filters.

of a target. Thus, when zero Doppler frequency ground clutter is present, it prevents the detection of any target whose Doppler frequency is in filter zero between plus and minus $1/8T$. Because of the Doppler ambiguity associated with the PRF values usually used for search radars, the target Doppler frequency can be approximated as a random variable uniformly distibuted between zero and $1/T$. Therefore, for this case, the probability that ground clutter prevents target detection equals 1/4, or $1/N$. Thus, generally, for an N-pulse transmission and processor, the probability of target detection is less than $1-1/N$ or $(N-1)/N$. Similar conclusions follow for nonzero Doppler clutter.

1.7.5 Noncoherent Summing

Implementation of the Doppler filter bank uses coherent integration. The summation of the returns from different pulse transmissions is made at a point prior to envelope detection, so the output depends upon the amplitudes and phases of successive pulses. Coherent summation causes the output to be zero for certain nonzero target velocities and also causes an SNR loss at other target velocities. If clutter is not present, and so the ability to reject clutter is not necessary, the complexity and SNR loss can be avoided by summing after the envelope detector. This is termed noncoherent integration. The conceptual block diagram for noncoherent summing is identical to that of Figure 1.2. The only difference is that the circuit input is now obtained from a point after the envelope detector. Note that unlike the Doppler filter bank, only a single circuit is required because noncoherent integration is not sensitive to Doppler frequency. The penalty for the equipment simplification is that the peak increase of SNR is less than that achieved by the Doppler filter bank. As will be discussed in detail later, the output SNR of the Doppler filter bank is almost equal to that of the optimum processor. The loss of noncoherent integration relative to the performance of the optimum coherent signal processor depends upon the number of pulses processed. The loss in decibels is approximately equal to

$$L_{\text{dB}} \approx \begin{cases} 2\log N & 1 \leq N \leq 10 \\ 2 + 4\log(N/10) & 10 \leq N \leq 1000 \end{cases} \tag{1.43}$$

1.7.6 Some Implementation Details

Doppler processing techniques can, in theory, be implemented using either analog or digital techniques. In practice, because of component limitations, Doppler filter banks are implemented primarily using digital techniques. In addition, though delay line cancelers are implemented using either analog or digital techniques, as a general rule, the digital implementations have superior clutter cancellation capabilities.

In analog implementations, filters are implemented using R, L, C components. Theoretically, these filters and the required delays can be implemented at an IF frequency. However, it is very difficult to implement the delay line at IF by delay of electromagnetic energy because of the high velocity of electromagnetic energy. Specifically, the delay through a cable or waveguide is approximately 1 ns per foot of length. Thus, if the radar PRF equals 1000 Hz, so that a delay of 1000 μs is required, the implementation requires 1×10^9 feet of cable. This is excessive and cannot be implemented. Instead, the output of an IF amplifier is demodulated to video and the video is used to modulate acoustic energy propagating through an acoustic medium such as a quartz line. As the velocity of acoustic energy is much lower than that of electromagnetic energy, the acoustic delay line is an implementable length. At the end of the delay line, the acoustic energy is remodulated to electromagnetic energy for further signal processing. However, because the demodulation and remodulation tend to induce errors, this technique is usually not capable of achieving very high clutter cancellations. The exact numeric value of the achievable cancellation is component sensitive, which in turn is sensitive to the particular calender year and cost of implementation. The calender year is significant as new components are continuously being developed. Cost is significant because of component tolerances. As an example, the actual delay of a delay line and values of the R, L, C components depends upon the temperature, so high cancellation is achieved only if good temperature control is implemented. Better temperature control can usually be achieved by increasing costs.

Digital implementation is implemented by converting the continuous time-continuous value waveform to time-sampled discrete-valued waveforms using synchronous demodulation as shown in Figure 1.20. The circuit operates by mixing the input with a sinusoidal oscillator whose frequency equals the IF

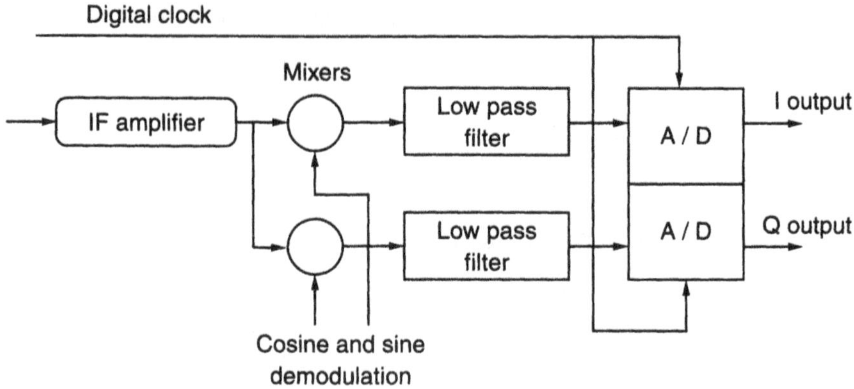

Figure 1.20 Synchronous demodulation.

frequency. Since mixing is approximately the same as multiplying, trigonometric expansion shows that this produces a component whose passband is centered at twice the IF frequency and also a component centered at zero frequency. The twice frequency term is rejected by the lowpass filter following the mixer. The mixing is done with both a cosine and sine (a 90-degree phase shift) version of the oscillator because an IF waveform is described by both an amplitude modulation and a phase modulation. If only a single-channel mixer is used, the IF waveform cannot be reconstructed, so there is a loss of information. It can be shown that the output of the two-channel mixing can be manipulated to reproduce the amplitude and phase modulation, so syncronous demodulation does not cause any information loss. After lowpass filtering, to remove the twice IF frequency component, both channels are converted to digital samples by a pair of analog-to-digital (A/D) converters. The digital clock determines the rate of digital sampling. The clock frequency must at least equal the two-sided bandwidth of the IF waveform. As will be shown later, the detectability of targets increases as the frequency of the sampling clock increases. However, the equipment complexity, and therefore cost, increases approximately linearly with clock frequency. Thus, there is a potential trade-off between equipment cost and target detectability.

The digital implementation of a delay line canceler or Doppler filter bank is performed using a special purpose digital computer. The delay operation is trivial because it simply depends upon read/write timing to memory. Long delays are no more difficult to implement than short delays. Filtering is done using standard digital filtering algorithms. The digital implementation has significant increased cost compared to an analog implementation but is capable of achieving cancellations that are considerably in excess of that achievable by analog techniques. As one example of the potential improvement, it was noted that the digital implementation processes two channels because the IF waveform is described by both amplitude and phase modulation. However, the analog implementation of the delay line canceler is done using a single channel partially because of the impossibility of matching the delay length of two acoustic delay lines. The use of a single channel causes an SNR loss of approximately 3 dB.

The digital implementation is subject to implementation errors just as is the analog one, but the errors occur in different ways. One example of errors is that the mixers are not perfect multipliers, so in addition to the video and twice IF frequency waveform, other terms appear. Some of these other terms cause components at baseband, distorting the desired signal. These distortion-causing terms can be reduced by careful mixer design, as an example, by using balanced mixers. Further distortion occurs because it is not possible to completely reject the second, and higher, harmonic terms that are generated at the output of the mixer. These terms can only be rejected if the rejection filters had infinitely sharp skirts and if the signal waveform was truly bandlimited. Neither of these conditions is true and thus contribute to the incomplete rejection. Another potential problem is that the I channel and Q channel processing channels are not exactly matched. That is, the post mixer filters are not identical, the A/D converters are not identical, and the "sinusoid" and "cosinusoid" oscillators are not precisely 90 degrees out of phase. Nonidentical post-mixer filters usually have differing gain and phase ripples across the filter's passband. All of these effects combine to limit the achievable clutter cancellation, and care is required to ensure that the channel matches are as close as possible to the ideal values.

1.8 Resolution and Measurement Accuracy

The resolution cell is a multidimensional concept. Its coordinates can be range, azimuth angle, elevation angle, and Doppler frequency, for example. Range resolution is depicted by the sketches of Figure 1.21. The left-hand sketch depicts the envelope detector output for a single isolated target neglecting thermal noise. The nominal size of the range cell depends upon the transmitted pulse width for a simple pulse or the compressed pulse width for a nonsimple pulse (pulse

Figure 1.21 Range resolution waveforms.

coding such as linear frequency modulation) transmitted waveform. The right-hand sketch depicts the output waveform for two closely spaced targets. Only a single peak appears. The middle sketch is for two targets that are approximately seperated by the compressed pulse width. Two distinct peaks are visible. These targets are deemed to be resolved and in two adjacent range cells. For a simple pulse of duration τ, the size of the range cell, R_τ, is

$$R_\tau = (c/2)\tau \tag{1.44}$$

and the pulse width is related to the bandwidth, B, as

$$\tau = 1/B \tag{1.45}$$

For a compressed pulse, the range resolution is

$$R_\tau = (c/2)B = (c/2)\tau_{\text{COMPRESS}} \tag{1.46}$$

The preceding concept of range resolution is classic and potentially misleading. Thus, note that it is obvious that the waveforms of the left-hand part of Figure 1.21 are not identical to the right-hand waveform so that in the absence of noise it is clear that the right-hand waveform cannot be caused by a single target. In general, if the SNR is sufficiently high, it is theoretically possible to resolve two targets at arbitrarily small range separations. However, as will be shown later, the required SNR values are very large and correspond to those for detection probabilities of 0.999 or 0.9999. The resolutions given by (1.44) and (1.45) are approximately applicable for the SNR values that correspond to detection probabilities of 0.7 to 0.95.

Measurement, or location accuracy, and resolution are similar in the sense that both depend upon the additive noise. In the absence of noise it is theoretically possible to resolve two arbitrarily closely spaced targets. In the presence of noise the resolution depends upon the SNR and the transmitted bandwidth. Resolution is essentially identical to measuring the differential range of the two targets. Similarly, the portion of the range location error due to noise depends upon the SNR and bandwidth. The location error is commonly modeled as a random variable. It is assumed that on average the range measurement is the true target range, so the average error equals zero. The standard deviation of the measurement error, σ_τ, is

$$\sigma_\tau \approx (1/B)/\sqrt{(\text{SNR})} \tag{1.47}$$

These equations can be generalized to other domains. Thus, an angle resolution cell, either azimuth or elevation, is approximately equal to the beamwidth. The standard deviation of the angle measurement accuracy, σ_θ, is

$$\sigma_\theta \approx \theta_{\text{BW}}/\sqrt{(\text{SNR})} \tag{1.48}$$

where θ_{BW} is the beamwidth in either azimuth or elevation. Note that angle resolution and measurement accuracy, like range resolution and measurement accuracy, approximate concepts. In either case, if the SNR is sufficiently high, two arbitrarily closely spaced targets can be resolved and accurately located. For the Doppler domain and an N-pulse transmission, the cell resolution size equals PRF/N. Usually Doppler frequency is not measured because it is highly ambiguous. If it is measured, the measurement accuracy is given by an equation similar to (1.48).

1.8.1 Ambiguity

Ambiguity refers to the concept that a reflector at one location reflects a waveform that, due to a deficiency of the radar, is interpreted by the radar signal processor as caused by a reflector at a different location. The word "location" is used to emphasize that the ambiguity can be in one of many location coordinates. One example is the angle ambiguity that is implied by the antenna pattern shown in Figure 1.5. The regions of the antenna pattern are the high-gain main beam and the sidelobes. The antenna is used by steering the main beam to a specified angle location. If a target is detected, the target's angle is *assumed* to be the steered angle. If, however, the received energy is caused by a reflector of very large cross section at a sidelobe angle, the reflector's angle is incorrectly reported as the main beam-steered angle. This is an angle ambiguity. Note that the reflector can be "detected" at many different angles and, therefore, has many angle ambiguities because the antenna is scanned and different sidelobes illuminate the reflector.

Similarly, there is a range ambiguity associated with the use of pulse compression waveforms. As will be indicated in more detail later, pulse compression waveforms are used to simultaneously achieve long detection range and high-range resolution (a small range cell duration). However, the defect of the waveform is that the compressed pulse has sidelobes along the range axis. The sketch of Figure 1.5, with a change of axis label, shows the output versus range of a typical pulse compression waveform.

There are several potential problems associated with range sidelobes. First, similar to the effect with angle ambiguity, is that a target at one range can be reported at many ranges if its cross section is sufficiently large so that the sidelobe levels are above the detection threshold. Another is that the sidelobe response of a large target can mask the presence of a close-by small target.

Range sidelobes are a problem for pulse compression waveform radars but not for conventional pulse transmission radars. Note that when the transmit/receive waveform is a simple pulse, the envelope detector output is nonzero over a time interval approximately equal to twice the transmit pulse width and there are no sidelobes outside this time interval. This time period is the approximate size of the time-range cell. For a pulse compression waveform, the envelope detector output is also nonzero over a time interval of approximately twice the transmit pulse width. However, the transmit pulse width is much greater than the compressed pulse width (the reciprocal of the bandwidth). For identical bandwidth radars, the twice transmit pulse width duration for a pulse compression waveform is much greater than that of the simple pulse radar. There are sidelobes over this interval, and these sidelobes can cause many range ambiguities.

A different form of range ambiguity occurs because of the transmission of repetitive pulses. As indicated previously, a benefit of transmission of an N-pulse train is that this increases the detectability of far-range small cross-section targets. However, this advantage is coupled with the disadvantage of target range ambiguity. As a specific example, assume that the pulse repetition interval is equal to 2000 μs. This equals the time required for two-way transmission to a range of 300 km. If a target is detected at a measurement range of 200 km, in theory, the target could actually be at a range of 100 plus 300 (equals 400 km), or, generally, 100 km plus any integral multiple of 300 km. One way of demonstrating this is via examination of the display of a plan-position-indicator (PPI). The PPI shows target position by a polar coordinate range-azimuth display as per Figure 1.22. The display operates by starting a sweep from the center of the display at the time instant of a pulse transmission. The angular direction of the sweep is that of the antenna's pointing angle. The sweep angle rotates synchronously with the

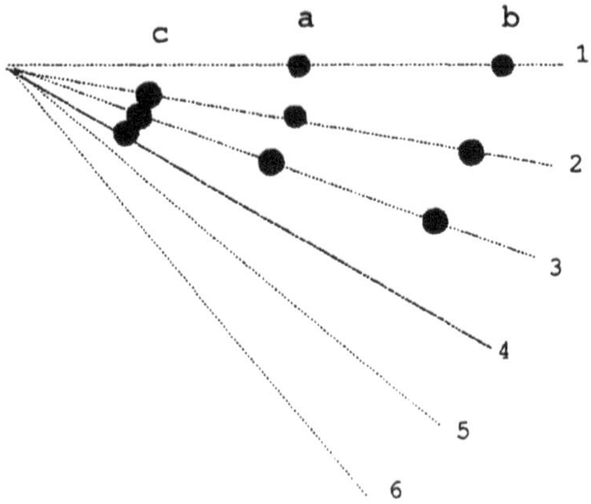

Figure 1.22 Plan position indicator (PPI) display.

antenna pointing angle. The display sweeps out in range at a rate such that the sweep is at the maximum instrumented range when the next pulse is about to be transmitted. The sweep is intensity modulated by the envelope detector output. The figure shows six consecutive sweeps. The plot indicates a total of six sweeps and three targets. The targets are all centered at the steering angle denoted by sweep two. The filled-in dots (blips) denoted as (a) are for a near-range target. The first indicated blip is for sweep one. The target is detected at an angle prior to its actual angle because the antenna gain is appreciable at angles away from the beam peak. Similarly, the last indicated blip is for sweep three. Target (b) is at a further range than target (a). Target (c) is at a still further range and its range is greater than the instrumented range. The two-way time delay for the reflection return exceeds the interpulse period. Since the delay exceeds the interpulse period, the first blip appears on sweep two rather than on sweep one. Note that all the blips for this target appear on the wrong sweep. The range measurement for targets (a) and (b) will be correct, and the angle measurements will correctly indicate that these targets are at the angle corresponding to that of sweep two. However, the range of target (c) will be incorrectly measured as a much shorter value than is correct and the angle will be incorrectly measured as that corresponding to sweep three. Targets whose range exceeds that defined as the maximum instrumented range are termed range ambiguous, or second-time-around, targets. It is also possible to have third- or fourth-time-around targets.

Figure 1.23 is a plot of the unambiguous range versus PRF. Some specific examples are given in Table 1.2.

The range ambiguous response is not due to the signal process but is caused by the repetitive waveform transmission. Figure 1.24 illustrates the problem for a Doppler filter integrator. The top sketch indicates successive pulse transmissions

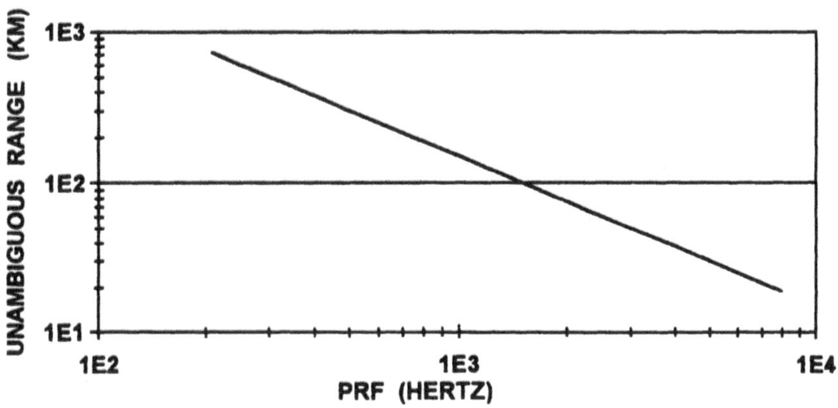

Figure 1.23 Unambigous range.

Table 1.2
Unambiguous Range

PRF (Hz)	PRI (ms)	Range (km)	Range (n. mi.)
300	3333	500	270
1000	1000	150	81
10000	100	15	8.1

at intervals of T. The next sketch indicates the returns from two targets such that the time delay of reception are τ_1 and τ_2, respectively. Note that the time delays are such that τ_1 is less than T whereas τ_2 is greater than T. The transmitted pulses are shown dashed. The next sketch indicates a conceptual implementation of the signal processor used to combine the returns from a 4-pulse transmission waveform. As discussed earlier, a significant feature of the signal processor is the tapped delay line with tap spacing equal to the pulse repetition interval and the summation. This allows the returns from each range cell due to each of the transmitted pulses to be added. The gain values at each tap are discussed in Section 1.7.3 and 1.7.4. The last sketch shows the waveform at the summer output versus time. During the first period of duration T, the output consists of only the return from the near-range target due to transmit pulse 1. The output during the second period of duration T is the return from the far-range target due to transmit pulse one and the sum of the returns of the close-range target from transmit pulses one and two. Note that, referenced to the time of the second pulse transmission, the return due to the far-range target appears at an apparent closer range than the actual near-range target. The summed returns in the third and fourth intervals are similar. The summed returns in the fourth interval are decreased for the near-range target but continue to increase for the far-range target. The summations continue to decrease for successive time intervals, but this is not important because this time period is not monitored. The output from the summer during the $3T$ to $4T$ period is used to make target present/absent decisions. The intrinsic assumption is that the returns due to transmission of the four pulses have been properly summed during this time period. It can be seen that this allows a far-range target to be detected ambiguously at the wrong range.

A PPI display can be made for this type of processor by generating synthetic video. Specifically, the output of the integrator during the period of $3T$ to $4T$ could be used directly to intensity modulate the sweep and an operator would make target present/absent decisions. More usually, the detection decisions are made digitally. The integrator output is compared to a detection threshold voltage. The output of the comparator is zero if the integrator output is less than the threshold and a maximum voltage if it exceeds the threshold. The comparator

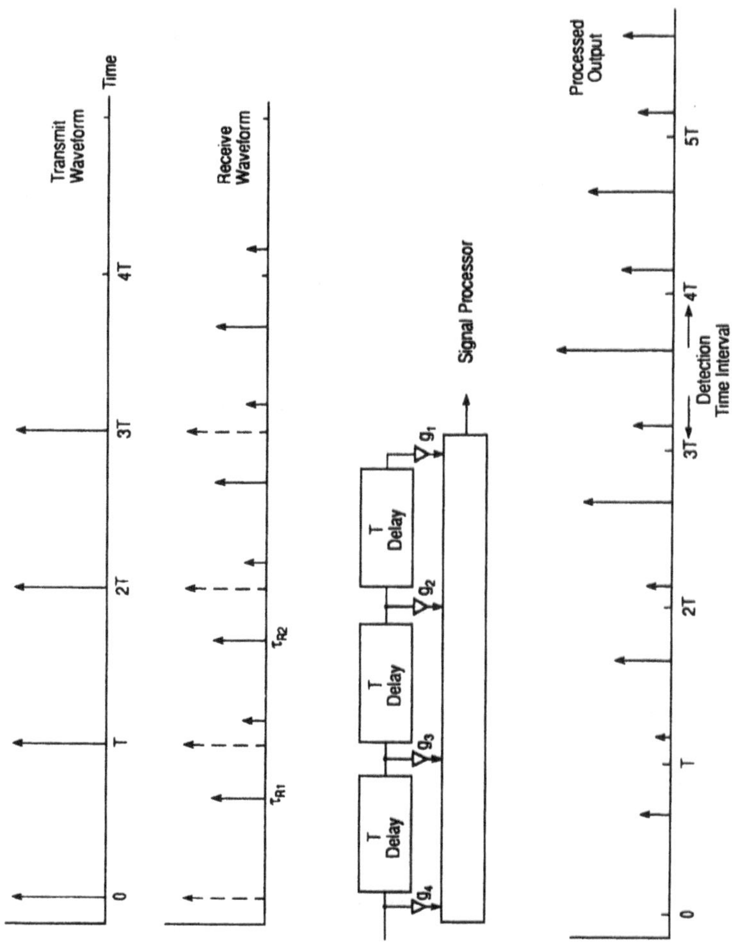

Figure 1.24 Repetitive pulse train waveform and range ambiguities.

output is used to intensity modulate the PPI display. Also, additional alphanumeric data can be generated and displayed. The target angle can be measured by curve fitting a parabola to the individual amplitudes of the returns for the four or more pulses. The parabola peak is used to derive the angle of arrival.

1.8.2 Doppler Ambiguity

To reiterate, a multiple-pulse waveform is used because it increases the transmit energy and thus the maximum detection range. The discrimination between targets and clutter requires long dwell times. Thus, a multiple-pulse waveform also enhances the discrimination of clutter and targets. However, as discussed previously, the multiple-pulse waveform causes range ambiguity. Also, as seen in the discussion of the delay line canceler and Doppler filter bank, the filter responses are periodic and the maximum unambiguous Doppler frequency equals the PRF. Thus, there is an identical response to a reflector whose Doppler frequency equals f_{D1} or $f_{D1} \pm 1/T$ or $f_{D1} \pm 2/T$ etc. Since ground clutter has a large cross section at zero Doppler frequency, that implies targets with Doppler frequency of $\pm 1/T$ or $\pm 2/T$, for example, cannot be detected because their reflection is obscured by that of the ground clutter. Also, if there is weather clutter of a large cross section at a Doppler frequency of f_W, that implies targets with Doppler frequency of $f_W \pm 1/T$ or $f_W \pm 2/T$, for example, cannot be detected because their reflection is obscured by that of the weather clutter. Examples of the Doppler ambiguity are given in Table 1.3.

The first row lists parameters for a typical ground-based search radar. Since the unambiguous velocity is only 29.1 knots and typical aircraft velocities can vary between zero (for a crossing trajectory) and hundreds of knots, it can be seen that Doppler is very ambiguous for a ground-based search radar and is not a useful parameter for data processor tracking. The other rows are parameters that are more typical of tracking radar and have reasonably large unambiguous velocities. However, from the previous section, it can be seen that high PRFs are associated with highly ambiguous range. Thus, there is a trade-off. High PRF

Table 1.3
Unambiguous Doppler

F_0 (GHz)	PRF (Hz)	V_{UNAMB} (m/sec)	V_{UNAMB} (knots)
3	300	15	29.1
3	10000	500	971.3
10	10000	167	323.8

increases the unambiguous Doppler band, but low PRF increases the unambiguous range.

1.9 Transmit Waveforms

The transmit waveform of first generation microwave radar all used repetitive simple pulse waveforms. This was primarily caused by the limitations of the only available transmitter tube, the magnetron. This device was gated on and off to generate a sinusoidal pulse. Current transmitters are usually based on the klystron oscillator. This device can be amplitude and phase modulated. The advantage of the more complex modulation is that the pulse duration and the waveform bandwidth can be independently controlled. The waveform and voltage spectrum of the simple pulse are shown in the sketches of Figure 1.25. The pulse has a duration of τ. The spectrum is centered at the carrier frequency of f_0 and has its maximum value at f_0. The nulls on either side of the peak are at $f_0 \pm 1/\tau$. For this waveform, the range resolution is approximately equal to the pulse duration, τ.

To obtain an increased detection range, with constant dwell time and surveillance volume, for example, it is necessary to transmit with increased average power. If all other waveform parameters are kept constant, this can be done by increasing the transmitted peak power. However, there is a limit to the ability to generate peak power because of dielectric breakdown in the transmitter tubes and in the waveguides. The waveguides are often pressurized in order to maximize the achievable breakdown voltage. Alternately, the PRF could be increased. However, this increases the range ambiguity problem and, thus, the problem of second-time-around clutter. If the pulse duration is increased, the range resolution is degraded, so this is not a desirable method of increasing the maximum detection range. The preferred method to increase detection range without

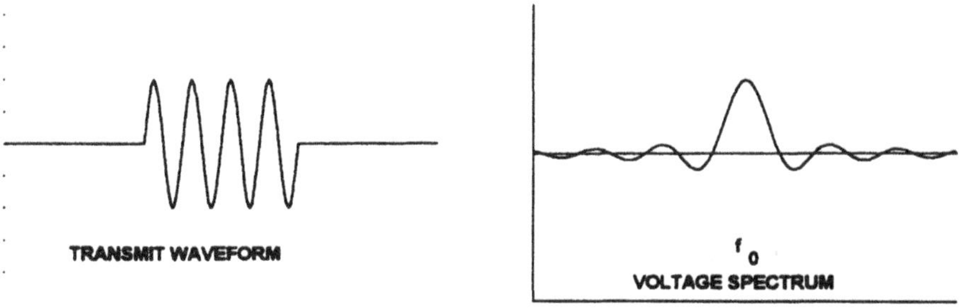

Figure 1.25 Gated sinusoid waveform and voltage spectrum.

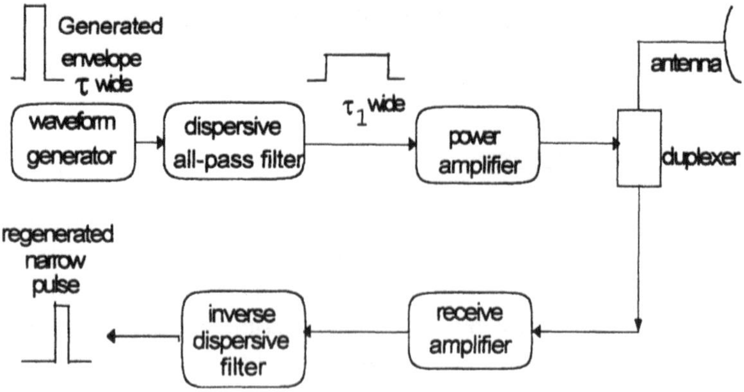

Figure 1.26 Pulse coding concept.

incurring dielectric breakdown or increased range ambiguity while maintaining the range resolution is to use pulse coding.

1.9.1 Pulse Coding Concept

The voltage spectrum indicated in Figure 1.25 shows that the major spectral components of the simple pulse are in phase. Therefore, these components add to a maximum value and generate the waveform peak. If some of these components are phase shifted before transmission, the waveform peak is reduced and will not cause dielectric breakdown. The complete transmit and receive concept is depicted in Figure 1.26. A narrow-width, high-energy pulse is generated. It is then passed through a phase dispersive filter. This filter changes the phase but does not change the amplitude of the spectral components. At the filter output the peak transmit power is low but the pulse is wider. The energy, and thus the maximum detection range, is high. On receive, an inverse dispersive filter is used, causing a phase shift that is the negative of that applied during transmission. Thus, the short-duration, high-resolution waveform is regenerated. Note that the range resolution is not equal to the transmitted pulse width but is much narrower. The range resolution is determined by the width of the initially generated pulse. Because the width of this pulse is not maintained at various points of the implementation but the bandwidth is maintained, it is convenient to express the range resolution in terms of the waveform bandwidth as $(c/2)/B$, as given in (1.3).

1.9.2 Candidate Wavefoms

The previous discussion of a conceptual generation of a phase-coded waveform does not necessarily result in a waveform with desirable properties. As an example,

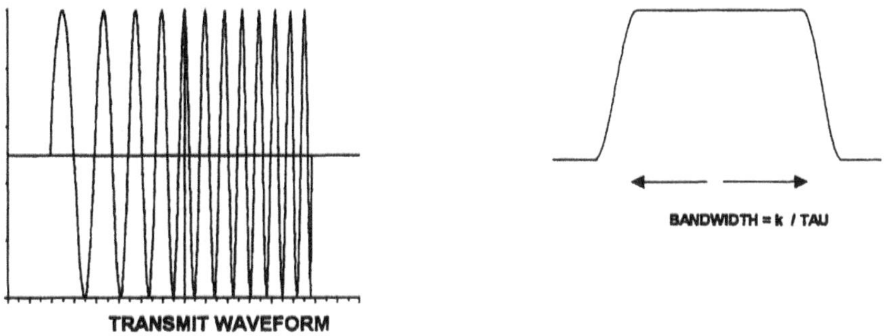

Figure 1.27 Linear frequency modulation waveform.

a desirable waveform should have a constant amplitude envelope voltage in order to generate the maximum possible transmit energy within the peak power constraints of the output amplifier and waveguide. Phase dispersion across the pulse spectrum does not necessarily accomplish this. A better procedure is to use phase changes in time. One common method of doing this is to use continuous phase change as illustrated by the waveform at the left of Figure 1.27. It can be seen that the time between zero crossings is continuously decreasing. This indicates that the instantaneous frequency is continuously increasing. For this waveform, the instantaneous frequency increases linearly with time and the waveform is termed chirp or linear frequency modulation (LFM). The waveform is of duration τ, but the bandwidth does not equal $1/\tau$ as it does for a simple pulse. The spectral form is complicated. An approximation to the spectrum is shown by the right-hand sketch of Figure 1.27. The spectrum is centered at the carrier frequency f_0, and its bandwidth is a multiple of $1/\tau$. The spectrum is approximately rectangular. The equation for the time waveform, $e(t)$, is

$$\begin{aligned} e(t) &= A\cos[\theta(t)] = A\cos[2\pi(f_0 t + kt^2/2)] \\ \theta(t) &= 2\pi(f_0 t + kt^2/2) \end{aligned} \tag{1.49}$$

where A is the constant envelope value. The first portion of the equation, $\theta(t)$, emphasizes that the waveform involves only phase modulation and, as can be seen, the phase varies quadratically with time. Since the instantaneous frequency, f_{inst}, is proportional to the time derivative of the phase,

$$f_{inst} = (1/2\pi)(d\theta/dt) = f_0 + kt \tag{1.50}$$

it can be seen that the instantaneous frequency varies linearly with time. For a pulse width of duration τ, the minimum instantaneous frequency equals f_0 and

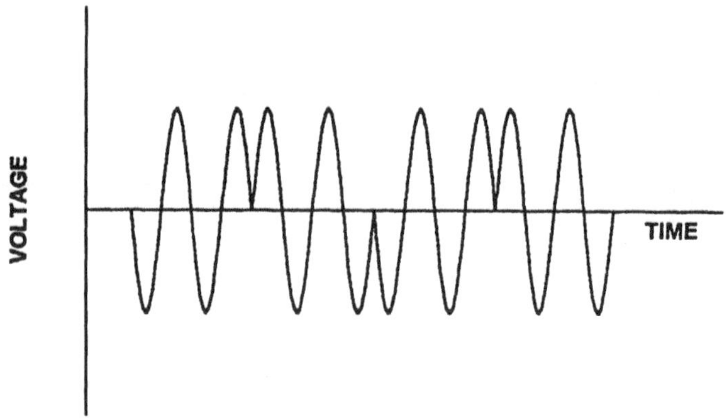

Figure 1.28 Phase coded waveform.

the maximum instantaneous frequency equals $f_0 + k\tau$. The deviation equals $k\tau$, which is approximately equal to the waveform bandwidth.

The disadvantage of this and other pulse compression waveforms is that the compressed waveform has range sidelobes with all the poor properties of sidelobes that have previously been discussed.

Another type of pulse compression waveform is termed discrete phase code, an example of which is shown in Figure 1.28. The generic waveform is made up of segments that are τ in duration. At the start of each segment, the phase can change by 180 degrees. The change can be designed to occur either deterministically or randomly. The waveform bandwidth is approximately the reciprocal of τ, but the total waveform duration is much much greater than τ; so the bandwidth and transmitted energy are also decoupled. There are many possible codings, and the preferred choice is made on the basis of a low-range sidelobe design.

The LFM waveform and the phase-coded waveform differ primarily because of the respective range sidelobes. The receiver for an LFM can be designed to achieve low sidelobes just as an antenna can be designed to achieve low sidelobes. There is no known method of reducing the sidelobes associated with a phase-coded waveform. The achievable sidelobes for the LFM waveform are much lower than that usually attained with a phase-coded waveform, but these low sidelobes are achieved by accepting an SNR loss that degrades the probability of target detection.

1.9.3 Multiple Subcarrier (Frequency Diversity) Waveforms

The detectability of a target in noise or clutter (reverberation) depends upon the SNR and the particular form of the target fluctuation characteristic. The fluctuation loss or increased SNR required for detection as compared to a non-fluctuating target is discussed in Section 1.6.3. As discussed, the fluctuation loss for a Swerling 1 target at a false alarm probability of $1. \times 10^{-6}$ is approximately 1.5 dB for a detection probability of 0.5 and 8 dB for a detection probability of 0.9.

The use of a multiple carrier frequency waveform to reduce the fluctuation loss is well known, and the theoretical improvement has been experimentally verified. This class of waveforms is denoted by some authors as frequency diversity and by others as frequency agility. A conceptual block diagram of a particular diversity waveform and processor is shown in Figure 1.29. As shown by Figure 1.29(a), each pulse of the pulse train is divided into K subpulses and each subpulse is at a separate carrier frequency f_1 through f_K. The transmit spectrum of a pulse is shown by Figure 1.29(b). The spectra are disjoint and

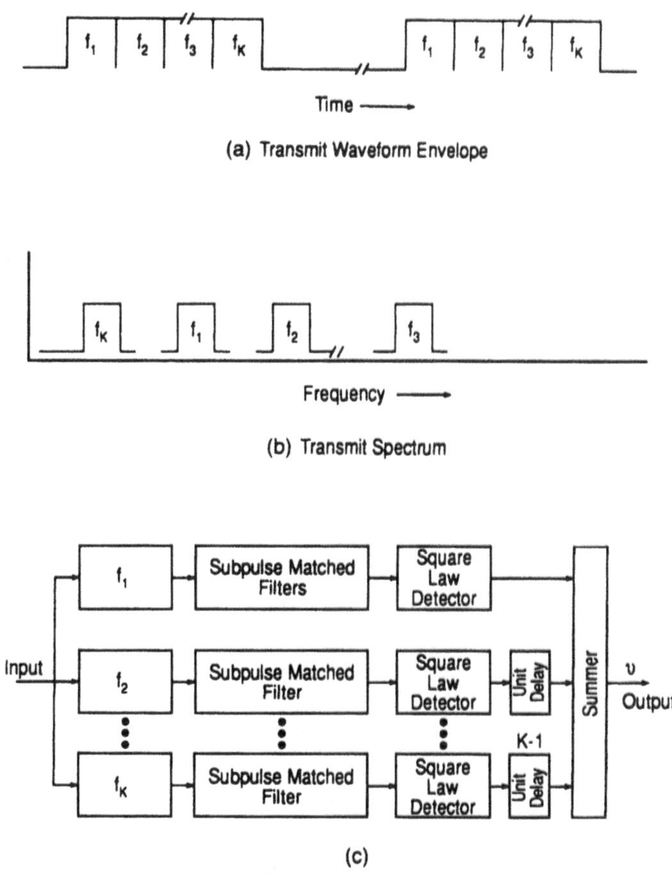

Figure 1.29 Conceptual block diagram of frequency diversity system.

are like parallel radar transmissions. A processor for the waveform is also shown. The processor consists of parallel channels with bandpass filters, pulse compressors, and envelope detectors for each subpulse. For a thermal noise environment, the outputs of the test cells of the individual channels are summed before target present/absent decisions are made. Each of the subpulses can be a coded waveform, and the codes can be different.

A qualitative indication that frequency diversity will decrease the fluctuation loss can be obtained by noting that the total return from a typical target consists of the superposition of the reflections from many individual scattering points. The magnitude of the sum depends upon the relative phasing of these reflections. For some phase conditions, the resultant magnitude is small and the detection probability is small. A change of transmit frequency changes the relative phasing of the reflections. It is improbable that a small target magnitude would occur on all of several different carrier frequencies and the fluctuation loss is decreased.

In addition to its effectiveness in improving target detection, diversity techniques have ECCM advantages against noise jammers. This occurs because in order for a noise jammer to be effective against this waveform, it is necessary that each channel be jammed. This usually requires the jammer to operate in a broadband mode, jamming over a bandwidth of hundreds or thousands of megahertz, because of the difficulty of establishing the carrier frequencies of each of the subchannels. Thus, the jammer is usually expending energy over portions of the band not occupied by the radar and this decreases the jammer's effectiveness. The technique also has improved ECCM capability against false target jammers for several reasons. First is that, just as for the noise jammer, it is necessary to jam each channel. Second is that when the radar transmits a coded pulse, the jammer must attempt to transmit the same coded pulse or the radar receiver substantially attenuates the jammer's false target jamming. When each subpulse is differently coded, the false target jammer's difficulty is substantially increased. In addition, the waveform can improve target tracking by reducing the effect of target glint. Glint refers to the phenomenon that the apparant location of the target return drifts over the target from pulse to pulse depending upon the relative phases of the returns from the target's highlights.

The disadvantage of the technique is that of implementation cost. The implementation is essentially that of K parallel radar with almost K times the cost.

1.10 Tracking Radar Techniques

The surveillance and tracking radar systems' primary function is the prediction of future relative target positions. This can be used to guide an intercepting vehicle to the target's location or to guide an aircraft to an airport runway. The radar methodology is to detect and measure the present and past range and angle locations of targets and, as the usual target is moving, to implement a location prediction technique.

In order to maintain accurate surveillance of a region, it is usually necessary to use several radar to perform this function. Typically, one is a surveillance radar whose primary function is, as its name suggests, to monitor a large volume of space and perform initial target detection. However, the location accuracy obtained from its measurements is usually insufficient to guide an interceptor or land an aircraft. The need to perform rapid surveillance over large regions and to derive highly accurate measurements on targets' locations are almost mutually exclusive. (Another trade-off example.) The primary reason for this is that low error measurement requires long durations of data whereas rapid region surveillance is best done with short-duration measurements. To obtain more accurate location information, the surveillance radar alerts a tracking radar as to the presence of a potential intruder and furnishes an estimate of the intruder's position. The tracking radar monitors a single target and thus can obtain long-duration measurement data. However, just as the surveillance radar's location accuracy is too poor to guide an interceptor to an intruder's location, it is only marginally adequate to alert a tracking radar. That is, because of measurement and prediction errors, the surveillance radar's position estimate is only an approximation of the actual location of the target. Often the error of the surveillance radar's location estimate is large; simply steering the tracking radar to the predicted target location is insufficient to ensure that the target is within the tracking radar's antenna beamwidths. Thus, the tracking radar performs an initial acquisition phase where it searches a small region around the predicted target's location. Obviously, more accurate surveillance radar location information increases the probability that the tracking radar will be able to successfully acquire the target. Thus, the surveillance radar design requires many tradeoffs.

The transition from surveillance (acquisition) radar to the tracker is termed a "hand-off." The tracker initiation phase must be done simultaneously in range and angle. The tracking is done by servo systems operating independently in range and angle.

1.10.1 Range Tracking

The generic technique used for range tracking is termed split gate or early-late gate tracking. It is implemented via a servo system using the range contiguous early and late gates shown in the upper part of Figure 1.30. The servo derives an error signal to drive the late and early gates to a range that brackets the target's range. A typical block diagram is shown in Figure 1.31. The initial portion of

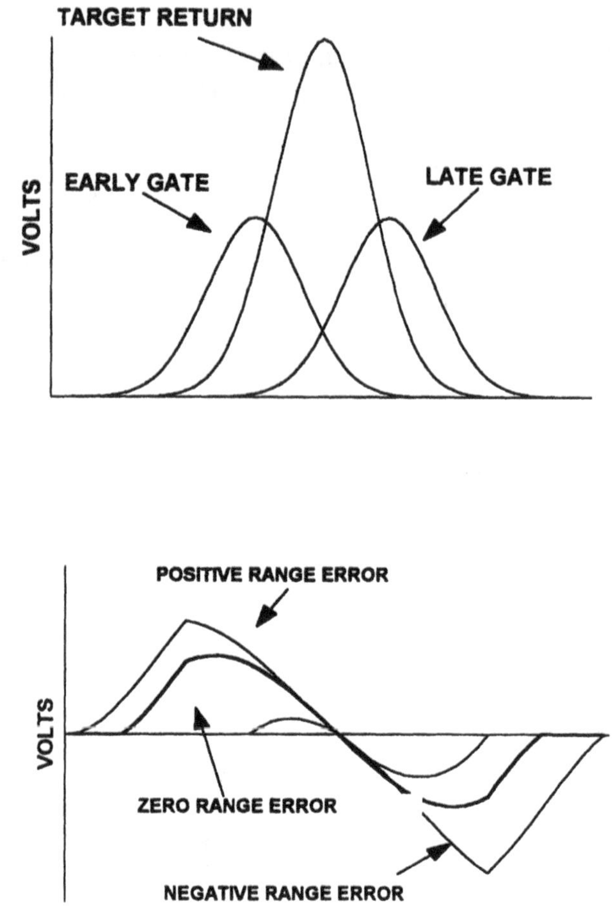

Figure 1.30 Early-late gate range tracking concept.

the block diagram is very similar to that of the surveillance radar. The differences occur primarily after the envelope detector where the range and angle tracking circuitry is implemented. Only range tracking is discussed and indicated in the block diagram. The angle tracking technique is similar in concept.

An example of a technique for generating the early-late range gates starts at the upper left of the block diagram. A track sweep voltage that varies linearly with range-time is generated and subtracted from a second voltage. As will be shown, the value of this second voltage also varies linearly with target range. As shown at the bottom of the figure, the track sweep is a repetitive waveform whose frequency equals the PRF. It is synchronized with the transmitted pulses and starts at zero voltage at the time corresponding to zero range. The subtracter output is fed to a circuit, termed a zero-crossing detector, that generates a pulse

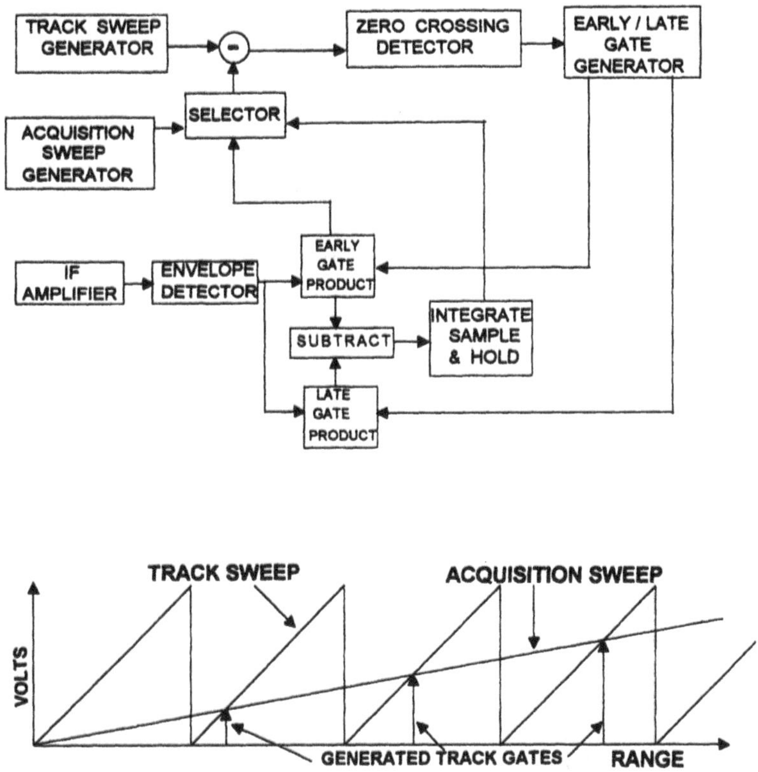

Figure 1.31 Range initiation and tracking block diagram.

when the input voltage is equal to zero. This pulse is differentiated to form positive and negative spikes. These spikes are fed to a pair of multivibrators that are triggered, respectively, by positive and negative spikes. The positive spike generates the early gate, and the negative spike generates the late gate. As will be discussed further, the pair of gates are generated at a time that brackets the target range when the servo error signal equals zero.

The voltage subtracted from the track sweep is generated via a servo loop. The error input to the servo is the difference of target energies contained within the early and late gates. The error signal is computed as the average of the difference of the product of the early gate and the target return and the product of the late gate and the target return.

The waveforms at the lower portion of Figure 1.30 are sketches of the difference of the products for three relative range conditions. These are the subtracter outputs of Figure 1.31. Note that the average value of the subtracter outputs is zero for no tracking error, positive when the the target range is less than the range of the gate's midpoint, and negative when the target range is greater than the range of the gate's midpoint. These subtracter outputs are averaged (integrated) and converted to a "DC" voltage by a special form of lowpass filtering. The special form of lowpass filtering, integrate sample and hold, is required because for most of the pulse repetition period the average output is only due to noise. Only the average output due to the target return should be used for tracking in order to obtain low tracking error. The sample and hold circuit samples the output only during the target return period, averages the waveform, adds the averaged value to prior values, and holds the result until the next PRI. The sample and hold circuit output voltage is the "second" voltage that is subtracted from the linear sweep voltage. The circuit tracks the range of the target since if the early-late gates are not centered on the target, the output voltage changes in the proper sense to correctly reposition the timing of the gates.

A typical range tracker block diagram, as shown in Figure 1.31, is more complicated than indicated by the prior discussion because it is necessary to position the early-late gates in the vicinity of the target range, range acquisition, before tracking can commence and because it is necessary to ensure that the error signal is not perturbed by noise that exists at ranges far from the target range.

The tracker block diagram is explained by first assuming that range acquisition is performed so that the gates generated by the early-late gate generator, at the upper right of the block diagram, are at approximately the target range. These gates and the output of the envelope detector are fed to the early and late gate multipliers. The products are subtracted and fed to an integrating sample and hold circuit. This circuit is a lowpass filter that is gated on only during the duration of the target return. The gating thus eliminates the noise output of the early-late gate subtracter that occurs at nontarget ranges. The subtracter output equals zero when the gates are centered on the target. The subtracter output is positive when the gates are leading the target and negative when the gates are lagging the target. When the input to an integrator is zero, its output value does not change. When the gates are not centered on the target, the output of the integrating sample and hold circuit increases if the gates occur at a range that is less than the target range and decrease if the gates occur at increased range. During track, the integrating sample and hold output is subtracted from the track sweep voltage generated by the circuit shown in the upper left of the block diagram. The frequency of the tracking sweep is the PRF, and its output varies linearly from zero to a maximum positive value. This maximum value equals the output value of the integrate sample and hold circuit when the target is at maximum range. The subtracter output is fed to the zero-crossing detector, and this circuit generates a pulse when the input voltage passes through zero. Thus, during track a gating pulse at the target range-time is generated at the PRF. Note that the range of this generated pulse increases and decreases as the target range increases or decreases since the output voltage of the integrate sample and hold changes with target range.

The range acquisition is accomplished by subtracting the acquisition sweep from the track sweep generator rather than subtracting the integrate sample and hold circuit output. The output of the early gate multiplier is used for selection. When the target is not being tracked, this output is zero and the acquisition sweep is selected. When the target is in track, the output is nonzero and the integrate sample and hold circuit output is selected. Both track and acquisition are linear sweeps varying from zero to the same positive voltage. They differ only as to sweep rate. The frequency of the track sweep equals the PRF whereas that of the acquisition sweep is much slower. The subtraction of the acquisition sweep from track sweep is depicted at the lowest portion of the figure. It can be seen that a zero subtraction value occurs at increasing range with increasing time, so the acquisition range gate occurs at increasing range.

The relative sweep frequencies of the acquisition and track sweeps are not shown to scale in the figure. As an approximation to the required acquisition sweep frequency, note that in order not to miss the target, the range gate should step approximately one range cell or less per PRI. For a waveform PRF of 1000 Hz, the PRI equals 1000 μs. If the waveform bandwidth equals 1 MHz, so that there are 1000 range cells within the maximum instrumented range, the acquisition sweep period equals at least $1 \times 10^6 \mu s$ (1 sec). In addition, this value is predicated on detecting the target using only a single pulse. In order to increase the detection range, it might be desirable to decrease the acquisition sweep frequency. If as an example, 8-pulse integration is desired, the sweep period should be increased to 8 sec. Thus, range acquisition can take a long time even without considering the additional complication due to angle tracking.

Also, note that the detection range during track is usually much greater than that during acquisition. The effective number of pulses integrated during track equals approximately the waveform PRF multiplied by the bandwidth of the integrate sample and hold filter. This bandwidth is chosen so as to be able to track the target's expected maneuvers in terms of rate of range changes. Usually, this bandwidth is relatively small and tens or hundreds of pulses are integrated.

1.10.2 Tracking Radar Angle Tracking Techniques

Several types of angle tracking techniques have been implemented. The main two techniques that are currently in use are conical scan and monopulse. The single-dimension version of conical scan is that the tracking is implemented by

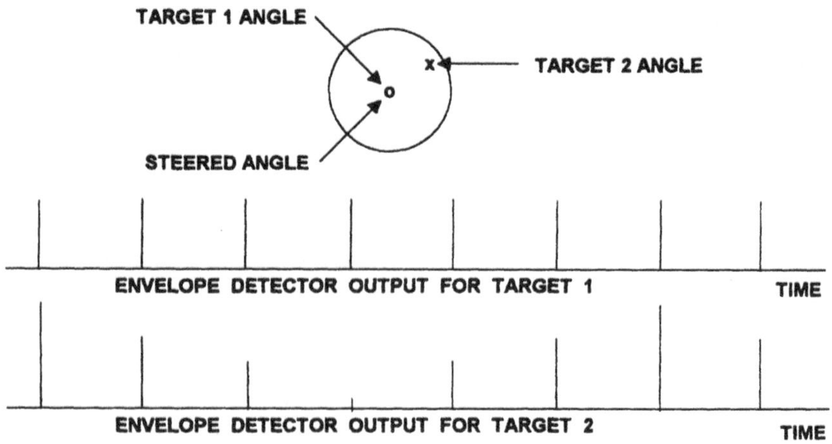

Figure 1.32 Conical scan technique.

steering the antenna slightly to the left of the antenna boresight and then slightly to the right and comparing the reflected target energy received during the two periods. The energy differential is used as the input to a servo to reposition the antenna boresight so that the antenna is steered to the target angle. This concept is identical to that of range tracking using the early and late gates. Two-dimensional angle tracking could be done by combining an up-down steering in addition to the left-right resteering. Instead, the antenna is steered in a conical path (nutation) about the boresight at a constant number of revolutions per second (typically 30 rps). The upper portion of Figure 1.32 depicts the conical scan geometry. The circle depicts the locus of the center of the antenna beam. The center point, denoted by the "o," is the point such that the antenna gain is constant independent of where the beam is steered on the circle. The point "o" is denoted as the antenna boresight. The antenna gain to a target depends upon the target angle relative to the boresight and the timing of the conical scan period. The envelope detected output is shown for two target locations. The first target location is at the antenna boresight. The second target's angular location relative to the antenna boresight is denoted by the "x." The vertical lines represent the output pulses and the pulses occur at the target's range. When the target is at the beam boresight, the antenna gain in the target direction is constant during the scan period and, as shown by the first trace of the figure, all the pulses have the same amplitude. When the target is not at the beam boresight, the antenna gain in the direction of the target varies during the scan period. As shown by the second trace, the amplitude varies at a sinusoidal rate induced by the nutation of the antenna. The target's angle relative to boresight can be deduced from the time of occurrence of the maximum pulse voltage. The ratio of the maximum voltage to the minimum voltage indicates the target's angular displacement from the antenna boresight.

A defect of this technique is that the pulse voltages vary not only as a function of the angular position of the target but also as a function of target reflectivity. As will be discussed in greater detail later, the reflectivity of most aircraft targets can vary by orders of magnitude over periods of hundreds of milliseconds in a nonpredictable manner. As the conical scan technique acquires its angle tracking data over comparable time periods, the reflectivity variation sometimes causes large errors in the angle tracking. A method of avoiding these errors is the monopulse technique. As the name suggests, the monopulse technique acquires its angle information on a single pulse rather than the multiple pulses of the conical scan technique.

The single-dimension introduction to the conical scan technique considered sequentially steering the antenna to the left and then the right of the boresight in order to measure the target's relative angle. The monopulse technique operates similarly but uses simultaneous beams rather than sequential beams. Since it measures the relative target angle based on the information of a single pulse, it is less sensitive to reflectivity fluctuations. It is not totally insensitive to the fluctuations because the target is not a point reflector but is distributed over space and, as the target reflectivity fluctuates, the apparent center of reflectivity moves. In addition, though it seems totally nonintuitive, the center of reflectivity can move completely outside the object's physical dimensions.

The monopulse technique is implemented using multiple-feed antennas. There are three variants of the technique. Though, in practice, the technique is used to track in both elevation and azimuth, the discussion will usually be in terms of single angle tracking. One variant, termed amplitude monopulse, is shown at the left of Figure 1.33. It uses a single antenna, shown as a parabolic reflector, and two offset feeds. Because the feeds are offset from the center of the

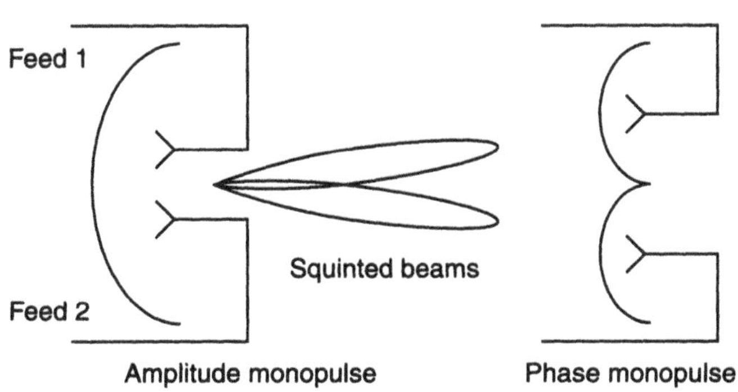

Figure 1.33 Amplitude and phase monopulse antennas.

dish, the two feeds create two nominally identical beams that are steered to either side of the perpendicular. The beams are called squinted beams because their amplitude patterns are identical, except for the steered angle. The feeds are closely spaced and nominally have the same phase center. Because of the (approximate) common phase center, a reflected waveform can differ in amplitude as received by the two beams, but the phases are identical. The amplitude difference of the two beams depends upon the reflector's spatial angle. Thus, the received waveforms of the two beams are subtracted and the subtracter output is used as an error signal for a servo system that drives the antenna until the target angle is at the crossover of the beams. The tracking can be extended to both elevation and azimuth angles using four feeds with two of them in the plane perpendicular to the paper.

Another monopulse variant is the phase monopulse depicted at the right of the figure. This technique uses two abutted antennas with a feed for each antenna. This generates two beams whose amplitude patterns are identical, and overlapping, but with differing phase centers. Thus, a reflected waveform can differ in phase as received by the two beams, but the amplitudes are identical. The phase difference of the two signals depends upon the reflector's spatial angle. Therefore, the received waveforms of the two beams are phase compared and the phase difference is used as an error signal for a servo system that drives the antenna until the target angle is at the crossover of the beams. The tracking can be extended to both azimuth and elevation angles using four feeds, with two of them in the plane perpendicular to the paper.

The third variant combines the two techniques as a phase-amplitude monopulse. Its advantage is that it is implemented with two feeds, rather than four, for two-dimensional angle tracking.

1.10.3 Track-While-Scan Data Processing

Phased array radar are usually used in a TWS mode. That is, the data processor is a computer system and it is used to maintain simultaneous track on many vehicles. The two distinct phases of the tracking process are initiation and maintenance. Initiation refers to the step of target detection and assigning a track channel to the vehicle. Maintenance, or track update, is the process of combining newly measured detection data, angle and range, with a current track. Since many tracks are being performed simultaneously, new data must be associated with the correct track and then the track is updated with this data. If the newly measured data is associated with the wrong track, the track of two different vehicles is degraded. Erroneous association is especially probable when the tracks of two vehicles cross. Also note that if the radar is not measuring an elevation angle, two vehicles can have crossing trajectories even when they are physically displaced by altitude differences of thousands of feet. Thus, radar design aspects that are seemingly far removed from the data processor can have significant impact on it.

Tracking is performed under the assumption that the target's trajectory will be continued at the same velocity and acceleration, for example. Each track uses logic based on the estimated multidimensional velocity and acceleration to predict the next target measured position. A new position measurement is compared to the set of predicted positions of all tracks in order to perform association. One possible association method is that the associated track is that with the least difference between the current measurement and the prediction set. Another more restrictive method is to associate the measurement with an established track only if the error between the measurement and the predicted position is small and "small" needs to be defined. TWS association and tracking is a complicated system trade-off. First, as mentioned several times previously, it is mandatory that false tracks do not overload the data processor. However, it is important that all targets are properly tracked. These two requirements cannot be simultaneously satisfied. Any method that tends to eliminate false tracks also tends to eliminate true tracks. Second, there are many possible methods of performing the various tracking tasks. Each has its own advantages and disadvantages, and it is difficult to choose between them.

False tracks are initiated in at least two different manners. One is a track caused solely by noise. Second is a track caused by a target maneuver. When there is a target maneuver, the newly measured target position can be significantly displaced from the predicted position. If the association algorithm is set up to accept a large error between prediction point and measurement point, it becomes easy to erroneously associate crossing tracks. If the association algorithm requires a small error between prediction and measurement, the maneuver causes a nonassociation and a new track is begun. Thus, there will be two tracks for the target. One track assumes that the previous trajectory is continued, but that detection did not occur on this scan. A second is based upon the vehicle track's trajectory after the maneuver. This is termed a bifurcated track and demonstrates that the tracking algorithms must develop a method of dropping old, no longer valid, tracks. As an example, one procedure is to drop any track where the target has not been detected for the previous, say, five scans. Clearly the number of nondetection scans for track dropping is a significant trade-off area. If the number is set very high, old tracks are carried for a large number of scans and the size of the tracking computer must grow to avoid track overload. However, if the number is set low, there is an increased danger of dropping proper tracks when the target is undetectable due to transient low cross section or clutter competition, for example. Some algorithms and trade-offs are discussed in the next section.

1.10.4 Track-While-Scan Track Initiation

Track initiation stategies (algorithms) are essentially a trade between properly initiating tracks on targets and not initiating tracks on noise. The example of Section 1.4 showed that if the signal processor detects targets with a false alarm probability of $1. \times 10^{-6}$ an average of 8.4 false targets are delivered to the data processor per scan. Decreasing the number of false target declarations by decreasing the false alarm probability decreases the detection probability. An alternate procedure to decrease the number of false tracks initiated due to noise is to initiate track only if a target is detected on two successive scans, or three successive scans, or two out of the three most recent scans, for example. This substantially decreases the probability of initiating tracks on noise with only a small decrease in the probability of initiating tracks on targets. This section develops the properties of some initiation algorithms.

Track initiation is a random event that depends upon the per-scan probability of detection, p, and the number of target illuminations (scans past the target). The important parameters are the particular scan on which track is initiated, denoted as k_{int}; the probability of initiating track on the kth scan, denoted as $P_{\text{II}}(k)$; and the probability of track initiation on or before the kth scan, denoted as $p_{\text{I}}(k)$. The simplest algorithm initiates track based upon a single detection, the 1 of 1 algorithm. For this algorithm, the probability of track initiation on the first scan, $P_{\text{II}}(1)$, equals p. Track is initiated on the second scan if the target is not detected on the first scan but is detected on the second scan. The probability that track is initiated on the second scan, $P_{\text{II}}(2)$, equals $(1-p)p$. Thus, $p_{\text{I}}(2)$ equals the sum of the probability of initiation on the first scan plus the probability that initation occurs on the second scan, which equals $p+(1-p)p$. Similarly, $p_{\text{I}}(3)$ equals $p_{\text{I}}(2)$ plus $P_{\text{II}}(3)$. This latter event occurs only if the target is not detected on scan one or two but is detected on scan three. The probability of this event is $(1-p)^2p$. Thus, the probability of initiation by scan three equals $p+(1-p)p+(1-p)^2p$, which can be regrouped as $p[1+(1-p)+(1-p)^2]$. This can be recognized as the start of a geometric series. Thus, the probability of track initiation by the kth scan $p_{\text{I}}(k)$ equals

$$p_{\text{I}}(k) = p(1-p)^{k-1} \tag{1.51}$$

This equation is evaluated and graphed in the top portion of Figure 1.34 for several values of per-scan probability of detection. It can be seen that the initiation probability approaches unity for every value of detection probability. This implies that every resolution cell is eventually initiated by noise, so it is mandatory that the tracking algorithm incorporate a method of dropping tracks!

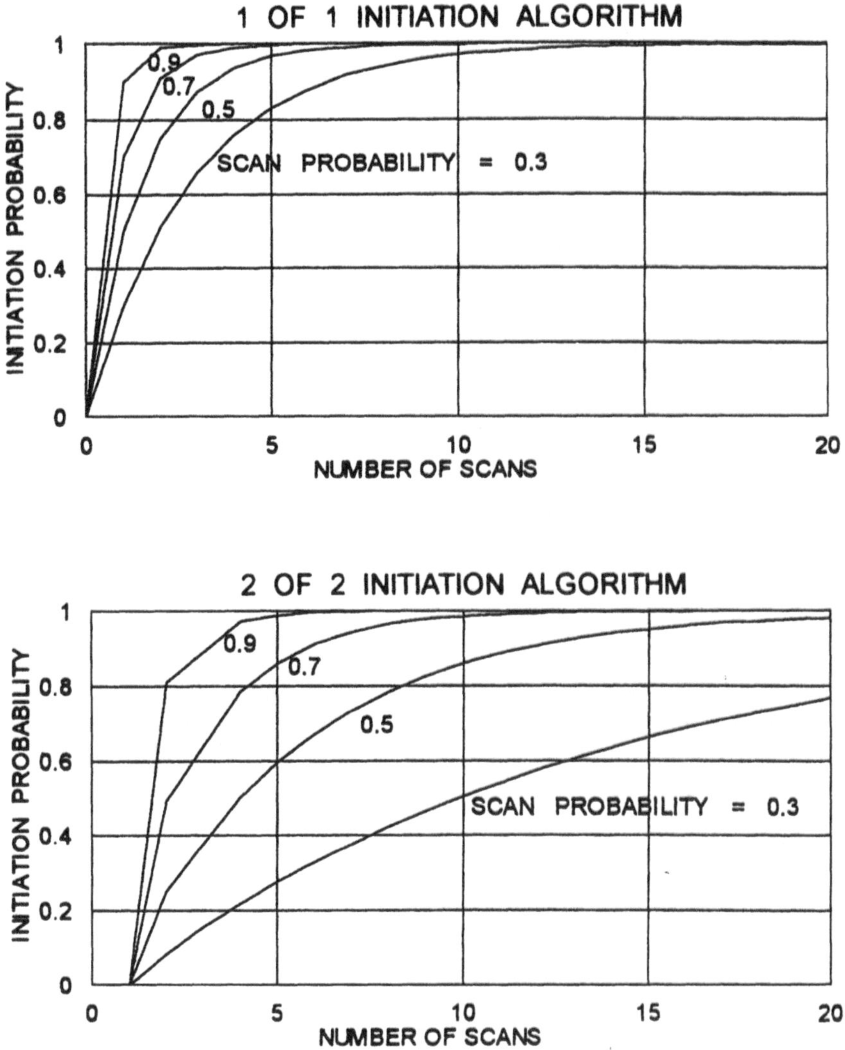

Figure 1.34 Probability of track initiation algorithms.

The probability of initiation, by noise or target, is decreased by considering alternate algorithms. One alternate is that track is initiated only when a target present declaration occurs on two consecutive scans, the 2 of 2 initiation algorithm. For this algorithm, the probability of track initiation equals zero on the first scan; equals p^2 on the second scan; and equals $2p^2 - p^3$ by the third scan. The last occurs because a track is established by scan three in several distinct ways:

1. Detect on all three scans;
2. Detect on scans one and two but not on scan three;
3. Detect on scans two and three but not on scan one.

The probability of event (1) equals p^3. The probability of event (2) equals $p^2(1 - p)$, where $(1 - p)$ is the probability of nondetection. The probability of event (3) is the same as that of event (2). The probability of initiation by scan three is the sum of the probabilities of the three events, which equals $2p^2 - p^3$. Similar computations can be performed to compute the probability of initiation by an arbitrary scan number. However, as will be shown in a later section, the computations are much simpler when performed using state matrix equations. The initiation probability for this algorithm is plotted at the bottom of Figure 1.34. Comparing the two algorithms, it can be seen that the 1 of 1 algorithm initiates faster but at the expense of more false noise tracks. Also, though the initiation probability increases monotonically with per-scan probability of detection, it is not easy to specify a minimum acceptable probability of detection. Specifying a minimum acceptable detection probability for track initiation is necessary for radar design since this value specifies a minimum acceptable SNR, which in turn determines the required radar parameters such as power and antenna aperture for a desired angular coverage, update rate, and maximum range. As will be discussed, the alternate measure of performance based upon k_{int} is a better indicator of relative performance of the various algorithms.

The particular scan on which track is initiated is a random variable. Because the probability density function of this random variable is complicated, it is convenient to compute the average initial track initiation scan only. Denote it as k_{AVE}. Then k_{AVE} can be computed as

$$k_{\text{AVE}} = \sum_{k=1,\infty} k P_{\text{II}}(k) \tag{1.52}$$

For the 1 of 1 algorithm, it can be shown that

$$k_{\text{AVE}} = 1/p \tag{1.53}$$

where p denotes the per-scan target declaration probability, either true target detection, p_{D}, or the false alarm probability, p_{F}. For a false probability of 1×10^{-6}, the average number of scans until track initiation of a specific resolution cell by noise alone equals $1. \times 10^6$ scans. For the previously considered example of 8.4×10^6 resolution cells, this evaluates to an average of 8.4 false tracks initiated per scan.

For the 2 of 2 algorithm, it can be shown that

$$k_{\text{AVE}} = (1 + p)/p^2 \qquad k_{\text{AVE}} \approx 1/p_{\text{F}}^2 \tag{1.54}$$

and for the same 8.4×10^6 resolution cells, the number of false tracks is reduced to 8.4×10^{-6} per scan. Thus, the change of initiation algorithm substantially alleviates the false track initiation problem, but target track initiation is degraded. The relative target track initiation of the two algorithms can be compared using the ratio of (1.54) and (1.53)

$$(k_{\text{AVE}} : 2 \text{ of } 2)/(k_{\text{AVE}} : 1 \text{ of } 1) = (1 + p)/p \tag{1.55}$$

For a per-scan target detection probability of 0.5, the ratio evaluates to 3. Thus, on average, the 2 of 2 algorithm requires three times as many scans to initiate track on a target as does the 1 of 1 algorithm. If the per-scan target detection probability equals 0.25, the ratio evaluates to 5. For the 2 of 3 algorithm, it can be shown that

$$\begin{aligned} k_{\text{AVE}} &= (1 + 2p - p^2)/(2 - p)(1 + p) \\ k_{\text{AVE}} &\approx 1/2p_F^2 \end{aligned} \tag{1.56}$$

Thus, the 2 of 3 algorithm initiates false tracks at twice the rate of the 2 of 2 algorithm, but this rate is still quite small. The relative target track initiation of the two algorithms can be compared via the ratios of (1.56) and (1.53). For a per-scan target detection probability of 0.5, the ratio evaluates to 7/9. A more complete relative evaluation of several track initiation algorithms is given by the curves of Figure 1.35. In addition, the figure contains a tabulation of the average number of scans required to initiate on noise for a false alarm probability of $1. \times 10^{-3}$.

In general, initiation algorithms are termed as *M* of *N*. The 2 of 2, 2 of 3, and 2 of 4 algorithms are examples of *M* of *N* algorithms with *M* equal to 2 and *N* ranging between 2 and 4. From the graphed results it can be seen that track initiation occurs earlier as *N* increases. However, from the tabulated results it can be seen that the number of scans until initiation by noise also decreases as *N* increases. This result is true in general and again illustrates the trade-off between high probability of target detection and low probability of false detection.

Comparing the curves of average initiation time versus per-scan probability of detection, it can be seen that there is a knee in the curves. Initiation times increase rapidly if the probability of detection is less than about 0.4 and is at about its minimum possible value if the probability of detection exceeds about

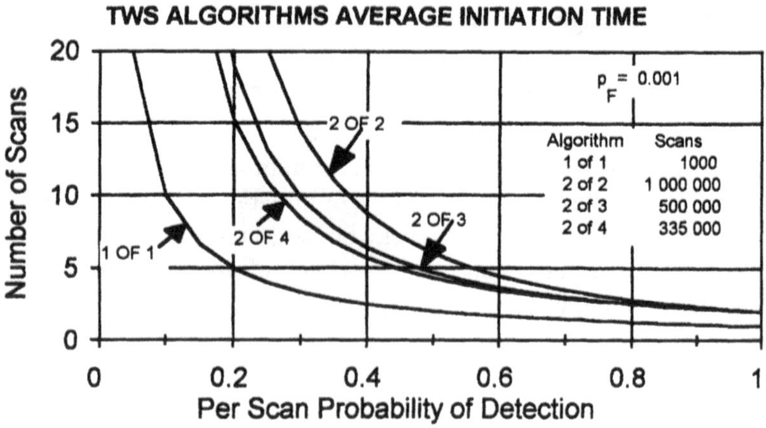

Figure 1.35 TWS algorithms average initiation time.

0.7. Thus, the minimum acceptable detection probability for radar design is approximately 0.7.

The previous results are approximations because of two significant simplifying assumptions. One is that the per-scan probability of detection is a constant value for all scans. This is approximate because target motion causes a range change and the approximation error tends to increase with increased target velocity. A second problem is that the analyses assume that if a target present decision is made for a particular resolution cell on a scan, it is adequate to monitor only that cell on successive scans. Usually the data processor needs to also associate detections in adjacent cells for successive scans to account for target motion. The size of the increased acceptance region depends upon the maximum expected target velocity. The main effect of the increased window is that the probability of false track initiation is increased. For the 2 of 2 algorithm, the probability of false initiation increases by a factor approximately equal to the number of resolution cells in the acceptance window.

1.10.5 Track-While-Scan Track Maintenance

Tracking is commonly performed using an algorithm denoted as α, β tracking. Though the tracking is usually done in three dimensions, for simplicity, it is described in terms of a single dimension, x.

After the $n - 1$ scan, the set of all previously measured data is used to predict the target locations for the nth scan for all targets in track. One use of the set of predicted positions is to associate n scan target measurements with the correct track. It is also used in the tracking algorithm. Denote the predicted position for the nth scan as $x_{\mathrm{P}}(n)$ and the position measurement on the nth scan as $x(n)$. Since the measurement is made in the presence of noise, the actual target position almost always differs from the measured position. The old measurements are combined with the new measurement to estimate the actual position of the target. The estimate of the true position (the smoothed position) on the nth scan is denoted $x_{\mathrm{S}}(n)$. The tracking algorithm is based on a constant velocity assumption. Thus, it is assumed that the target motion is due only to a constant velocity, and the measurements are combined as

$$\begin{aligned} x_{\mathrm{S}}(n) &= x_{\mathrm{P}}(n) + \alpha[x(n) - x_{\mathrm{P}}(n)] \\ v_{x\mathrm{S}}(n) &= v_{x\mathrm{S}}(n-1) + \beta[x(n) - x_{\mathrm{P}}(n)]/T_{\mathrm{U}} \end{aligned} \tag{1.57}$$

where $v_{x\mathrm{S}}(n)$ denotes the smoothed velocity for the nth scan, $v_{x\mathrm{S}}(n-1)$ denotes the smoothed velocity for the prior scan, and T_{U} denotes the data update period. The α, β algorithm is extended to three dimensions using (1.57) separately for the x, y, and z dimensions. A further extension is to the α, β, γ algorithm. This assumes the target motion is described by a constant acceleration trajectory so that a third smoothing equation for acceleration is added to (1.57).

A complicating factor for the tracking algorithm is that the assumption of constant target velocity or acceleration seems reasonable for a Cartesian coordinate system but the radar measurements are made in the polar coordinates of ρ, θ, ϕ. Thus, trigonometric coordinate conversion is required and this can be computationally intensive.

An additional aspect of track maintenance is the track association and track dropping problems. The usual method of association is to center a three-dimensional acceptance gate on each predicted target position. If a new measurement is within a gate, it is associated with the track. If a new measurement is not associated with any track, a new track is initiated. The size of the acceptance gates is a trade-off area. Small gates are required to decrease the potential problems induced by targets on crossing trajectories. However, small gates have an increased tendency to miss maneuvering targets and thus create bifurcated tracks. A compromise is to use small gates initially and increase the gate size if a measurement does not associate with any track. However, a new question then becomes, what are the proper sizes of the small and large gates?

One way of deciding whether to drop a track is to associate a quality number with each track. The value of the quality number depends upon how frequently and how well measurements are associated with the track. As an example, if a new measurement is associated with a track via a small gate, the quality number is increased by two units. If a new measurement is associated via a large gate, the quality number is increased by one unit. If no measurement is associated with the track, the quality number is decreased by one unit. When the quality number decreases below a specified value, the track is dropped.

1.11 Summary and System Trade Areas

The primary function of a radar system is to use position measurements to predict future target locations. The types of radar used to achieve this objective are search, track, and TWS. The radar system operates by monitoring a volume, usually by repetitively scanning the volume, and making a target present or absent decision for each resolution cell on each scan. A major limitation on the design of the radar is the finite number of available track channels used to track and perform position prediction. To avoid overloading the available track channels, a maximum allowable false alarm probability for each resolution cell is of the order of $1. \times 10^{-6}$. To achieve satisfactory scan-to-scan target tracking, the surveillance radar should be designed for a probability of target detection of approximately 0.7 at its maximum range.

The maximum detection range depends primarily upon the average transmit power, the track update (scan) period, the antenna area, the azimuth and elevation interrogation angles, and the target reflectivity and fluctuation characteristics. The maximum range depends indirectly on the transmitted pulse width, PRF, and carrier frequency, for example. In a thermal noise environment the maximum range varies as the fourth root of the radar parameters; in a noise jamming environment, it varies as the square root of the parameters.

Several different types of the transmit waveforms can be used. Multiple carrier frequency waveforms can be used to decrease the fluctuation loss. Pulse phase modulation, such as LFM, is used for independent control of the transmit energy (pulse duration) and the range resolution (signal bandwidth).

The fluctuation loss occurs because two targets with the same average radar cross section can have substantially different detection probability. The difference in detectability between a nonfluctuating target and a fluctuating target is termed the fluctuation loss. The value of the loss increases with the value of detection probability. The loss can be reduced using a frequency diversity waveform.

The maximum detection range depends upon the average transmit power and one of the parameters that affects the average power is the pulse width. For a simple pulse, the range resolution cell width is a linear function of the pulse width. Thus, increasing the transmit average power by increasing the pulse width causes an increased range resolution cell and decreased range resolution. Increased average transmit power without degrading the range resolution can be achieved using pulse coding such as LFM.

Often, it is necessary to be able to detect targets in the presence of clutter. The primary methods of doing this exploit the Doppler characteristics of moving targets and the polarization characteristics of the reflected energy. The goal of Doppler processing is to attenuate the clutter while not significantly attenuating the target. One form of processor is the delay line canceler (MTI). It works well against ground clutter but poorly against wind-blown clutter such as rain. A better, though much more expensive, processor is the Doppler filter bank. The Doppler bank is an approximation to an optimum processor. The optimum processor increases the per-pulse SNR by a factor of N for an N-pulse train. A poor property of all Doppler processors is that they have blind speeds. These occur at radial velocities such that the Doppler frequency is an integral multiple of the PRF. The blind speeds are a significant problem especially because sometimes a target's trajectory is such that the target is not competing with clutter in particular regions. For these regions, using a Doppler processor can change an easily detectable target into a nondetectable target. This problem can be avoided using a noncoherent integrator in parallel with the Doppler processor. Doppler processors add the multiple pulse returns before envelope detection. The noncoherent integrator adds the multiple pulse returns before envelope detection and thus are not sensitive to the target Doppler. However, noncoherent processing has a loss because its SNR gain for an N-pulse train is less than N.

1.11.1 System Trade Areas

There is no best solution! The design of a radar is a constant trade-off of achievable properties. It usually happens that changing a design variable to improve one desired property degrades another desired property. A prime example is that increased maximum range can usually be achieved using parallel channels. However, this entails increased cost, which is undesirable. Another example is that the surveillance volume can be sequentially scanned with a small frame time to improve tracking of maneuvering targets, but this degrades clutter rejection, which is improved by long dwell times.

Angle scanning can be done using a single narrow beam, with azimuth beamwidth approximately equal to the elevation beamwidth, using either a continuous scan or a step scan mode. Alternately, multiple parallel narrow beams can be used on receive. This tends to increase dwell time without increasing frame time and therefore can improve clutter rejection by Doppler processing. However, implementation cost increases. Somewhat similar increased dwell time without increasing frame time can be achieved using a fan beam where the elevation beamwidth is much greater than the azimuth beamwidth. This avoids the increased cost of parallel receive beams, but the received clutter cross section is often increased because of the increased footprint of the beam on the ground or on the weather clutter.

In addition to the pulse modulation choices, another important waveform variable is the PRF. Low PRFs have a large unambiguous range but a small unambiguous Doppler region. High PRFs have a small unambiguous range but a large unambiguous Doppler region.

There are many trade-offs for the TWS data processor. One is the choice of algorithm used for target initiation. Different algorithms have differing probabilities of initiating false tracks and initiating true tracks and the number of scans required to initiate a track. The choice of algorithm is made commensurate with the size of computer used to implement the data processor. Bigger, more expensive computers always achieve improved performance. A similar choice, and its effect on the required computer size, is the size of acceptance windows used on successive scans to account for target motion during track initiation. Too small a window rejects tracks and too large a window causes many false tracks.

Range resolution depends upon the transmitted signal bandwidth. Angular resolution depends upon the antenna size. Increased resolution allows better resolution of closely spaced targets and also tends to decrease the clutter cross section. Decreasing the resolution cell size can be done by increasing the bandwith and/or antenna size. Either increases implementation costs.

All resolutions also depend upon the SNR. For very high SNR, arbitrarily closely spaced targets can be resolved using so-called super-resolution techniques. Super-resolution is related to measurement accuracy. Range measurement accuracy varies inversely with bandwidth and SNR. Similarly, angle accuracy varies directly with the beamwidth and inversely with the SNR.

There are several types of potential ambiguities. Some are induced by sidelobes. Examples are angle ambiguity due to the antenna sidelobes and range ambiguity due to the sidelobes of pulse compression waveforms. Multiple pulse waveforms are used to increase the maximum detection range by increasing the average transmit power and to improve the discrimination against clutter by increasing the dwell time on target. The defect of the technique is that it causes a range ambiguity due to second (or third, for example) time around returns from far-range targets. Far-range targets are incorrectly identified as near-range targets. Second-time-around clutter can also cause increased clutter rejection problems due to the increased clutter cross section. In addition, as will be discussed later, sometimes there are advantages to using varying PRF waveforms for clutter rejection since the varying PRF significantly reduces the blind speed problem. However, second-time-around clutter cancellation is difficult unless the PRF is constant.

Angle tracking is done by either conical scan or monopulse techniques. Conical scan has increased angle tracking error due to pulse-to-pulse target cross-section variation. Monopulse is a more expensive implementation.

1.12 Bibliography

The following chapters discuss most of the topics of this chapter in more detail. Also, there are many books that expand significantly on the material presented in this introductory text. Some of these additional references are listed.

Barton, D. K., *Modern Radar System Analysis*, Norwood, MA: Artech House, 1988.

Cook, C. E., and M. Bernfeld, *Radar Signals*, New York: Academic Press, 1967.

DiFranco, J. V., and W. L. Rubin, *Radar Detection*, Englewood Cliffs, NJ: Prentice-Hall, Inc., 1968.

Levanon, N., *Radar Principles*, New York: John Wiley & Sons, 1988.

Nathanson, F. E., *Radar Design Principles*, New York: McGraw-Hill Book Co., 1969.

Nitzberg, R., *Adaptive Signal Processing for Radar*, Norwood, MA: Artech House, 1992.

Papoulis, A., *Signal Analysis*, New York: McGraw-Hill Book Co., 1977.

Skolnik, M. I., Ed.-in Chief, *Radar Handbook*, New York: McGraw-Hill Book Co., 1970.

Skolnik, M. I., *Introduction to Radar Systems*, New York: McGraw-Hill Book Co., 1980.

Whalen, A. D., *Detection of Signals in Noise*, New York: Academic Press, 1971.

2

Waveforms and Matched Filtering

A general expression for a real-valued waveform, denoted $e(t)$, is

$$e(t) = r(t)\cos[2\pi f_0 t + \theta(t)] \tag{2.1}$$

where $r(t)$ is the envelope modulation and $\theta(t)$ is the phase modulation. Both $\theta(t)$ and $r(t)$ are real valued and the latter is strictly positive. Often analyses are simplified by considering complex notation. This is based on the relationships

$$\begin{aligned} \cos(\phi) &= [\exp(j\phi) + \exp(-j\phi)]/2 \\ \sin(\phi) &= [\exp(j\phi) - \exp(-j\phi)]/(2j) \\ \exp(j\phi) &= [\cos(\phi) + j\sin(\phi)] \end{aligned} \tag{2.2}$$

Substituting the first equation of (2.2) into (2.1) gives

$$\begin{aligned} e(t) &= r(t)\{\exp(j[2\pi f_0 t + \theta(t)]) + \exp(-j[2\pi f_0 t + \theta(t)])\}/2 \\ v(t) &= r(t)\exp\{j[2\pi f_0 t + \theta(t)]\} \end{aligned} \tag{2.3}$$

Note that $e(t)$ is real valued but $v(t)$ is complex valued and $e(t)$ is one-half of the real part of $v(t)$. From $v(t)$ it is convenient to define $s(t)$ as

$$s(t) = v(t)\exp[-j(2\pi f_0 t)] = r(t)\exp[j\theta(t)] \tag{2.4}$$

and $s(t)$ is termed the complex envelope. The complex envelope and complex notation are useful because it is usually simpler to analyze the effect of a processor on $s(t)$ and then deduce the processor's effect on $e(t)$. The definition of the complex envelope via (2.4) is only approximately correct. The exact definition is based on Hilbert transforms but the approximation is satisfactory for almost all radar waveforms.

2.1 Fourier Transforms

The analysis of waveforms and processors, their properties, and the derivation of optimum processing is partially based on Fourier transform theory. Some aspects of the theory are reviewed in this section. The limits of all integrals are from minus infinity to plus infinity unless otherwise indicated.

The transform is defined for finite energy signals. That is, for the signal $x(t)$, the integral $\int |x(t)|^2\, dt$ must be finite. This contrasts with the Fourier series where the signals are of infinite energy but finite power. The series and transform can be related by limiting operations involving delta functions.

The fundamental integral relations defining the Fourier transform are

$$\begin{aligned} X(f) &= \int x(t)\exp(-j2\pi ft)\, dt \\ x(t) &= \int X(f)\exp(-j2\pi ft)\, df \end{aligned} \tag{2.5}$$

The first integral equation is usually thought of as $x(t)$ is a time function and $X(f)$ is its transform into the frequency domain. $X(f)$ is termed the voltage spectrum, often abbreviated simply as the spectrum of $x(t)$. The second equation is the inverse transform that obtains the time function from the frequency function. As an example, consider the pulse waveform of time duration τ centered on zero time. It is defined as

$$\text{Rect}(t/\tau) = \begin{cases} 1 & |t| \le \tau/2 \\ 0 & |t| > \tau/2 \end{cases} \tag{2.6}$$

Applying (2.5) gives

$$X(f) = \int \exp(-j2\pi ft)\, dt$$

with integral limits of $-\tau/2$ and $\tau/2$. This can be evaluated as

$$X(f) = \tau \sin(f\pi\tau)/(f\pi\tau) \tag{2.7}$$

This is the sin x/x function graphed in Figure 1.5. As indicated by the figure, the spectrum of (2.7) has values for negative frequency. This represents an important difference between the concept of the transform frequency and frequency of the real-valued time function $e(t)$. The latter is a real-valued sinusoidal waveform and the value of the frequency parameter is only positive. The Fourier transform is a set of cisoid functions of the form $\exp(-j2\pi ft)$ and the frequency of this function can be positive or negative. Also note that the spectrum of Figure 1.5 is symmetric and centered on zero frequency.

An important property of the Fourier transform is that the transform of a time-delayed waveform equals the original transform multiplied by $\exp(-j2\pi ft_0)$, where t_0 denotes the time delay. This can be proved by direct substitution into (2.5). Thus, for the time function $x(t-t_0)$, the transform $X'(f)$ is

$$X'(f) = \int x(t-t_0)\exp(-j2\pi ft)\,dt \tag{2.8}$$

and a change of integration variable from t to t' equals $t-t_0$ gives

$$X'(f) = \exp(-j2\pi ft_0)X(f) \tag{2.9}$$

which proves the theorem. A corollary is that the inverse transform of the frequency-shifted spectrum, $X(f-f_0)$, is the original waveform multipled by $\exp(j2\pi f_0 t)$. These results are summarized as

$$\begin{aligned} x(t) &\Leftrightarrow X(f) \\ x(t-t_0) &\Leftrightarrow \exp(-j2\pi ft_0)X(f) \\ X(f-f_0) &\Leftrightarrow \exp(j2\pi f_0 t)x(t) \end{aligned} \tag{2.10}$$

and the two-sided arrow denotes the transform relationship.

The spectrum of $\mathrm{Rect}(t/y)\cos(2\pi f_0 t)$ can easily be derived using these results. First note that the waveform can be rewritten as $\mathrm{Rect}(t/\tau)[\exp(j2\pi f_0 t)+\exp(-j2\pi f_0 t)]/2$. From (2.7) and (2.10) the spectrum equals

$$\begin{aligned} Y(f) = \tau\{&\sin[(f-f_0)]\pi\tau/[(f-f_0)\pi\tau] \\ &+\sin[(f+f_0)]\pi\tau/[(f+f_0)\pi\tau]\} \end{aligned} \tag{2.11}$$

This can be recognized as a spectrum consisting of two sin x/x functions. One is centered at f_0 and the other is centered at $-f_0$.

Another important theorem is that the transform of a real-valued function has an even real part and an odd imaginary part. The proof is obtained by substituting in the first equation of (2.5) and conjugating both sides while recognizing that the waveform is real. Thus, for $e(t)$ of (2.1)

$$E^*(f) = \int e^*(t)\exp(-j2\pi ft)\,dt \tag{2.12}$$

Noting that $e(t)$ is real so that $e^*(t)$ equals $e(t)$ and substituting the decomposition of $E(f)$ as the sum of a real part $E_R(f)$ and an imaginary (or quadrature) part $E_Q(f)$ gives

$$\begin{aligned} E_R(f) &= E_R(-f) \\ E_Q(f) &= -E_Q(-f) \end{aligned} \tag{2.13}$$

Note that the squared magnitude of a spectrum is always real and for this spectrum it is also even. This can be proved using (2.5). The time function that is the inverse transform of the squared magnitude of a spectrum is

$$R(t) = \int |E(f)|^2[\cos(2\pi ft) + j\,\sin(j2\pi ft)]\,df \tag{2.14}$$

Since $E(f)$ is even, the product of $|E(f)|^2$ and the sine function is an odd function and integrates to zero. Thus, $R(t)$ is real and even. Similarly, if $e(t)$ is purely imaginary, $R(t)$ is real and even. This is proved by noting that a purely imaginary function equals j times a real function. The voltage spectrum is then j times the spectrum of the real function. The magnitude spectrum is the magnitude spectrum of the real function, and (2.14) again is applicable.

Figure 2.1 is a sketch of an example spectrum of a real-valued waveform. The first trace shows the real portion of the spectrum. It consists of two disjoint portions. One is centered at f_0 and the other portion is centered at $-f_0$. Note that it is an even function of frequency but note that the spectral portion at the carrier frequency is not necessarily even with respect to f_0. The next trace shows the imaginary portion of the spectrum. It also consists of two portions centered at plus and minus f_0 and is an odd function. The next trace shows the magnitude of the spectrum $E(f)$. It is obtained as the square root of the sum of the squares of the real and imaginary portions. The magnitude spectrum is the square root of the energy spectrum. The next trace is the positive frequency portion of the magnitude spectrum. For radar waveforms this is approximately equal to the magnitude of the spectrum of $v(t)$ of (1.59). It is exactly equal if the spectrum of each portion of $E(f)$ is strictly bandlimited, that is, if the tails of the positive portion of the spectrum in the vicinity of f_0 do not extend

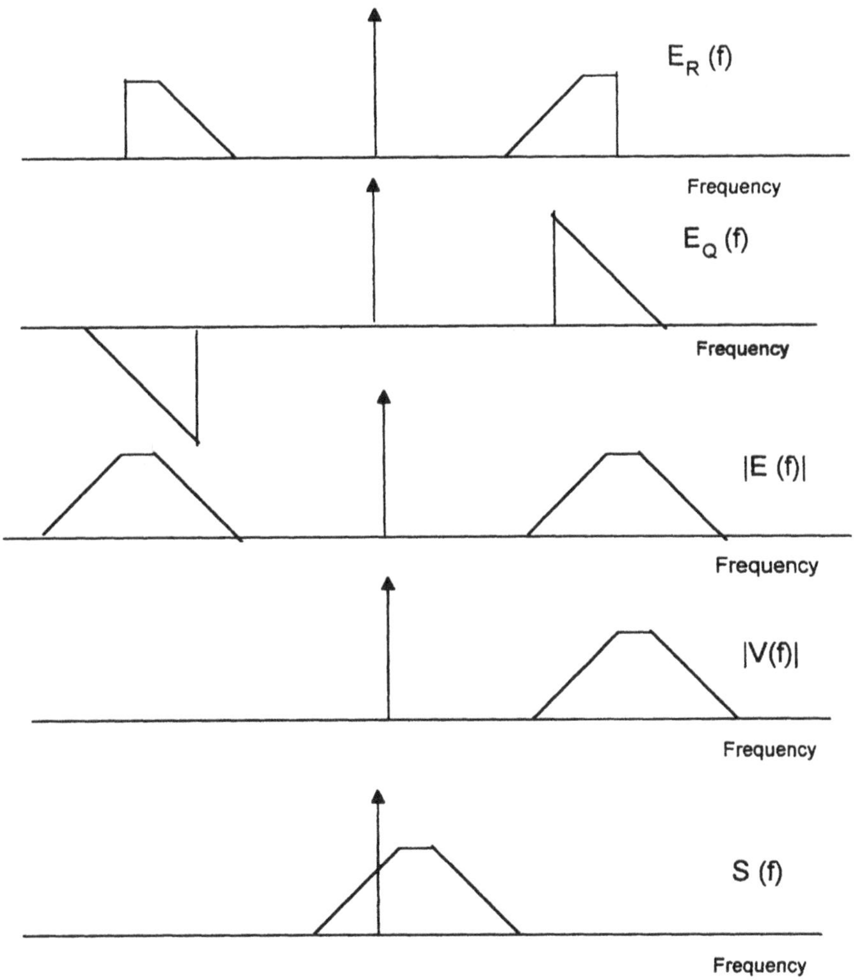

Figure 2.1 Spectra of real waveform and its complex envelope.

into the negative frequency region and similarly that the tails of the negative portion of the spectrum do not extend into positive frequency. Since the bandwidth of radar waveforms almost always is much less than the carrier frequency, the radar waveform is almost bandlimited and this approximation is usually quite good. When, as is sometimes true for communication signals, the waveform is not approximately bandlimited, it is necessary to use Hilbert transform theory to derive the spectrum, $V(f)$. The last trace is the spectrum of the complex envelope $s(t)$. The spectral relation between $v(t)$ and $s(t)$ from (2.4) and (2.10) is

$$V(f) = S(f - f_0) \tag{2.15}$$

where $S(f)$ is termed a baseband signal and its carrier frequency is zero.

2.1.1 The Autocorrelation Function

The autocorrelation function is defined as

$$R_x(\tau) = \int x(t)x^*(t - \tau)\, dt \tag{2.16a}$$

and is a function of the time delay, τ. It will be shown that the inverse transform of the magnitude squared of the spectrum, given in (2.14), is the autocorrelation function. Specifically, for an arbitrary time function, $x(t)$, with transform $X(f)$

$$R_x(\tau) = \int |X(f)|^2 \exp[j(2\pi f\tau)]\, df \tag{2.16b}$$

The proof is obtained from (1.72a) via

$$\mathcal{F}[R_x(\tau)] = \int \left[\int x(t)x^*(t - \tau)\, dt\right] (-j2\pi f\tau)\, d\tau \tag{2.17}$$

and $\mathcal{F}[\ ]$ denotes the Fourier transform operation. Interchanging orders of integration to integrate on τ and making a change of variable gives

$$\mathcal{F}[R_x(\tau)] = X^*(f) \int x(t)\exp[-j2\pi ft]\, dt = |X(f)|^2 \tag{2.18}$$

An alternate proof is also of interest because it introduces the delta function and its integral property.

2.1.2 The Delta Function

The delta function, $\delta(t)$, is not a true function but is termed a generalized function. It is defined in terms of an integral property. For an arbitrary function, $f(t)$ the integral property is

$$\begin{aligned} f(0) &= \int f(t)\delta(t)\, dt \\ f(t_0) &= \int f(t)\delta(t - t_0)\, dt \end{aligned} \tag{2.19}$$

where $f(0)$ and $f(t_0)$ denote the value of the $f(\ )$ function at arguments of zero and t_0, respectively. The delta function and its integral properties may be thought of in terms of a limiting process. Thus consider the function $(1/\tau)\text{Rect}(t/\tau)$. It is a pulse of width τ and amplitude $(1/\tau)$, so it has a unit area independent of the value of τ. Then as τ approaches zero the error of the integral approximation

$$\int f(t)(1/\tau)\,\text{Rect}(t/\tau)\,dt \approx f(0) \tag{2.20}$$

approaches zero. Thus, in the limit, the delta function can be thought of as a function that is zero everywhere except at the origin, tends to infinity at the origin, and has the quoted integral property.

From (2.19) a corollary integral property of the delta function is

$$I(f) = \int \delta(t) \exp[-j2\pi ft]\,dt \tag{2.21}$$

and $I(f)$ is termed the identity function and equals one for all frequencies. The inverse transform relation then gives

$$\delta(t) = \int \exp[j2\pi ft]\,df \tag{2.22}$$

Consider the inverse transform of $|X(f)|^2$

$$q(\tau) = \int X(f)X^*(f)\exp[j2\pi f\tau]\,df \tag{2.23}$$

Substituting the first equation of (2.5) for $X(f)$ and its conjugate for $X^*(f)$ gives

$$q(\tau) = \iiint x(t)x^*(t')\exp[j2\pi f(\tau + t' - t)]\,df\,dt\,dt' \tag{2.24}$$

Integrating with respect to f gives

$$q(\tau) = \iint x(t)x^*(t')\delta[t' - (t - \tau)]\,dt\,dt' \tag{2.25}$$

and integrating with respect to t' gives

$$q(\tau) = \int x(t)x^*(t - \tau)\,dt \tag{2.26}$$

which is equal to (2.16a), the definition of the autocorrelation function. Thus, this is alternate proof that the magnitude squared of a spectrum and the autocorrelation function are a transform pair and serves as a demonstration of the use of delta functions.

2.2 Matched Filter Processing

After frequency translation to a convenient IF frequency, the remaining portion of the receive channel prior to the envelope detector is a linear filter. The design of the optimum filter characteristic is analyzed in this section. The design is based on the theorem that for a thermal noise environment the probability of detection is maximized using the filter that maximizes the output SNR. This theorem is used to derive the form of the optimum filter and to determine the optimum output SNR.

The input to the filter is a scaled, time-delayed version of the complex envelope of the transmit signal, $s(t)$. The Fourier transform of $s(t)$ is denoted as $S(f)$. Without loss of generality, the scaling factor and time delay may be neglected. The complex envelope of the filter's impulse response is denoted as $h(t)$. Its transfer function is denoted as $H(f)$. The filter's output voltage as a function of time is given by

$$e_0(t) = \int H(f)S(f)\exp(j2\pi ft)\,df \tag{2.27}$$

In general, this voltage varies with time. Denote t_M as the time instant when the magnitude of the output voltage is its maximum value. This time is of interest because the maximum magnitude causes the highest probability of detection.

The noise is assumed to be white, which means its power density is constant over all frequency and N_0 denotes the one-sided noise power density. The filter's output noise power, denoted p_{noise}, is

$$p_{\text{noise}} = (N_0/2)\int |H(f)|^2\,df \tag{2.28}$$

The output SNR at time instant t_M is defined as

$$\begin{aligned}\text{SNR} &= |e_0(t_M)|^2/p_{\text{noise}} \\ &= \left|\int H(f)S(f)\exp(j2\pi ft_M)\,df\right|^2 \Big/ \left[(N_0/2)\int |H(f)|^2\,df\right]\end{aligned} \tag{2.29}$$

A pertinent theorem is the Schwarz inequality. It states that for arbitrary functions $A(f)$ and $B(f)$,

$$\left| \int A(f) B^*(f)\, df \right|^2 \le \int |A(f)|^2\, df \int |B(f)|^2\, df \qquad (2.30)$$

and that equality is attained for $A(f)$ equal to $g\,B(f)$, where g is any nonzero scalar. For

$$A(f) = H(f) \qquad B^*(f) = S(f)\exp(j2\pi f t_M)$$

$$\text{SNR} \le \int |H(f)|^2\, df \int |S(f)|^2\, df \Big/ \left[(N_0/2) \int |H(f)|^2\, df \right] \qquad (2.31)$$

$$\text{SNR} \le \int |S(f)|^2\, df/(N_0/2)$$

$$\text{SNR}_{\text{MAX}} = \int |S(f)|^2\, df/(N_0/2)$$

These last two equations state that for any filter the output SNR is less than or equal to the ratio of the signal energy to the noise power density (noise energy). The maximum SNR is attained when the filter transfer function equals $g\,S^*(f)$, where the effect of the time instant of the maximum is incorporated into the arbitrary constant, g. Thus, the transfer function of the optimum filter that maximizes the target detection probability equals the complex conjugate of the transmitted waveform. Also note that using the inverse Fourier transform gives the impulse response of the optimum filter, given by

$$\begin{aligned} h_{\text{OPT}}(t) &= g^* s^*(-t) \\ H_{\text{OPT}}(f) &= g\,S^*(f) \end{aligned} \qquad (2.32)$$

The spectrum of the output voltage when using the optimum filter equals $|S(f)|^2$, and the output voltage is the inverse transform of $|S(f)|^2$. From (2.18) the output voltage when using the optimum filter equals the autocorrelation function of the complex envelope of the radar waveform. An alternate expression for the output voltage is obtained from the convolution equation by substituting (1.77) and changing variable of integration

$$e_{\text{OPT}}(t) = \int h(x) s(t-x)\, dx = \int h(t-y) s(y)\, dy = \int s^*(t'-t) s(t')\, dt' \qquad (2.33)$$

This is often rewritten as

$$e_{\mathrm{OPT}}(\tau) = \int s(t)s^*(t-\tau)\,dt \tag{2.33a}$$

to emphasize that the output voltage is the autocorrelation function.

2.3 Matched Filtering Examples

This section gives some examples of the matched filter output. Equation (2.33a) requires the integration of the multiplication function $s(t)$ with a delayed replica $s^*(t-\tau)$. Often solving the equation is easily done graphically; some of the examples considered here are solved this way.

2.3.1 Simple Pulse Example

The first example is for a pulsed sinusoid waveform of duration τ. The complex envelope is equal to the Rect function of width τ given by

$$s(t) = \mathrm{Rect}(t/\tau) \tag{2.34}$$

From (2.32) the impulse response of the optimum filter is the conjugate of the time-reversed signal. When, as is true for this example, $s(t)$ is real and even, $h(t)$ equals $s(t)$. This example is shown in Figure 2.2. The complex envelope of the transmit signal, $s(t)$ is shown in trace A. The zero-delay complex envelope of the impulse response of the optimum filter is shown in trace B. The output voltage for zero time is given by the area of the product of trace A and B. This voltage value is shown at the zero time instant of the last trace. Trace C shows the impulse response for a delay of $\tau/4$. The integrated product of traces A and C is smaller than the zero delay and for Rect functions the output decreases linearly between zero time and the $\tau/4$ time. The linear change is also shown by the last trace. Similarly, trace D is for a delay of $\tau/2$ and causes a further linear decrease. The output voltage goes to zero and remains zero when the delay is equal to or greater than τ. Thus, the output voltage for positive time has been determined. Since $s(t)$ is real and even, it is known from (2.14) that its autocorrelation function is even and the negative time portion of the output voltage is the reflection of the positive time portion.

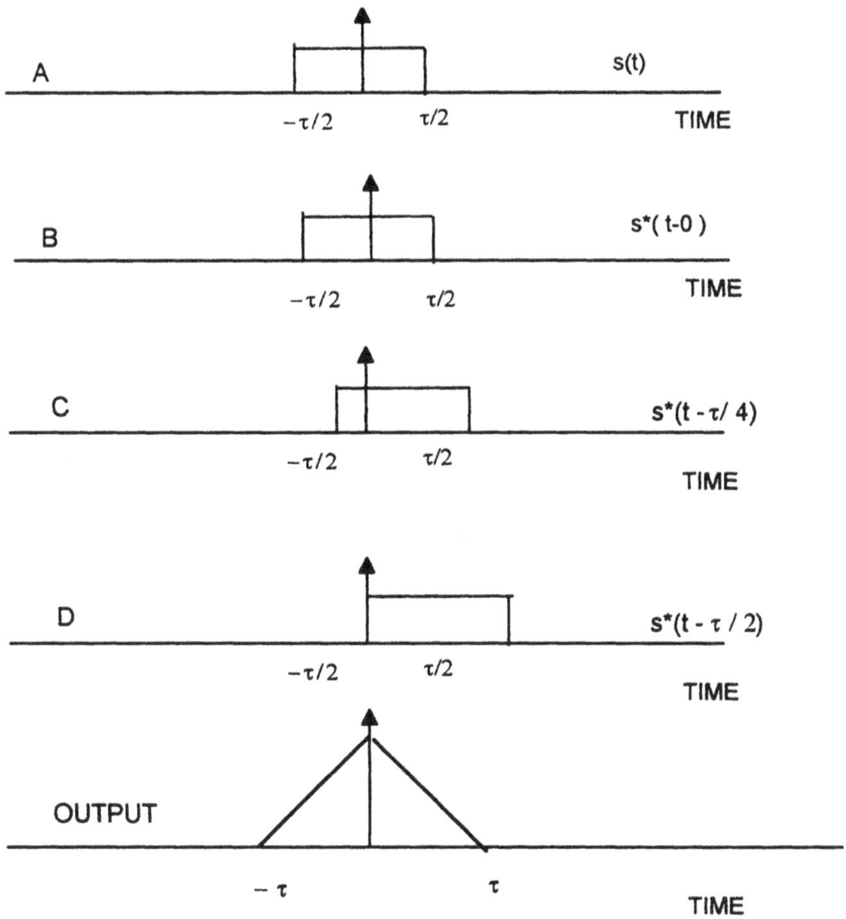

Figure 2.2 Matched filter output for Rect function signal.

2.3.2 Right Triangle Example

The complex envelope of the signal of duration τ starting at time t_0 is defined as

$$s(t) = 1 - (t - t_0)/\tau \qquad t_0 \leq t \leq t_0 + \tau \tag{2.35}$$

The impulse response of the optimum filter is the time reversal of the complex envelope of the signal

$$h(t) = 1 + (t + t_0)/\tau \qquad -(\tau + t_0) \leq t \leq -\tau \tag{2.36}$$

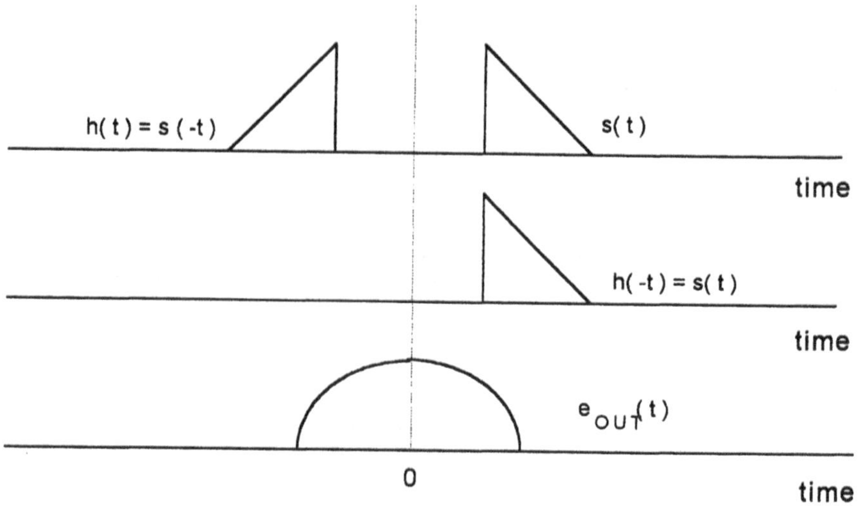

Figure 2.3 Matched filter output for triangle signal.

The example is sketched in Figure 2.3. The first trace shows the signal and the optimum impulse response. The second trace shows the time-reversed impulse response with zero delay. The last trace shows the voltage output, which equals the autocorrelation function.

2.3.3 sin *x*/*x* Type Spectrum Example

The signal spectrum is

$$S(f) = [\sin(\pi f B)/(\pi f B)] \exp[j\theta(f)] \tag{2.37}$$

so the transfer function of the optimum filter is

$$H(f) = [\sin(\pi f B)/(\pi f B)] \exp[-j\theta(f)] \tag{2.38}$$

and the spectrum of the filter's output voltage is

$$E_0(f) = [\sin(\pi f B)/(\pi f B)]^2 \tag{2.39}$$

The output voltage is the transform of $E_0(f)$ and may be evaluated as the triangle function of (2.40).

$$\begin{aligned} e_0(t) &= \text{Tri}(tB) \\ \text{Tri}(tB) &\equiv 1 - (|t|B) \qquad 0 \le |t| \le 1/B \\ &\equiv 0 \qquad |t| \ge 1/B \end{aligned} \tag{2.40}$$

The two-sided total width equals $2/B$, and the range resolution is commonly defined as equal to $1/B$.

2.4 The LFM Pulse

The waveform was called a chirp to contrast it with the "bang" of the simple pulse. It is one of the unit amplitude phase-modulated waveforms whose phase is given by $\exp[j\theta(t)]$. The complex envelope of the LFM waveform of duration τ is given by

$$s(t) = \exp(j\pi kt^2) \qquad 0 \le t \le \tau \tag{2.41}$$

so $\theta(t)$ equals πkt^2 and k is a positive constant that determines the bandwidth. The instantaneous frequency is equal to

$$f_{\text{instantaneous}} = (1/2\pi)[d\theta/dt] = kt \tag{2.42}$$

and the maximum instantaneous frequency, Δ, the frequency deviation from the carrier, is

$$\Delta = k\tau \tag{2.43}$$

The LFM waveform is described and analyzed in great detail in the report by Klauder, J. R., et al., "The Theory and Design of Chirp Radars," *BSTJ*, Vol. 39, July 1960, pp. 745–808. They showed that the matched filter output is given by

$$e_{\text{MF}}(t) = \text{Tri}(t/\tau)[\sin \pi\Delta t(1 - |t|/\tau)]/[\pi\Delta t(1 - |t|/\tau)] \tag{2.44}$$

The compressed pulse width is approximately $1/\Delta$, and in the vicinity of the main lobe the output is well approximated as a $\sin x/x$ type function.

The spectrum is given by

$$S(f) = \int \exp(j\pi k t^2)\, dt \tag{2.45}$$

where the limits of the integral are from zero to τ. The solution of (2.45) is given in terms of Fresel integrals. Using asymptotic expansions for the Fresnel integrals, it can be shown that the spectrum is approximately constant in the region close to zero and its relative magnitude is -6 dB at $\Delta/2$. For the tail region of $f > \Delta/2$, an approximation valid for the $\Delta\tau$ product greater than about 20 is

$$|S(f)| \approx (1/\pi\Delta)\{(2f/\Delta)/[(2f/\Delta)^2 - 1]\} \qquad f > \Delta/2 \tag{2.46}$$

which may be further approximated as

$$|S(f)| \approx 1/(2\pi f) \qquad f \gg \Delta/2 \tag{2.47}$$

Thus, in the region of the skirts, the spectral decay rate is identical to that of a single pole filter. The approximate spectrum magnitude is indicated in Figure 2.4.

This spectrum magnitude is often further approximated as

$$|S(f)| \approx \mathrm{Rect}(f/\Delta) \tag{2.48}$$

The phase of the spectrum is approximately quadratic, as given by

$$\Phi(f) \approx (2\pi f)^2/4\pi k \tag{2.49}$$

so

$$S(f) = \mathrm{Rect}(f/\Delta)\exp[j\Phi(f)] \tag{2.50}$$

The transfer functions of the matched filter and the spectrum output voltage are

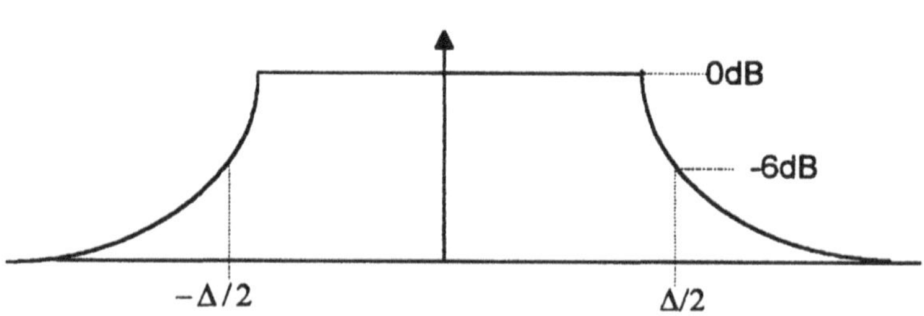

Figure 2.4 Approximate spectrum for LFM waveform.

given by

$$H(f) = \text{Rect}(f/\Delta)\exp[-j\Phi(f)]$$
$$E_{\text{OUT}}(f) = \text{Rect}(f/\Delta) \tag{2.51}$$

and as the inverse transform of a Rect function is a sin x/x type function, the matched filter output is approximated as

$$e_{\text{MF}}(t) \approx c\,\sin(\pi\,\Delta t)/\pi\,\Delta t \tag{2.52}$$

and c denotes an unspecified amplitude. Note that (2.52) is an approximation of (2.44).

Equations (2.44) or (2.52) imply that the compressed pulse has relatively high-range sidelobes. Later material discusses the use of weighting functions to reduce the sidelobes and are usually based upon the form of (2.52). Though the approximation is conceptually valid, it should be applied cautiously. These techniques are not applicable for sidelobes that are lower than about −45 dB. This is partially because the spectrum magnitude close to zero frequency is not constant but has small ripples. It is also because the spectral tails are going down relatively slowly.

2.4.1 LFM Pulse Doppler Tolerance

For some applications it is necessary to transmit a high-energy waveform using a long-duration waveform. If the target has a high radial velocity and the transmitted waveform is a simple pulse, the large Doppler shift induced by the target causes the received waveform to shift totally out of the matched filter passband. To prevent this, it is necessary to implement a more complicated receiver that encompasses the expected Doppler shift. As shown in this section, an LFM waveform is more tolerant of Doppler shift and so the receiver implementation is simpler. However, as also shown, the simplicity comes at the cost of an uncertainty of the target range.

As an example of the problem, assume that the target is a stealth type aircraft and it is necessary to transmit a pulse width of 1000 μs in order to develop the energy required for target detection. One difficulty with a waveform of this duration is that the receiver of a monostatic radar is gated off until after the end of the pulse transmission. Thus the radar cannot detect targets at a range less than approximately 150 km unless separate transmit and receive antennas are used. This effect is independent of whether the transmit waveform is a simple pulse or a coded pulse.

If a simple pulse is used, a second problem is that the matched filter bandwidth is approximately equal to 1 kHz. For a target velocity of 300 m/s, approximately Mach 1, the Doppler shift for a carrier frequency of 3 GHz is 6 kHz.

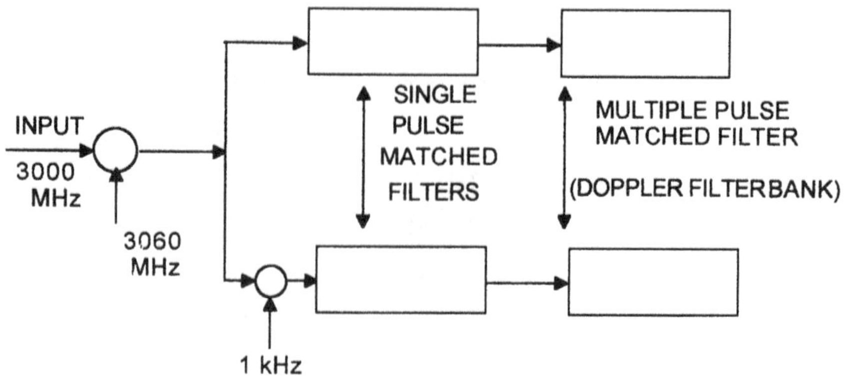

Figure 2.5 Simple pulse receiver for high Doppler targets.

Thus, the received waveform is completely shifted out of the passband of the matched filter receiver. Figure 2.5 shows the conceptual block diagram of a receiver that uses parallel channels to receive the Doppler shifted waveform. Since the effect of trajectory geometry will cause the radial velocity to span the range from zero to Mach 1, it is necessary to use a sufficient number of channels to span the expected Doppler range. For simplicity, the sketch shows only two channels and the second channel has its bandpass response shifted by 1 kHz by an offset oscillator. Note that the receiver performs multiple pulse Doppler processing after the single pulse-matched filtering.

A major difference between a simple pulse transmission and an LFM pulse for this application is that the bandwidth of an LFM waveform with a pulse width of 1000 μs is much greater than 1 kHz. One common characterization of an LFM waveform is its time-bandwidth product (with the semantically inconsistent notation of "the BT product"). A typical LFM pulse has a BT product of 64, which implies a bandwidth of approximately 64 kHz for this application. Since the bandwidth is much greater than the expected target Doppler shift, the single channel-matched filter should be adequate, so the receiver for the LFM waveform is much simpler than that required for the simple pulse waveform. However, the receiver simplicity comes at the cost of uncertainty about the range of the target.

To derive this effect, note from (2.49) and (2.50) that the spectrum of the transmitted LFM waveform, $S(f)$, and its matched filter, $H(f)$, are approximately

$$\begin{aligned} S(f) &= \mathrm{Rect}(f/\Delta)\exp[j\alpha f^2] \\ H(f) &= (1/\Delta)\,\mathrm{Rect}(f/\Delta)\exp[-j\alpha f^2] \\ \alpha &= (\pi/k) = (\tau/\pi\Delta) \end{aligned} \tag{2.53}$$

For a positive Doppler shift of f_D, the reflected waveform $S_R(f)$ is

$$S_R(f) = \text{Rect}[(f - f_D)/\Delta] \exp[j\alpha(f - f_D)^2] \tag{2.54}$$

The output voltage, e_{out}, is the inverse transform of the product of $H(f)$ and $S_R(f)$, and the magnitude can be evaluated as

$$|e_{out}| = [1 - (f_D/\Delta)]\, |(\sin\{\pi(\Delta - f_D)[t - \alpha f_D]\})/\{\pi(\Delta - f_D)[t - \alpha f_D]\}| \tag{2.55}$$

This is the product of $[1 - (f_D/\Delta)]$ and a $\sin x/x$ type function. Its peak value is attained at t equal to αf_D, and the peak value equals $[1 - (f_D/\Delta)]$). For the simple pulse waveform and a receiver using a filter matched to the transmitted waveform, the magnitude of the output is approximately equal to zero at a Doppler shift equal to $1/\tau$, the reciprocal of the transmitted waveform duration. From (2.55) the magnitude of the output voltage for this Doppler frequency and the LFM waveform equals $1 - 1/(\Delta\tau)$, and for the typical waveform BT product of 64 this equals 0.98. Thus, the output magnitude is almost unaffected by the Doppler shift. However, the time of the peak is $(\tau/\pi\Delta) f_D$. The differential peak time for a Doppler frequency of zero and $1/\tau$ equals $1/\pi\Delta$, and since the range cell size equals $1/\Delta$, the shift equals 0.32 range cells. The preceding example was for a target velocity of approximately Mach 1. If the target speed is Mach 5, the magnitude of the peak equals 0.92 and the time shift of the peak increases to 1.6 range cells.

The peak shift as a function of target Doppler is referred to as range-coupling; that is, two targets at the same range but with different Dopplers peak at different times. Further, two targets at different ranges and different Dopplers can have their peak magnitude at the same time instant. An approximate explanation of the phenomenon is illustrated by the bottom sketch of Figure 2.6.

The sloping lines in the top portion of the sketch represent the instantaneous frequency of four targets. The three targets depicted by the solid lines are for targets at different ranges and zero Doppler. The dashed line represents a target at the same range as the center solid line but a higher Doppler. The time instant of the peak response of the compressed pulse for the solid line targets occurs at the end of the frequency sweep depicting the return target. These peak responses occur at times commensurate with the relative ranges of the targets. Note that if the Doppler shift of the dashed line target is sufficiently high, its instantaneous frequency looks very much like that of the early-range target. The matched filter response is at an early range but decreased amplitude since its instantaneous frequency range is not identical to that of the matched filter.

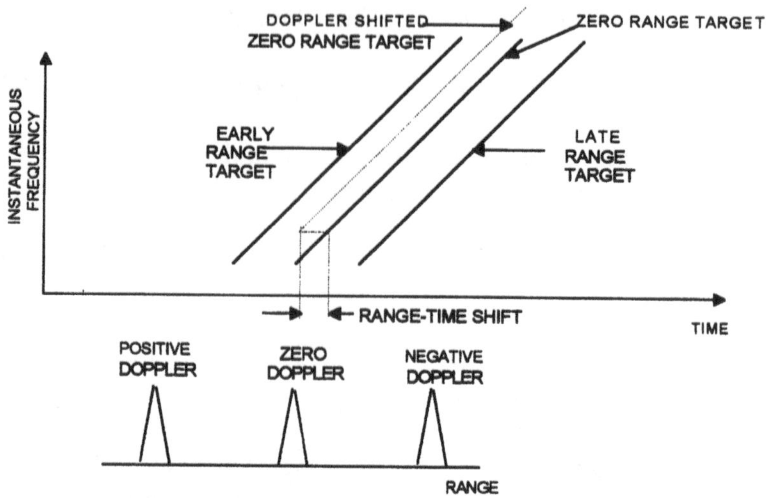

Figure 2.6 LFM range-Doppler cross coupling.

Similarly, if the Doppler shift of the dashed line target is sufficiently low, its instantaneous frequency looks very much like that of the late-range target. Based on this, the bottom portion of the figure shows the magnitude of the output of the matched filter for three different targets. Though identified as positive, negative, and zero Doppler, they could be at the same range but different Dopplers, at the same Doppler but different ranges, or at a combination of ranges and Dopplers.

From Figure 2.6, it may be deduced that the range-time delay, t_D, caused by a Doppler shift of f_D is given by

$$t_D/(1/\Delta) = -f_D\tau \tag{2.56}$$

The negative sign indicates that a positive Doppler causes the apparent range of the target to decrease. The left-hand side of (2.56) is the range shift, expressed by the number of range resolution cells, induced by the Doppler shift. If f_D is only a fraction of the nominal Doppler resolution of $1/\tau$, the range shift is the same fraction of a range resolution cell. Since, however, the Doppler shift can be an appreciable fraction of Δ, the range shift can be many range resolution cells.

2.4.2 LFM Generation Techniques

There are several methods of generating an LFM waveform. One technique is based upon the time delay versus frequency characteristic of the LFM waveform. Time delay, t_D, is the frequency derivative of the spectral phase and from (2.49) is given as

$$t_D = -(d/df)[\Phi(f)]/2\pi = -f/k \tag{2.57}$$

Thus, the time delay is approximately a linear function of frequency. The generation technique based upon this property excites a dispersive delay line with a very short pulse (an impulse approximation). The bandwidth of the delay line is designed to equal the bandwidth of the desired LFM waveform. If the delay-versus-frequency characteristic of the delay line is approximately linear, the output waveform closely approximates the theoretical LFM signal. A delay line with the inverse delay characteristic can be used to compress the received waveform.

Initially, delay lines with the desired characteristic were not available. An approximation to this generation technique is shown in Figure 2.7. The upper left shows the waveform generator. It consists of a bank of filters with cascaded delays. As indicated by the sketch in the lower left of the figure, the bandwidth of each filter equals B and the center frequency of each filter is offset by B. The total bandwidth of the generated waveform equals B times the number of filters. Each channel is delayed before summing, and the delay increases as

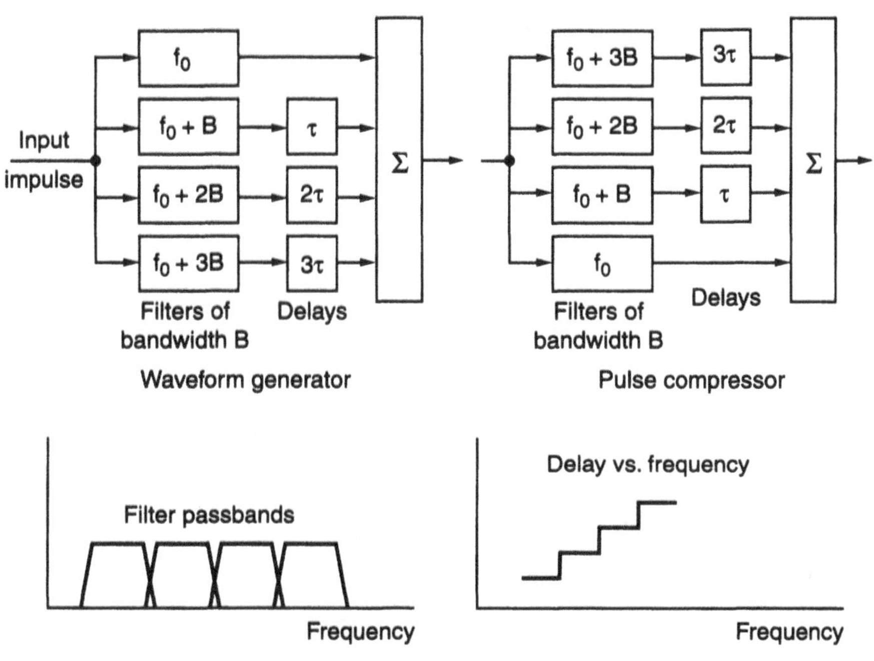

Figure 2.7 Parallel filter LFM generation technique.

the center frequency increases. The overall delay-versus-frequency characteristic of the generated waveform is indicated by the lower right sketch. It is a staircase approximation to the linear variation of (2.57). Thus, the generated waveform is an approximation to an LFM waveform. The approximation is improved using more filters that encompass the same total bandwidth.

The pulse compressor is shown in the upper right-hand sketch. It is similar to the generation technique. The major difference is that the delay variation versus frequency is the negative of the generation delay variation.

Some modern day radar are implemented using digital techniques. Digital generation is more complicated than the dispersive delay line but easily allows for bandwidth and code diversity. Multiple bandwidths are useful for a TWS radar since it can be desirable to use relatively small bandwidths for the surveillance function and relatively large bandwidths for tracking or identification. Also, code diversity can be an important antijam because it makes it more difficult for a repeater jammer to insert false targets into the radar's signal processor.

A conceptual block diagram of a digitally based LFM generation technique is shown in Figure 2.8. It is based upon generating a phase-coded waveform using the relationship

$$s(t) = \cos(\omega t)\cos[\theta(t)] - \sin(\omega t)\sin[\theta(t)] = \cos[\omega t + \theta(t)] \quad (2.58)$$

where $\theta(t)$ is the time-varying phase deviation from the carrier frequency. The initial box of Figure 2.8 generates a digital quadratic voltage that is proportional to the LFM waveform's instantaneous phase deviation $\theta(t)$. This can be done by reading successive words from digital storage or via an algorithmic computation. From (2.41) the voltage, $\theta(t)$, is

$$\theta(t) = \pi k t^2 \quad (2.59)$$

Next the voltage is trigonometrically converted, yielding the two voltages $\cos[\theta(t)]$ and $\sin[\theta(t)]$. The conversion is usually done by a table look up.

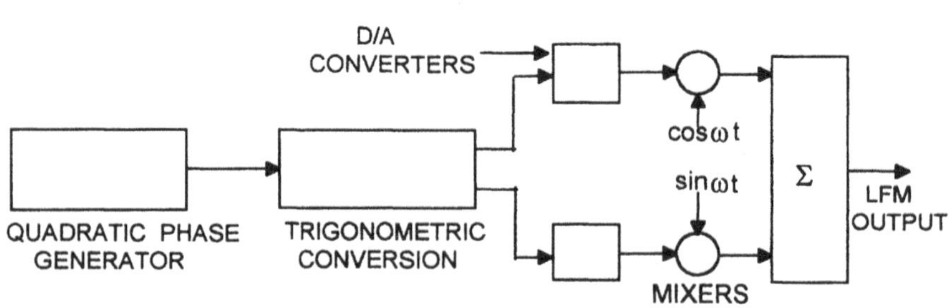

Figure 2.8 Digital LFM waveform generation.

This does not require much memory since if the phase is carried as an eight-bit word, the table consists of only 256 pairs. Because the phase word size is finite, there is a phase error in the generation that causes noise-like range sidelobes after compression. The word size is chosen large enough so that other factors limit the achievable sidelobe level. After trigonometric conversion, each word is passed through an A/D converter and then a mixer where the carrier frequency is inserted.

2.4.3 Biphase-Coded Waveforms

The phase of the LFM waveform changes continuously with time. The biphase waveform only has two discrete phases; an example is shown in Figure 1.28 and for convenience is repeated as Figure 2.9.

The generic waveform is made up of segments that are τ in duration. At the start of each segment, the phase can change by 180 degrees. The waveform bandwidth is approximately the reciprocal of τ, but the total waveform duration is much greater than τ. As shortened notation, the waveform is denoted by the phase of adjacent segments and for Figure 2.10 the waveform is denoted as $(0, \pi, 0, \pi)$. The sketch of Figure 2.10 shows an alternate biphase code, its pulse compressor processor, and the processor's output voltage.

The code of this waveform is $(0, \pi, 0, 0)$. The matched filter impulse response is the conjugate time-reversed signal and thus is $(0, 0, \pi, 0)$. The output voltage from the matched filter can be determined by convolving the signal with the impulse response. The signal voltage values are obtained as the complex exponential of the appropriate phase and therefore has values of plus or minus one. It is convenient to perform the convolution via Table 2.1. Time point one

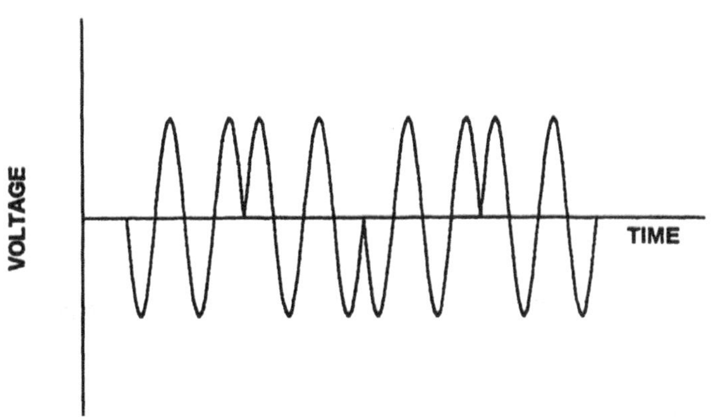

Figure 2.9 Biphase coded waveform.

Table 2.1
Biphase code convolutions

											Output voltage
$h(t)$			1	1	−1	1					Output voltage
$s(t)$: Time 1	1	1	−1	1							1
Time 2		1	1	−1	1						0
Time 3			1	1	−1	1					−1
Time 4				1	1	−1	1				4
Time 5					1	1	−1	1			−1
Time 6						1	1	−1	1		0
Time 7							1	1	−1	1	1
$h(t)$			−1	1	−1	1					Output voltage
$s(t)$: Time 1	−1	1	−1	1							−1
Time 2		−1	1	−1	1						2
Time 3			−1	1	−1	1					−3
Time 4				−1	1	−1	1				4
Time 5					−1	1	−1	1			−3
Time 6						−1	1	−1	1		2
Time 7							−1	1	−1	1	1

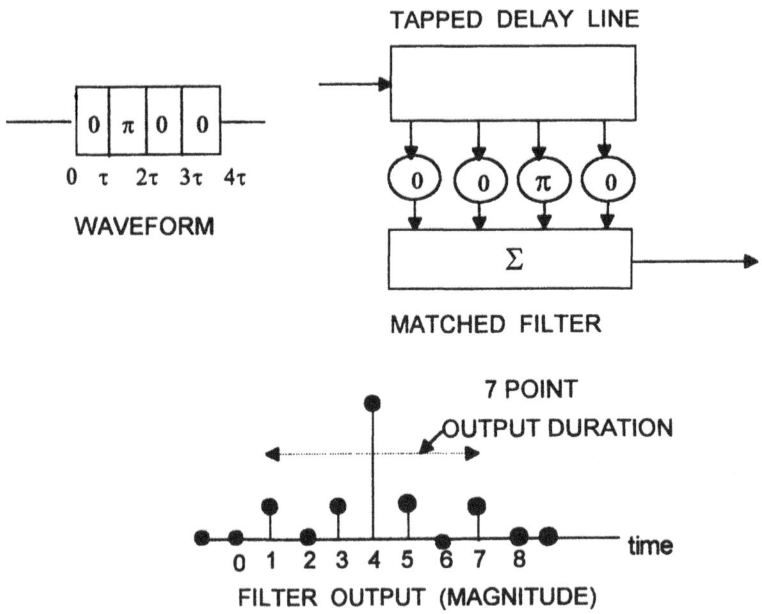

Figure 2.10 Biphase coded waveform and processor.

indicates the first time point when the signal starts to enter the delay line. Note that the signal voltage values are read from right to left. When the signal is K time units long, the duration of the output is $2K - 1$ units long. Here K equals 4, so the output is seven units in duration. For time point one, only a single time value of $s(t)$ and $h(t)$ overlap, so multiplication of $h(t)$ by $s(t)$ and summing give the output voltage as plus 1. For time point two, two time values of $s(t)$ and $h(t)$ overlap, so multiplying and summing result in an output voltage of 0. The output voltage value at other time values are obtained in a similar manner and displayed in Table 2.1. These output values are those depicted as the output for Figure 2.12. The peak amplitude of the output equals 4, which represents the pulse compression gain. Also note that the output has sidelobes and the magnitude of the sidelobes equals one-quarter of the peak magnitude.

The lower portion of Table 2.1 shows the results of convolving the original biphase signal, $(0, \pi, 0, \pi)$, with its matched filter impulse response. The peak magnitude equals 4, as it does for the previous example. However, the sidelobe level is much different and higher.

Any biphase code has a maximum peak value equal to the number of segments of the waveform. No code can have 0-level sidelobes everywhere. An "ideal" code would have a maximum sidelobe level equal to 1. There are some codes that have these properties. They are known as Barker codes. The $(0, \pi, 0, \pi)$ biphase code example is a member of the class of Barker codes and is termed a Barker 4 code. The remainder of the Barker codes are tabulated in Table 2.2 with their relative sidelobe levels. The largest sized Barker code that is known is a Barker 13. Thus, the maximum peak to sidelobe level for this class of code equals 13 or a relative range sidelobe level of approximately −22 dB. Often it is necessary to use waveforms with lower range sidelobe levels so that Barker codes have limited application. Though, as discussed later in detail, special weighting functions that can significantly reduce the sidelobes levels for an LFM waveform exist, there is no known method of reducing the sidelobes of Barker codes.

Table 2.2
Barker phase codes

Number of segments	Code	Sidelobe level (dB)
3	+ + −	−9.5
4	+ + − +	−12.0
4	+ + + −	−12.0
5	+ + + − +	−14.0
7	+ + + − − − + − − + −	−16.9
11	+ + + − − − + − − + −	−20.8
13	+ + + + + − − + + − + − +	−22.3

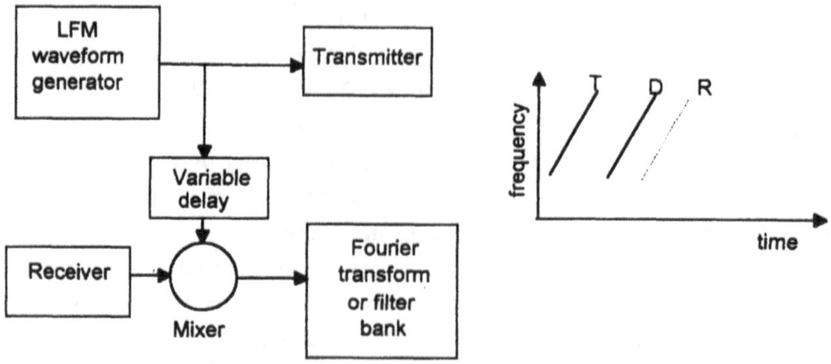

Figure 2.11 Very large bandwidth LFM generation and processing.

An alternate phase code that can theoretically achieve lower sidelobes is the pseudorandom code. For this code the phase reversals for adjacent segments are "random." It is not truly random because it is a fixed code whose reversals have been carefully selected to achieve relatively low sidelobes. Since it is noise-like, it has noise-like properties. For an N-segment code the average sidelobe level is of the order of $1/N$ though the peak sidelobes are higher. To prove this, note that the matched filter processor is like that shown in Figure 2.11. The code is not truly random, so the processor phase shift values are known and matched to the waveform. Define the voltage of each segment as either $+1$ or -1 and assume each voltage is random with probability of $\frac{1}{2}$ and statistically independent from segment to segment. The output voltage of the matched filter for an N-segment code has a peak value of N that occurs when the signal fills the delay line so that it is aligned with the proper phase shifts. For a shift of one segment, the phase shift values are not aligned with the corresponding waveform segment. The product of phase shift and voltage value causes the output for each segment product to equal $+1$ or -1 with $\frac{1}{2}$ probability for each value. The average value of the product for each segment equals 0. The square of the product for each segment always equals $+1$, so the variance equals 1. For signals, only the number of terms in the sum for the processor output depends upon the degree of misalignment between the signal and the processor phase shifts. The variance of the output, and this represents the noise sidelobes, ranges between $N-1$ and 1. Thus the largest sidelobe level equals $N-1$ and the peak energy output equals N^2 so that the sidelobe level is of the order of $1/N$. Note that -22-dB sidelobes can be obtained using a 13-segment Barker code but requires approximately a 158 segment pseudorandom code. Thus low sidelobes require long duration codes. This is sometimes undesirable because the minimum detection range-time equals the duration of the transmitted signal. A 1-MHz bandwidth waveform with -30-dB sidelobes has a duration of approximately 1000 μs, which gives a minimum range of 150 km.

In addition to biphase codes, it is theoretically possible to use tri-phase or quad-phase codes, for example. These have not found practical application.

2.5 Very Wideband LFM Waveforms

For some applications, the desired range resolution is of the order of a meter, so waveform bandwidths of hundreds of megahertz are required. It is difficult to generate and signal process waveforms of this bandwidth using the techniques previously discussed. Two alternate techniques that are easier to generate and process are discussed in this section.

2.5.1 Linear Frequency Sweep

The technique uses an oscillator whose generated frequency depends upon voltage. The controlling voltage is changed linearly and the generated frequency also changes linearly. This technique is most often used for very large bandwidth waveforms in a range-tracking mode. It is restricted to this application because of the difficulty of compressing the reflected signal. As indicated by the conceptual block diagram of Figure 2.11, the signal processor mixes a delayed replica of the generated LFM waveform with the reflected signal. The delay is set equal to the nominal delay of the tracked target. The right portion of the figure shows an instantaneous frequency representation of the transmitted signal, designated as T, the delayed transmitted signal, D, and the received signal, shown as the dotted R. The mixer output is a constant frequency pulse whose frequency depends upon the time delay between the target's actual range and its nominal range. The frequency of the pulse is analyzed using a Fourier transform processor or a filter bank. If there are several closely spaced targets, they are resolved by the frequency difference between the processor outputs. As an example, the technique could be used in a high-resolution mode to perform target identification by measuring the target length or more generally to identify characteristic protruding objects on the vehicle.

2.5.2 Bandspan Waveforms

Section 1.9.3 discussed a multiple subcarrier waveform where each of the subpulses is combined noncoherently after subpulse processing. As will be shown in this section, if the subpulses are combined coherently, the effective waveform bandwidth for range resolution equals the total span of the spectrum. For a specified range resolution this implementation is simpler than that of a single

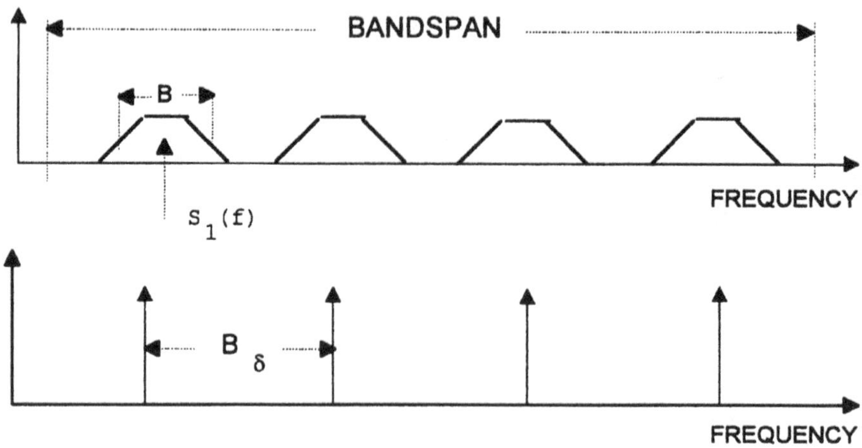

Figure 2.12 Bandspan waveform spectrum.

channel. If the sub-bands are noncontiguous, the implementation is further simplified. However, noncontiguous sub-bands cause range ambiguities, but for some applications this is not significant.

The spectrum of the bandspan waveform is depicted by the top sketch of Figure 2.12. This example consists of four noncontiguous spectra spaced at a frequency of B_δ. The spectrum of each sub-band, after matched filtering, is denoted as $S_1(f)$ with bandwidth B. The lower portion of the figure shows four delta functions, each centered at the center frequency of each sub-band. The spectrum of the upper sketch can be recognized as the convolution of $S_1(f)$ and an infinite train of these impulses and then the convolution is multiplied by a rect function in frequency. Specifically, denoting the total spectrum as $S(f)$,

$$S(f) = \left[S_1(f) \otimes \sum_{n=-\infty,+\infty} \delta(t - nB_\delta) \right] \mathrm{Rect}(f/B_{\mathrm{span}}) \tag{2.60}$$

where $\otimes$ denotes the convolution operation. The output time waveform is the inverse Fourier transform of $S(f)$. Evaluation of the inverse transform is simplified using several standard Fourier transform relations. One is that the transform of the convolution of two spectra equals the product of the transforms of the two spectra. That is, denoting the two spectra as $A(f)$ and $B(f)$ and their respective transforms as $a(t)$ and $b(t)$,

$$A(f) \otimes B(f) \Longleftrightarrow a(t)b(t) \tag{2.61}$$

where the two-sided arrow denotes the Fourier transform pair relation. Similarly, the transform of the product of two spectra equals the convolution of the transforms of the two spectra

$$A(f)B(f) \Longleftrightarrow a(t) \otimes b(t) \tag{2.62}$$

To evaluate (2.60) using (2.61) it is necessary to demonstrate that the Fourier transform of an infinite train of impulses with time spacing of T is an infinite train of impulses in the frequency domain with spacing $1/T$ and vice versa.

The Fourier transform of an N-impulse finite train, $E(f)$, is

$$E(f) = \sum_{n=1,N} \delta(t - nT)\exp[-j2\pi ft] \tag{2.63}$$

and from (2.19) this evaluates to

$$E(f) = \sum_{n=1,N} \exp[-j2\pi nfT] \tag{2.64}$$

This is a geometric series whose magnitude evaluates to

$$|E(f)| = |[\sin(\pi NfT)]/[\sin(\pi fT)]| \tag{2.65}$$

This function is sketched in Figure 1.17. It peaks at all integer multiples of $1/T$; the peak magnitude equals N; and the bandwidth of each peak is approximately $1/N$. As N approaches infinity, the time waveform approaches an infinite-duration impulse train and its transform approaches an impulse train in frequency with $1/T$ spacing. Thus, the transform of the convolution of $S_1(f)$ and an infinite impulse train is the transform of $S_1(f)$ sampled at time instants that are integer multiples of $1/B_\delta$. This is then convolved with the transform of $\text{Rect}(f/B_{\text{span}})$, which equals $\sin(\pi B_{\text{span}}t)/(\pi B_{\text{span}}t)$. The resulting time function is sketched in Figure 2.13 as the solid lines. As can be seen, the time output is range ambiguous. The time spacing between compressed pulses is equal to $1/B_\delta$. The time duration of each compressed pulse is approximately $1/B_{\text{span}}$, so the range resolution is commensurate with the spectral bandspan. The envelope modulating the amplitude of the compressed pulses, shown as the dashed line, is the transform of $S(f)$ and has an approximate duration of $1/B$. If this waveform is used to resolve highlights on an object whose length is less than a range-time of $1/B_\delta$, the range ambiguities are not important. Also, if the environment consists of multiple objects whose range-time spacing is greater than $1/B$, the "sidelobes" of the responses from the multiple objects are nonoverlapping and do not cause

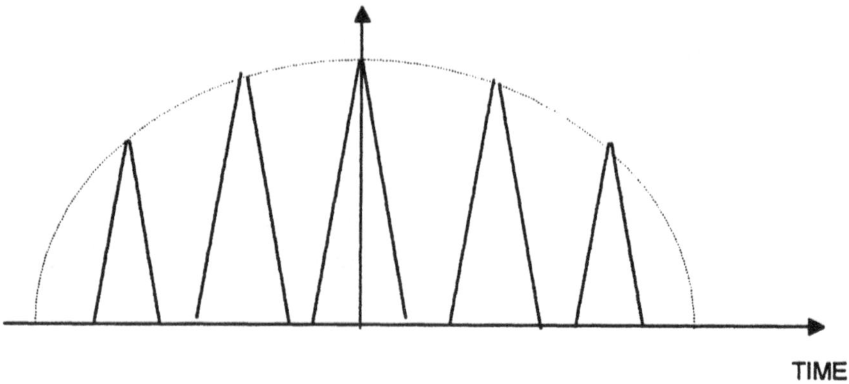

Figure 2.13 Bandspan waveform matched filter output.

ambiguities. Thus, the waveform bandwidth parameters can be chosen as a function of the expected target size and spacing such that the potential ambiguities do not cause trouble.

2.6 Waveform Summary

There are essentially three types of pulses: the simple pulse, the LFM pulse, and discrete biphase codes.

The simple pulse is a constant amplitude, constant carrier frequency waveform. Its major characteristic is that the pulse width determines both the transmit energy and the range resolution. This is often an undesirable coupling of important system characteristics. Its benefit, as compared to other pulse waveforms, is that it does not have any range sidelobes. Another defect of this waveform is that it can require an implementation with multiple IF channels in order to accommodate high Doppler targets when using long-duration pulses.

The LFM, linear frequency modulation or chirp pulse, has a constant envelope like the simple pulse but its instantaneous frequency varies linearly. The instantaneous frequency of an up-chirp LFM increases from the carrier frequency to the carrier frequency plus a deviation. That of the down-chirp LFM decreases. The deviation is approximately equal to the signal's bandwidth. Its major characteristic is that the pulse width determines the transmit energy but not the range resolution. It requires more complex waveform generation and processing equipment. Also, it has relatively high range sidelobes. The sidelobe level can be reduced by weighting techniques. But the processor using weighting is not a matched filter and therefore the resultant SNR is less than that achieved by a matched filter. The SNR loss increases monotonically as the sidelobes decrease.

The waveform has a very different ambiguity diagram than that of the simple pulse. There is significant range-Doppler coupling. The good aspect of this property is that multiple IF channels are not needed to process large Doppler shift targets. The bad aspect of this property is that the measured range of a target is usually not its actual range as the Doppler shift perturbs the measured location. The waveform can be implemented by digital techniques that start by generating a digital quadratic function. It is then easy to change codes on a pulse-to-pulse basis and achieve improved antijam capability against false target repeater jammers.

The biphase-coded waveform has a constant envelope and constant carrier frequency. It is implemented using 180-degree phase reversals of the carrier at a rate equal to the transmitted bandwidth. Similarly to the LFM waveform, the total pulse duration is independent of the range resolution. The waveform has relatively high range sidelobes, and there is no known method of weighting to reduce the sidelobe level. Its ambiguity diagram is similar to that of the simple pulse. Therefore, it also requires multiple IF channels to process very high Doppler shifts. The waveform can use code agility on a pulse-to-pulse basis and achieve an antijam capability against false target repeater jammers.

3

Antennas

Antennas are used both to transmit and receive electromagnetic energy. An important property of antennas is given by the reciprocity theorem. This states that an antenna has the identical gain pattern when used either in transmit or receive modes. The analysis of an antenna is done for the mode that simplifies the analysis.

The first analysis will determine the farfield pattern of line and planar antennas excited by a current flow and establish that the antenna field pattern is given by the Fourier transform of the illumination pattern.

3.1 Line Antenna

The electric field intensity $F(P)$ at a point P with coordinates (x_p, y_p, z_p) excited by a differential element located on the y-axis is

$$k_1 i(y) \exp(j2\pi r_p/\lambda)/r_p \, dy$$

where k_1 is a constant, $i(y)$ is the current of the differential element, and r_p denotes the distance from the current element to P. For a line antenna of length D aligned along the y-axis with current $i(y)$, the field intensity is given by the integral

$$F(P) = k_1 \int_{-D/2}^{D/2} [i(y)/r_p] \exp(j2\pi r_p/\lambda) \, dy \tag{3.1}$$

The expression is difficult to evaluate in general but an approximation, valid for the Fraunhofer, or farfield, region simplifies the equation. The simplification applies when the distance to P is such that

$$r_p > D^2/\lambda = \lambda/(\theta_{\text{beamwidth}})^2 \tag{3.2}$$

and the relation to the beamwidth is obtained from (1.12) or (1.13). As an example for a 3-GHz carrier frequency, 10-cm wavelength, and 1-degree beamwidth, the farfield region is for ranges that exceed 360 meters. If the 1-degree beamwidth is obtained for a carrier frequency of 30 MHz, the farfield region is for ranges that exceed 36 km.

The difficulty of evaluating (3.1) is due to the effect of r_p. Its influence is both in the amplitude term and in the phase term. In the farfield region, the approximation that r_p equals r, where r is the distance from the origin to P and equals $[x_p^2 + y_p^2 + z_p^2]^{1/2}$ is valid for the amplitude effect as $[i(y)/r]$ but not for the phase effect. To derive an approximation that is valid for the phase term note that

$$\begin{aligned} r_p &= \left[x_p^2 + (y_p - y)^2 + z_p^2\right]^{1/2} = \left[x_p^2 + y_p^2 + z_p^2 + y^2 - 2yy_p\right]^{1/2} \\ &= [r^2 + y^2 - 2yy_p]^{1/2} = r[1 + (y^2 - 2yy_p)/r^2]^{1/2} \approx r[1 - yy_p]/r^2 \end{aligned} \tag{3.3}$$

since y_p is much greater than y. Since y_p equals $r\sin(\theta)$, where θ is the angle measured from the normal to the antenna,

$$r_p \approx r - y\sin(\theta) \tag{3.4}$$

Thus

$$F(\theta) = [k_1 \exp(j2\pi r/\lambda)] \int_{-D/2}^{D/2} i(y) \exp[-j2\pi y \sin(\theta)/\lambda]\, dy \tag{3.5}$$

so the farfield field intensity is proportional to the Fourier transform of the illumination function.

3.1.1 Uniformly Illuminated Line Antenna

From (3.5) and previously derived Fourier transform relationships, when $i(y)$ is constant (uniform illumination) the electric field is proportional to a $\sin x/x$ function. Equation (3.5) evaluates to

$$F(\theta) = G_0^{1/2}\{\sin[\pi(D/\lambda)\sin(\theta)]/[\pi(D/\lambda)\sin(\theta)]\} \tag{3.6}$$

where G_0 equals the peak gain at θ equals zero, and for small θ, the gain pattern in the vicinity of the mainlobe is approximately

$$G(\theta) \approx G_0\{\sin[\pi(D/\lambda)\theta]/[\pi(D/\lambda)\theta]\}^2 \tag{3.7}$$

3.1.2 Properties of sin x/x Functions

Since the sine of 0 equals 0, the value of $F(\theta)$ is undefined at θ equal to 0. Using a series approximation for the sine function it can be shown that the function approaches unity as θ equals zero. Thus, considering the more general function

$$f(x) = \sin x/x \approx [x - x^3/3! + x^5/5! - + \cdots]/x) \approx 1 - x^2/3! + x^4/5! \tag{3.8}$$

As x approaches 0, $f(x)$ approaches its maximum value of unity, and the maximum value of (3.7) equals G_0. From (3.7) the first null of $G(\theta)$ occurs for $(D/\lambda)\theta$ equal to unity. Thus, the peak-to-null beamwidth equals

$$\theta_{\text{beamwidth}} = \lambda/D \tag{3.9}$$

This is approximately equal to the two-sided -3-dB beamwidth and the basis for the beamwidth definitions of (1.12) and (1.13).

3.2 Planar Antenna

Similar to (3.1) the equation for the electric field insensity for a planar antenna is

$$F = k_2 \iint [i(y, z)/r_p] \exp(j2\pi r_p/\lambda)\, dy\, dz$$

for a rectangular current sheet of dimension D_y and D_z in the yz-plane and where the limits of integration are $\pm D_y/2$ and $\pm D_z/2$. A similar approximation for the amplitude-dependent term is that r_p is approximately equal to r and for the phase-dependent term

$$r_p \approx r[1 - (y_p y + z_p z)/r^2] \tag{3.10}$$

The relation between spherical coordinates and Cartesian coordinates is

$$z_p = r \sin(\phi) \qquad x_p = r \cos(\phi)\cos(\theta)$$

$$y_p = r \cos(\phi)\sin(\theta) \qquad \left[x_p^2 + y_p^2\right]^{1/2} = r \cos(\phi)$$

so

$$r_p \approx r\{1 - [yr\ \cos(\phi)\sin(\theta) + zr\ \sin(\phi)]/r^2\}$$
$$= r - y\cos(\phi)\sin(\theta) - z\ \sin(\phi) \tag{3.11}$$

and the farfield intensity is

$$F = k_3 \int\int i(y, z)\exp\{-j2\pi[y\ \cos(\phi)\sin(\theta)+z\ \sin(\phi)]/\lambda\}\,dy\,dz \tag{3.12}$$

Often the illumination function is separable as a product. That is,

$$i(y, z) = i(y)i(z) \tag{3.13}$$

so

$$F = k_3 \int i(y)\exp\{-j2\pi[y\ \cos(\phi)\sin(\theta)]/\lambda\}\,dy \times \int i(z)\exp\{-j2\pi[z\ \sin(\phi)]/\lambda\}\,dz \tag{3.14}$$

Thus, the electric field is separable as the product of the individual illumination functions.

3.3 Uniformly Illuminated Planar Antenna

A constant amplitude, constant phase illumination function is seperable. The antenna gain is the product of two functions of the form of (3.6) and equals

$$G(\phi, \theta) = G_0\{\sin[\pi D_y \cos(\phi)\sin(\theta)]/\lambda \sin[\pi D_z \sin(\phi)]\lambda)] \times([\pi D_y \cos(\phi)\sin(\theta)]/\lambda[\pi D_z \sin(\phi)]/\lambda)^{-1}\}^2 \tag{3.15}$$

In the vicinity of the mainlobe

$$G(\phi, \theta) = G_0\{\sin[\pi D_y\theta]/\lambda \sin[\pi D_z\phi]/\lambda/[\pi D_y\theta]/\lambda[\pi D_z\phi]/\lambda\}^2 \tag{3.16}$$

or the product of two sin x/x type functions.

Note that for both the line and planar antennas the gain pattern depends upon the wavelength. Thus the antenna pattern is frequency dependent, acts like a filter, and has a bandwidth. Also, since sin x/x type functions have relatively high sidelobes and high sidelobes cause ambiguity and other problems, alternate illumination patterns that synthesize lower sidelobe patterns are of interest. Both of these questions are addressed in later sections.

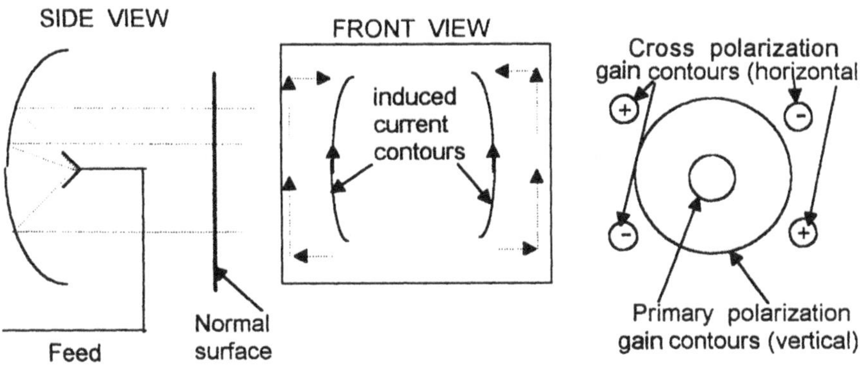

Figure 3.1 Parabolic reflector antenna.

3.3.1 Parabolic Reflectors

A current sheet illumination pattern cannot be directly implemented. The common method of approximating the desired antenna pattern is by using a reflector and feed as sketched in Figure 3.1. As indicated by the leftmost sketch, the transmitter power is connected to the antenna feed. The feed is a small antenna. The energy propagates from the feed toward the reflector and is then reflected. When the reflector's surface is parabolic and the feed is located at the focus of the parabola, the reflected rays are parallel. In addition, the total length of the rays from the feed point and the reflecting surface to a plane that is perpendicular to the axis of the parabola are all equal. Thus, the rays are all in phase at the normal surface, so the field at the normal plane is approximately that caused by an equiphase current sheet illumination function. Therefore the reflector antenna synthesizes a gain pattern similar to that derived in the preceding section. However, there are some differences.

One difference is due to the directive pattern of the feed. Because the feed's gain decreases as the angle from the feed increases, the illumination function at the edges of the reflector is lower than at the center. Also, the reflected rays are blocked by the feed and the pattern shows the effect of the blockage. In addition, antennas are commonly covered with radomes to protect the antenna from the effects of wind, snow, and ice. The radome causes attenuation of the transmitted energy.

Another difference is that, as indicated by the center sketch of Figure 3.1, the induced current contours are curved due to the curvature of the surface. The current can be decomposed into vertical and horizontal components. For this example the induced current is primarily vertical. The vertical component of the current causes a vertically polarized transmission and this is the primary, or intended, polarization. It causes the antenna pattern similar to that of the current sheet example of Section 3.2.

The currents' horizontal components cause transmission of a horizontally polarized transmission. As indicated in the middle portion of Figure 3.1, the direction of the horizontal current is toward the right for the upper left and lower right portions of the reflector. The direction is toward the left for the upper right and lower left portions of the reflector. The horizontal components change sign from the top to the lower portions of the reflector and are zero in the middle. The resultant antenna pattern due to the cross-polarization illumination has zero gain at the antenna boresight (where the primary polarization gain pattern is maximum) and four maxima in the region near the sidelobes of the primary polarization. The sense of the four maxima differs because the sense of the horizontal current components varies over the reflector area. This is depicted by the rightmost sketch of Figure 3.1. The maxima of the cross-polarization patterns are of the order of 25 dB to 35 dB lower than the primary polarization maximum gain. The effect of the cross-polarization patterns depends strongly on the application. For search radar the main effect is that it is difficult to reduce the antenna sidelobes to a level less than the cross-polarization sidelobes. For tracking monopulse radars, these sidelobes perturb the tracking by causing a wrong sense error signal. This effect can also be exploited to create a vulnerability to track breaking jammers.

3.4 Linear Phased Array

An N-element linear phased array is shown in Figure 3.2. It is convenient to analyze this configuration by assuming it is a receiving antenna. Thus, as indicated in the figure, a wavefront is impinging upon the array, inducing voltages in each array element; and the wavefront's angle of arrival, the angle between the normal to the wavefront and the normal to the array, is θ. Denote the gain pattern of the nth element as $g_n(\theta)$. The voltages induced in each element are $g_n(\theta)x_n$, where x_n is the voltage that would be induced in a unit gain antenna. The phase and amplitude controls shown in the figure are complex-valued gains (gain magnitude and phase shift) and are commonly referred to as weights. The array output is

$$e_{\text{out}} = \sum_{n=1,N} g_n(\theta) w_n^* x_n \tag{3.17}$$

and the conjugate is used to simplify succeeding equations. In general, the elements have arbitrary spacing and the spacing from the leftmost element to the nth element is d_n. The time delay of propagation from the leftmost element to the nth element, t_n, equals $d_n \sin\theta / c$. Denote x_1 as $x(t)\exp(j2\pi ft)$, where $x(t)$ is a complex envelope. Then

$$\begin{aligned} x_n &= x(t - t_n) \exp(j2\pi ft) \exp(-j\phi_n) \\ \phi_n &= 2\pi f t_n = 2\pi (d_n/\lambda) \sin\theta \end{aligned} \tag{3.18}$$

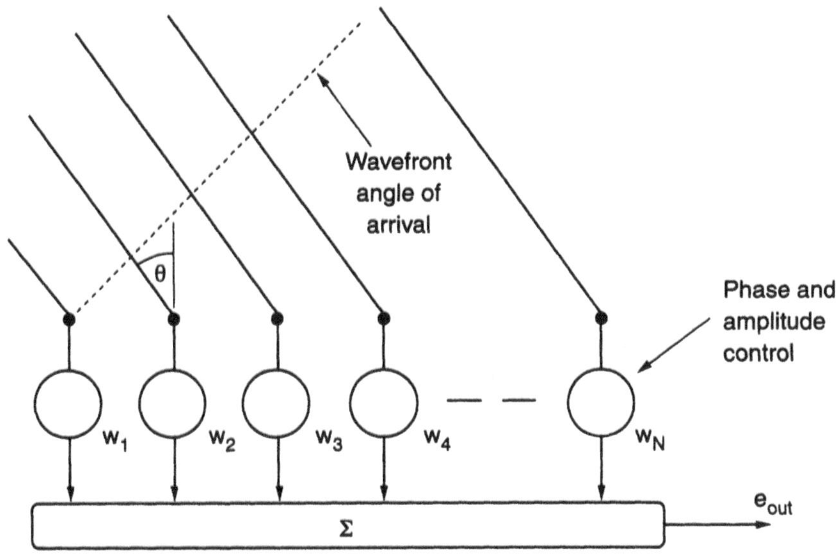

Figure 3.2 Linear phased array.

so

$$e_{\text{out}} = \exp(j2\pi ft) \sum_{n=1,N} g_n(\theta) w_n^* x(t - t_n) \exp[-j2\pi (d_n/\lambda) \sin\theta] \quad (3.19)$$

and the portion of (3.19) within the summation is the complex envelope of the output voltage.

The antenna gain is defined for the case that the incident waveform is a cisoid, so $x(t)$ equals the constant value A. Also, usually the array has constant interelement spacing of d, so $d_n = (n-1)d$ and the complex envelope of (3.19) reduces to

$$v_{\text{out}} = A \sum_{n=1,N} g_n(\theta) w_n^* \exp[-j2\pi (n-1) d/\lambda \sin\theta] \quad (3.20)$$

Note that just as for the line and planar antennas, the output voltage is obtained via a Fourier transform relation.

The antenna power gain, $G(\theta)$, equals $|v_{\text{out}}/A|^2$. Neglecting mutual coupling, the element gain does not depend upon the element location, so $g_n(\theta)$ equals $g_e(\theta)$. This simplifying assumption is useful in deriving general properties of arrays, but it must be used with caution because it leads to grossly erroneous results for adaptive arrays and other processing. With this assumption, the antenna gain is

$$
\begin{aligned}
G(\theta) &= \left| g_e(\theta) \sum_{n=1,N} w_n^* \exp[-j2\pi(n-1)d/\lambda \sin\theta] \right|^2 \\
G(\theta) &= \left| g_e(\theta) \sum_{n=1,N} w_n^* \exp[-j(n-1)\phi] \right|^2 \qquad (3.21) \\
\phi &= 2\pi(d/\lambda)\sin\theta
\end{aligned}
$$

The gain is the product of two factors. One is the element gain, $G_e(\theta)$; the other is the array factor gain, $G_a(\theta)$, defined as

$$
\begin{aligned}
G(\theta) &= G_e(\theta)G_a(\theta) \\
G_e(\theta) &= |g_e(\theta)|^2 \\
G_a(\theta) &= \left| \sum_{n=1,N} w_n^* \exp[-j(n-1)\phi] \right|^2 \qquad (3.22)
\end{aligned}
$$

The phase of the weights is used to steer the beam. The magnitude of the weights is used to control the sidelobe level. As previously done, it is convenient to constrain the weights such that the output noise power equals the noise power at an element. With this constraint the array factor equals the SNR gain from an element to the output.

In (3.17) let x_n be the per-element noise voltage. The output noise power due to $G_a(\theta)$ is

$$
p_{\text{out}} = E\left(\left| \sum_{n=1,N} w_n^* x_n \right|^2 \right) = \sum \sum_{n,m=1,N} w_n^* w_m E(x_n x_m^*) \qquad (3.23)
$$

where $E(\;)$ denotes the statistical expectation operator. Assume the element noise is identically distributed, independent from element to element (i.i.d.), has zero mean, and a variance (the element noise power) of p_n. Thus $E(x_n x_m^*) = 0$ unless $m = n$ and $= p_n$ when $n = m$, so (3.23) reduces to

$$
p_{\text{out}} = p_n \left(\sum_{n=1,N} |w_n|^2 \right) \qquad (3.24)
$$

and the output noise equals the input noise when $\sum |w_n|^2 = 1$. For uniform illumination the weights are equivalued, each equals $N^{1/2}$, and the array factor gain equals

$$G_a(\theta) = \left| \sum_{n=1,N} \exp[-j(n-1)\phi] \right|^2 \Big/ N^{1/2} \tag{3.25}$$

This may be recognized as a geometric series, so

$$\begin{aligned} G_a(\theta) &= N[\sin(N\phi/2)/N\,\sin(\phi/2)]^2 \\ &= N\{[\sin[N\pi(d/\lambda)\sin\theta]/N\,\sin[\pi(d/\lambda)\sin\theta]\}^2 \end{aligned} \tag{3.26}$$

At $\theta = 0$ the array gain, and therefore the SNR gain, equals N. This is identical to the improvement achieved when processing an N-pulse waveform using the optimal matched filter processor. The antenna is combining N spatial distributed signals in an optimal manner just as the optimal signal processor is combining N time distributed signals. For small values of θ, (3.26) can be approximated as

$$\begin{aligned} \sin[N\pi(d/\lambda)\sin\theta] &\approx \sin[N\pi(d/\lambda)\theta] \\ G_a(\theta) &\approx N\{\sin[N\pi(d/\lambda)\theta]/[N\sin[\pi(d/\lambda)\theta]]\}^2 \\ &\approx N\{\sin[N\pi(d/\lambda)\theta]/[N\pi(d/\lambda)\theta]\}^2 \end{aligned} \tag{3.27}$$

Thus, in the vicinity of the maximum gain angle, the antenna pattern is approximately equal to that of the uniformly illuminated current sheet derived in Section 3.2 with the substitution that the antenna dimension $D = Nd$. From (3.26) or (3.27) it can be seen that the nulls of the pattern in the vicinity of the main lobe occur for spatial angles that are integer multiples of λ/Nd. The sidelobe peaks occur approximately halfway between the nulls, and the relative levels can be evaluated as -13.5 dB and -17.9 dB, for example.

3.4.1 Grating Lobes

Note from (3.26) that there are maxima at $\theta = 0$ degrees and $\theta = 180$ degrees. In addition, the gain function attains a maximum whenever $(d/\lambda)\sin\theta$ equals an integer. From (3.21) this condition is equivalent to ϕ equal to multiples of π. Figure 3.3 shows a portion of the antenna pattern as a function of ϕ. The main lobe is indicated as the maximum at $\phi = 0$. The other maxima are termed grating lobes. The spatial angles of the grating lobes, θ_G, are at solutions of

$$\sin\theta_G = (k\lambda/d) \tag{3.28}$$

where k is a nonzero integer and thus depends upon the array's element to element spacing. There is no solution of (3.28) when the element spacing is less than the wavelength. There are grating lobes for element spacings greater than the

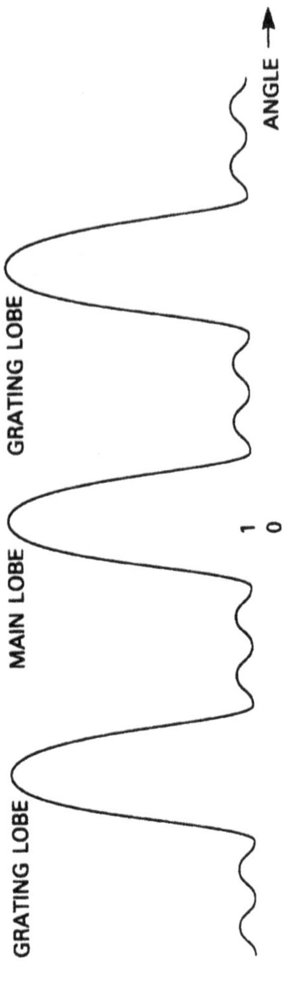

Figure 3.3 Linear array antenna pattern.

wavelength. Thus, the antenna pattern of the array is a strong function of the element spacing.

The cost of the array is approximately a linear function of the number of array elements, N. Because the array beamwidth depends upon the product of the number of array elements and the interelement spacing, large interelement spacing seems a low-cost method of achieving narrow antenna beamwidth. However, large interelement spacing causes grating lobes and grating lobes cause significant angle ambiguity on receive. Thus, there is a significant trade-off between grating lobes and antenna beamwidth. Beam steering also affects grating lobes so that usually the value of d/λ is set at approximately $\frac{1}{2}$.

3.4.2 Beam Steering

The previous section showed that when the incident plane wave emanates from a source, that is, at an angle θ, the complex envelope of the weighted induced voltage on the nth element, v_n, is

$$v_n = w_n^* \exp[-j2\pi(n-1)(d/\lambda \sin\theta)] \tag{3.29}$$

and for equal magnitude zero phase weights the complex envelope of the output voltage is

$$v_{\text{out}} = N^{-1/2} \sum_{n=1,N} \exp[-j2\pi(n-1)d/\lambda \sin\theta] \tag{3.30}$$

When $\theta = 0$, v_{out} equals its maximum possible value of $N^{1/2}$. As indicated by (3.26), the antenna power gain is less than N for other angles, representing an SNR loss. The loss can be eliminated by steering the beam to the wavefront's angle of arrival. Steering the beam to the angle of arrival θ_S is done by including phase shifts for the weight values. When

$$w_n = N^{-1/2} \exp[-j2\pi(n-1)(d/\lambda)\sin\theta_S] \tag{3.31}$$

the output

$$v_{\text{out}} = N^{-1/2} \sum_{n=1,N} \exp[-j2\pi(n-1)(d/\lambda)(\sin\theta - \sin\theta_S)]$$

and the antenna gain

$$G_a(\theta) = N\{\sin[N\pi(d/\lambda)\delta_S]/[N\sin[\pi(d/\lambda)\delta_S]]\}^2$$
$$\delta_S = \sin\theta - \sin\theta_S = 2\sin[(\theta_S - \theta)/2]\cos[(\theta_S + \theta)/2]$$

and for angles close to the steered angle

$$\delta = (\theta - \theta_S) \qquad (\theta_S + \theta)/2 \approx \theta_S$$

$$G_a(\theta) = N\{\sin[N\pi(d/\lambda)\delta\cos(\theta_S)]/N\sin[\pi(d/\lambda)\delta\cos(\theta_S)]\}^2 \quad (3.32)$$

Thus, the antenna pattern is similar to that derived in the previous section. In the vicinity of the steered angle, small δ, it is of the $\sin x/x$ type. The maximum power gain equals N. However, the steering causes the pattern to be modified. From (3.32) note that the effective width of the antenna equals $Nd\cos(\theta_S)$, and this is the projection of the total antenna width onto the steered angle of arrival. Thus, the effective antenna width depends upon the steered angle and, since the antenna beamwidth is proportional to the inverse of the antenna width, steering the beam off the broadside to the antenna causes the beamwidth to increase.

Another aspect of beam steering is that it affects the maximum interelement spacing that can be used without causing grating sidelobes. Grating lobes occur when the denominator of (3.32) equals zero. Thus, the grating lobe condition is

$$\sin[\pi(d/\lambda)(\sin\theta - \sin\theta_S)] = 0$$

For a specified steering angle, the maximum value of the argument of the sine function, argmax, is

$$\text{argmax} = \pi(d/\lambda)(1 + |\sin\theta_S|)$$

so a grating lobe never occurs if $(d/\lambda)(1 + |\sin\theta_S|)$ is less than unity. Thus, the condition for no grating lobes is

$$(d/\lambda) < 1/(1 + |\sin\theta_S|) \quad (3.33)$$

When the steered angle equals zero, the maximum value of d/λ that prevents grating lobes is unity; for a steered angle of 45 degrees, the maximum value of d/λ that prevents grating lobes is 0.59; for a steered angle of 90 degrees, the maximum value of d/λ that prevents grating lobes is 0.5.

3.4.3 Matched Filter Comparison

In the previous sections, the beam-steering weight values were chosen by intuitive means. Section 2.2 discusses optimum processing of waveforms by linear filters. Optimum filtering was defined as the linear filter that maximized the output SNR. An antenna can be considered to be a linear spatial filter where the filter spatial transfer function is determined by the antenna elements' weights. Therefore, a pertinent question is whether the intuitively chosen weight values are optimum or are there other weight values that can produce a larger output SNR? This section shows that the intuitively chosen weight values are the optimum values to maximize the output SNR.

From (3.21) the output signal energy, S_{out}, is

$$S_{\text{out}} = \left| \sum_{n=1,N} w_n^* \exp[-j(n-1)\phi] \right|^2 \tag{3.34}$$

Since the output noise power is proportional to $\sum |w_n|^2$, the output SNR is maximum when S_{out} is maximized under the constraint that $\sum |w_n|^2 = 1$. This maximization can be set up using Lagrange multiplier techniques. Defining the Lagrange multiplier as λ, we wish to find the set of weights $\{w_n\}$ to maximize Q

$$Q = S_{\text{out}} + \lambda \left(1 - \sum w_n w_n^*\right) \tag{3.35}$$

This is done by differentiating (3.35) with respect to both the real and imaginary parts of w_k and setting the two resulting equations to zero. An equivalent, and computationally simpler, technique that is valid for quadratics like that of (3.35) is to differentiate with respect to w_k^* while considering w_k to be an independent variable and setting the resulting equation to zero. The validity of this simplification can be proved by considering the general real-valued quadratic Q'

$$Q' = a\sum w_n w_n^* - b\sum u_n w_n^* - b^* \sum w_n u_n^* + c \tag{3.36}$$

where a, b, and c are arbitrary constants. Denote the real part of the kth weight as w_{kr} and the imaginary (or quadrature) part as w_{kq}. Then

$$\begin{aligned} Q' &= a\sum w_{nr}^2 + w_{nq}^2 - b\sum u_n(w_{nr} - jw_{nq}) \\ &\quad - b^* \sum u_n^*(w_{nr} + jw_{nq}) + c \\ \partial Q'/\partial w_{kr} &= 2aw_{kr} - bu_n - b^* u_n^* \\ \partial Q'/\partial w_{kq} &= 2aw_{kq} + jbu_n - jb^* u_n^* \end{aligned} \tag{3.37}$$

Setting the two partial derivatives to zero gives

$$\begin{aligned} w_{kr} &= [bu_n + (bu_n)^*]/2a \\ w_{kq} &= -j[bu_n - (bu_n)^*]/2a \end{aligned} \tag{3.38}$$

so

$$w_k = [bu_n + (bu_n)^*]/2a + j\{-j[bu_n - (bu_n)^*]/2a\} = bu_n/a \qquad (3.39)$$

Differentiating (3.36) with respect to w_k^* and assuming w_k is an unrelated variable gives

$$\partial Q'/\partial w_k^* = a\, w_k - b\, u_k$$

Setting this partial derivative to zero gives the same result as (3.39) and demonstrates the validity of the simplification.

The derivation is additionally simplified by rewriting (3.35) in vector notation. Thus, define the N-dimensional vectors W and U with nth component w_n and $\exp[-j(n-1)\phi]$, respectively, so that (3.35) equals

$$Q = W^{\mathrm{H}} U U^{\mathrm{H}} W + (1 - \lambda W^{\mathrm{H}} W) \qquad (3.40)$$

where the superscript H denotes the transpose conjugate. Define the vector Z with the nth component z_n defined as the nth component of the vector $UU^{\mathrm{H}} W$. Then (3.40), expressed as a summation, is

$$Q = \sum z_n w_n^* + \left(1 - \lambda \sum w_n w_n^*\right) \qquad (3.41)$$

Differentiating with respect to w_k^* gives

$$\partial Q/\partial w_k^* = z_n - \lambda w_k \qquad (3.42)$$

The vector of derivatives is then obtained from (3.42). Denote the vector of derivatives as $\partial Q/\partial W^{\mathrm{H}}$ so that

$$\partial Q/\partial W^{\mathrm{H}} = UU^{\mathrm{H}} W - \lambda W \qquad (3.42a)$$

Note that this vector of derivatives is identical to that obtained by directly differentiating (3.40) with respect to the vector W^{H}, and thus the direct differentiation is valid and simpler.

Setting (3.42a) to zero gives the result that the optimum W is a vector such that it transforms to a scalar multiple of itself. Vectors with this property are called eigenvectors. Thus, the solution of (3.42a) is based on the theory of eigenvectors. Section 3.4.3 is an introductory summary of some of the aspects of eigenvectors that are required to solve (3.42a). It is shown that the solution of (3.42a) is that the optimum weight is colinear with U; therefore, the nth weight component equals any constant times $\exp[-j(n-1)\phi]$. Thus, the optimum weight set equals the previously developed intuitive weight set.

3.4.4 Introduction to Eigenvectors

Algebraic equations of conic sections, ellipses, and parabolas, for example, can be difficult to interpret when the conic axes are not colinear with the coordinate axes. Coordinate transformations are used to simplify the interpretation of the conic's properties. Eigenvectors are special vectors that are useful because they can be used to obtain transformations that, like coordinate transformations, simplify the interpretation of equation properties. Also, often using eigenvector theorems allows an answer to be obtained essentially by inspection. This section summarizes some of the important properties of eigenvectors. Other aspects of eigenvectors are introduced later as required. In most cases the development does not include proofs.

The theory of linear equations can be used to develop the concept of matrix transformations. Specifically, for a vector X and a transformation matrix A a new vector Y is obtained as

$$Y = AX \tag{3.43}$$

Another type of transformation that is often of interest is an integral transformation. Thus, for a function $x(\,)$ and an integral kernal $a(\,)$ a new function $y(\,)$ is obtained as

$$y(f) = \int a(f, t)x(t)\,dt \tag{3.44}$$

Note that the Fourier transform is of the form of (3.44). The emphasis of this section will be on matrix transformations, though similar developments can be developed for integral transformations.

In general, the transformation matrix of (3.43) can be rectangular. However, this section considers only square matrices. With this restriction, when X is an N-dimensional vector, Y is also an N-dimensional vector in the same space. For any particular A there can be specific vectors X such that the transformation of X is a scalar multiple of X. These vectors are called eigenvectors or characteristic vectors. Specifically, an eigenvector is a nonzero vector X such that

$$AX = \lambda X \tag{3.45}$$

and λ is called the associated eigenvalue. Though eigenvectors cannot be the zero vector, eigenvalues can have the value of zero.

The number of eigenvectors depends upon the matrix A and is often difficult to determine. When A is restricted to the class of Hermitian matrices, defined by the property that for any matrix element a_{mn}

$$a_{mn} = a^*_{nm} \tag{3.46}$$

the transformation has N orthogonal eigenvectors and all the eigenvalues are real valued. An example of an Hermitian matrix is

$$A = \begin{pmatrix} 1 & 1+j & -3+2j \\ 1-j & 4 & 6-2j \\ -3+2j & 6+2j & -2 \end{pmatrix} \tag{3.47}$$

Note that the elements on the main diagonal are real and the elements bracketing the main diagonal are complex conjugates as per (3.46). An equivalent phrasing of (3.46) is that an Hermitian matrix equals its transpose conjugate, so A equals A^{H}.

The N eigenvectors constitute a basis for the space. A basis set means that any vector in the space can be written as a linear combination of the basis set. Thus, when the set $\{E_N\}$ is a basis, an arbitrary vector Z can be written as

$$Z = \sum_{n=1,N} \alpha_n E_n \tag{3.48}$$

In general, finding eigenvalues and eigenvectors is a tedious task. However, for some matrices, it is easy to determine the eigenvectors. Consider the Hermitian matrix R

$$R = U_1 U_1^{\mathrm{H}} \tag{3.49}$$

where U_1 is any N-dimensional vector, an $N \times 1$ matrix. Note that R is Hermitian since R^{H} equals R. Then

$$Y = RX = U_1 U_1^{\mathrm{H}} X = (U_1^{\mathrm{H}} X)\, U_1 \tag{3.50}$$

Since $(U_1^{\mathrm{H}} X)$ is a scalar, this demonstrates that if X is any vector that is not orthogonal to U_1, the transformed vector is colinear with U_1. In particular, when X equals U_1, Y equals $(U_1^{\mathrm{H}} U_1)\, U_1$. Thus, U_1 is an eigenvector with associated eigenvalue $(U_1^{\mathrm{H}} U_1)$. Also, any vector orthogonal to U_1 is an eigenvector with eigenvalue zero.

Therefore, the solution of (3.42a) is that the optimum weight is colinear with U, so the nth weight component equals $\exp[-j(n-1)\phi]$, the previously developed intuitive weight set.

Equation (3.49) is for the special case where the transformation is one-dimensional. That is, the transformed vectors all lie in a one-dimensional space. Another relatively simple case to analyze is for the two-dimensional Hermitian matrix transformation R' defined as

$$R' = U_1 U_1^{\mathrm{H}} + U_2 U_2^{\mathrm{H}} \tag{3.51}$$

where U_1 and U_2 are not colinear and the transformation of any X that is not orthogonal to both U_1 and U_2 results in a vector that is in the two-dimensional space spanned by U_1 and U_2. Thus, there are N-2 eigenvectors orthogonal to the space spanned by U_1 and U_2 with zero value eigenvectors. The eigenvectors of the two-dimensional space spanned by U_1 and U_2 can be found via the properties that they are linear combinations of U_1 and U_2 and mutually orthogonal.

3.5 The Maximum Detection Range of Multiple Beam Radar

Some radar are implemented using parallel beams in elevation. For a mechanically scanned antenna this can be done using multiple feeds clustered in elevation about the parabola focal point. This is sketched in the top portion of Figure 3.12. Each feed is connected to a duplexer, so the feed can be used for both transmit and receive. Feeds that are above the parabola's focal point create beams that are depressed in elevation angle; feeds that are below the focal point create beams that are elevated in angle. Phased arrays can create multiple beams using a parallel structure for transmit or receive that uses different sets of phase shifts for each beam that is created. A conceptual block diagram of the phase shift structure is shown in the bottom of Figure 3.4 for a three-beam system formed from four elements. Each element is weighted for sidelobe control. Then each weighted element is split into three parts and phase shifted. The weighted-phase elements are then summed to form the three beams.

Section 1.6.1 derived system equations for the maximum range of a radar in a noise environment. Section 1.7.1 derived system equations for the maximum range of a radar in a clutter environment. Both derivations assumed that the radar used a single beam to scan in azimuth and elevation. As will be discussed, the use of parallel beams modifies the angle cell dwell time. It is shown that the dwell time modification does not affect the maximum range in a thermal or jamming noise-limited environment but does improve it in a clutter-limited environment.

For ease of comparison the performance of two specific systems are compared. Assume both systems interrogate 15 elevation cells for each azimuth cell. System A does this using a single beam to interrogate all elevation cells. System B

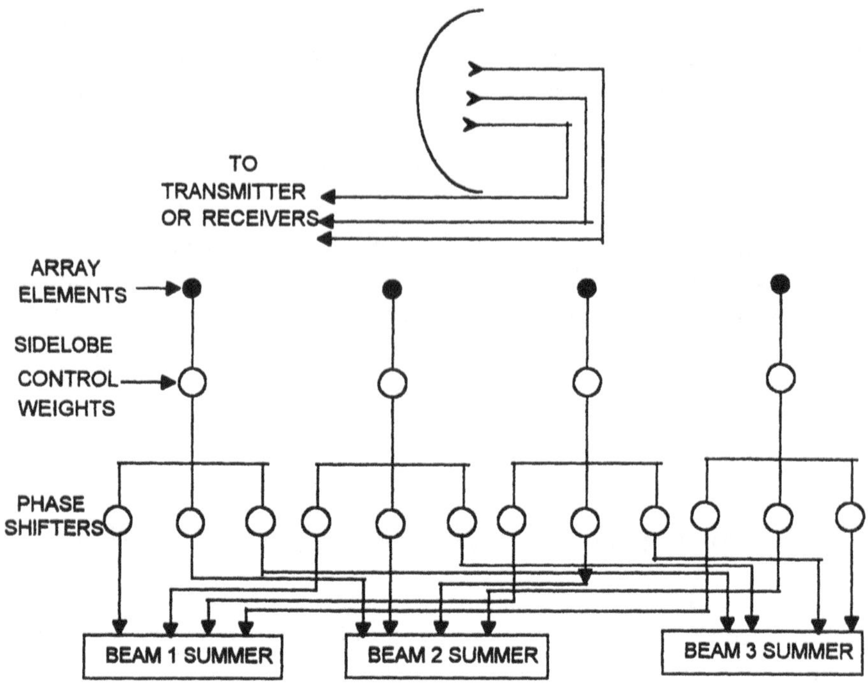

Figure 3.4 Antennas with multiple beams.

uses three parallel beams and each beam interrogates 5 elevation cells. For the same scan period (data update period) for each system, the dwell time of system B, and therefore the number of pulses transmitted for each cell, are 3 times greater than that of system A since only one-third of the number of angle cells are interrogated by each beam. Since (1.8) indicates the radar range in noise increases as a function of the number of pulses, it could appear as if the use of multiple beams increases the detection range in a noise-limited environment. This does not occur because the multiple feeds required to form the three beams on transmit split the transmit power so that system B transmits only one-third of the power per beam as does system A. The decreased transmit power exactly compensates for the increased dwell time so that the SNR, and therefore the detection range per angle cell are the same for both systems. However, system B does have better performance in a clutter-limited environment. This occurs because from (1.36) and (1.37) in a clutter-limited environment, the per-pulse ratio of the signal-to-clutter ratio does not depend upon the transmitted power. However, as discussed in Section 1.7.4, the detectability of a target in clutter increases with the number of pulses transmitted and processed per angle cell. Thus, the decreased transmit power does not compensate for the increased dwell time and the performance of system B is better than that of system A in a clutter-limited environment. Note that though the previous discussion assumed a three-beam multiple-beam system, the relative performance of it and the single-beam system is true for an arbitrary number of beams.

3.6 Sidelobe Reduction Weighting

Peak sidelobe levels of sin x/x type patterns usually are too high and cause various types of ambiguity. These high sidelobes are a consequence of using uniform illumination functions. There are several other illumination functions that are commonly used. One disadvantage of the alternate illumination functions is that uniform weights maximize the output SNR so that using the alternate illumination functions decreases the achievable SNR. Another disadvantage is that they increase the beamwidth and thus degrade the resolution and increase the clutter cross section due to the increased antenna pattern footprint. There is no best choice of illumination pattern. A trade must be made among the features of high SNR and good resolution.

As shown in previous sections, the farfield pattern of antennas can be obtained from the Fourier transform of the illumination pattern. An example of the use of this theorem is given by the sketches of Figure 3.5. The left sketch depicts the field pattern for uniform illumination. The vertical arrows indicate the weight values for each element of a linear array. The dashed rectangle is the equivalent illumination pattern of a continuously illuminated reflector antenna. The right-hand sketch depicts the resulting sin x/x type pattern.

An example of an illumination function that produces decreased sidelobes is shown in the left-hand portion of Figure 3.6. It shows a sampled

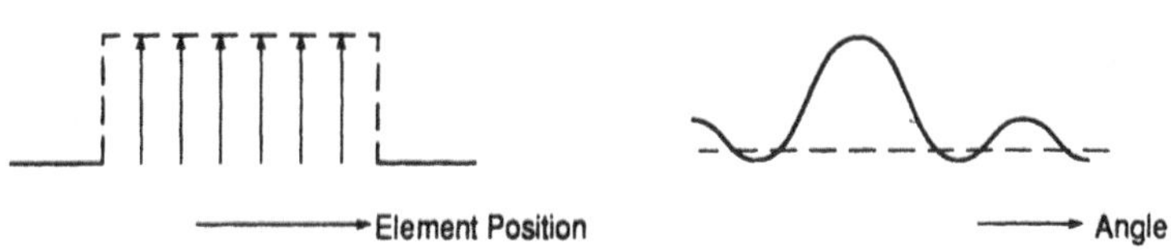

Figure 3.5 Uniform illumination function antenna pattern.

Figure 3.6 Cosine squared illumination function antenna pattern.

cosine-squared illumination function and its continous illumination approximation. The dashed curve of the right-hand sketch is the resulting antenna pattern. To derive the antenna gain pattern, note that the continuous illumination function for an antenna width of D is given by

$$\begin{aligned} I(y) &= \cos^2(\pi y/D)\mathrm{Rect}(y/D) = \{[1+\cos(2\pi y/D)]\ \mathrm{Rect}(y/D)\}/2 \\ &= [1/2 + \exp(j2\pi y/D)/4 + \exp(-j2\pi y/D)/4]\ \mathrm{Rect}(y/D) \\ &= I_1 + I_2 + I_3 \end{aligned}$$

$$\begin{aligned} \text{where}\quad I_1 &= (1/2)\ \mathrm{Rect}(y/D) \\ I_2 &= (1/4)\exp(j2\pi y/D)\ \mathrm{Rect}(y/D) \\ I_3 &= (1/4)\exp(-j2\pi y/D)\ \mathrm{Rect}(y/D) \end{aligned} \tag{3.52}$$

The illumination pattern has been decomposed into three functions and the Fourier transform of each has been previously developed. Thus,

$$\begin{aligned} f(y) = \mathrm{Rect}(y/D) &\Leftrightarrow F(\phi) = D\ \sin(\pi\phi D)/\pi\phi D \\ &\qquad\qquad \text{peak at 0, nulls at} \pm 1/D \\ f(y)\exp(j2\pi y/D) &\Leftrightarrow F[\phi-(1/D)] \quad \text{peak at } 1/D\text{, a null of } F(\phi) \\ f(y)\exp(-j2\pi y/D) &\Leftrightarrow F[\phi+(1/D)] \quad \text{peak at } -1/D\text{, a null of } F(\phi) \end{aligned} \tag{3.53}$$

The transform of I_1 is a $\sin x/x$ type function of relative amplitude 1/2; that of I_2 is a right-shifted $\sin x/x$ type function of relative amplitude 1/4; of I_3 is a left-shifted $\sin x/x$ type function of relative amplitude 1/4. These three functions are shown solid in the right-hand sketch. The antenna field pattern is the sum of the three functions and is shown by the dashed curve. Note that the sidelobes of the weighted antenna pattern are lower. An explanation of the decrease is that the two components of one-quarter magnitude have sidelobes that are oppositely phased to the one-half magnitude component sidelobes, so the sum of the sidelobes of the three components tend to cancel. Also note that the distance between the nulls either side of the peak of the weighted antenna pattern is twice as wide as those of the $\sin x/x$ function. This indicates that the beamwidth is wider, so the resolution is degraded. An additional degradation due to the weighting is a decreased output SNR. Though the magnitude of the decrease has not yet been evaluated, it is known to be a decrease since the optimum weight set has uniform magnitude and any other weight set produces a smaller output SNR.

A generalized illumination function is given by

$$I_a(x) = [a_1 + a_2 \cos(2\pi x/D)] \operatorname{Rect}(x/D) \tag{3.54}$$

This function includes the cosine-squared function and the Hamming weighting function as special cases. Hamming has lower sidelobes and less SNR loss than cosine-squared weighting. The values of a_1 and a_2 that minimize the largest sidelobe, the Hamming weights, are

$$I_H(x) = [0.54 + 0.46 \cos(2\pi x/D)] \operatorname{Rect}(x/D) \tag{3.55}$$

For a line array, from (3.34) the output energy at the steered angle is proportional to

$$S_{out} = \left| \sum_{n=1,N} w_n^* \right|^2 \tag{3.56}$$

From (3.24) the ouput noise power is proportional to

$$p_{noise} = p_n \sum_{n=1,N} |w_n|^2 \tag{3.57}$$

For continuous illumination functions, the summations can be replaced by integrals so that for the illumination function of (3.54) the output SNR is proportional to

$$\text{SNR} = \left[\int [a_1 + a_2 \cos(2\pi x/D)]\, dx \right]^2 \Big/ \int [a_1 + a_2 \cos(2\pi x/D)]^2\, dx \tag{3.58}$$

where the integrals are over the limits of $-D/2$ to $D/2$. This evaluates to

$$\text{SNR} = a_1^2/(a_1^2 + +a_2^2/2) \tag{3.59}$$

For cosine-squared weights $a_1 = a_2$ and (3.59) evaluates to a loss of 1.76 dB. For the Hamming weights of (3.55) the loss evaluates to 1.34 dB.

The other performance degradation for nonoptimum weights is that the beamwidth increases. The increase factor depends upon how the beamwidth is defined. From (3.3) the "natural" beamwidth was defined as the angle between the beam peak gain and the first zero adjacent to the beam peak. From the right-hand sketch of Figure 3.6 it can be seen that the "natural" beamwidth doubles

Table 3.1
Effects of Weighting

Function	dB SNR Loss	−3-dB Width (Ratio)	Max sidelobe
Uniform	0.00	$0.89^*(D/\lambda)$ (1.00)	−13.2 dB
Cosine squared	1.76	$1.46^*(D/\lambda)$ (1.65)	−31.7
Hamming	1.34	$1.30^*(D/\lambda)$ (1.47)	−42.8
Taylor, $\bar{n} = 8$	1.14	$1.25^*(D/\lambda)$ (1.41)	−40.0

for either Hamming or cosine-squared weights as compared to uniform weights. However, as tabulated in Table 3.1, the two-sided −3-dB beamwidth increases by a smaller factor. Which is the appropriate beamwidth increase criterion depends upon the application. If the radar application is to detect a target in distributed clutter, then computing the total cross section of the clutter using the −3-dB beamwidth gives a close approximation to the value computed by integrating over the beam pattern. But, if the application is to detect a target close to point clutter, using the −3-dB beamwidth is often inadequate. As an example, assume the point clutter radar cross section is 25 dB greater than that of the target and that an SCR of 10 dB is required for detection. When the peak of the beam is pointing at the target, the required SCR is obtained only when the angular separation between the target and point clutter is sufficient, so the relative antenna gain is less than −35 dB. Since this angle is close to the null angle, the resolution beamwidth should be considered the angle of the null rather than the −3-dB angle.

The "natural" beamwidth was introduced as a useful concept because of its ease of computation. For the weights of (3.54), the angles of the nulls do not depend upon the values of a_1 and a_2 and the "natural" beamwidth is twice that of the uniform illumination pattern. However, it is not always easy to compute the exact angles of the nulls. Thus, for the illumination function

$$I_b(x) = \{b_1 + b_2 \cos(2\pi x/D) + b_3 \cos[2\pi x(2/D)]\} \, \mathrm{Rect}(x/D) \qquad (3.60)$$

the location of the nulls depends upon the exact values of b_1, b_2, and b_3. This is because the $\sin x/x$ component due to b_3 peaks at the location of the first null for the Hamming weights of (3.55). Therefore, this angle is no longer a null angle. If all three values of b_m are positive, the first null moves to an angle further from the beam peak; if, however, b_1 and b_2 are positive and b_3 is negative, the first null moves to an angle closer to the beam peak. Usually, however, the "natural" beamwidth is approximately twice that of the uniformly illuminated pattern. A comparison of uniform, cosine-squared and Hamming weighting is given by Table 3.1.

It can be seen that Taylor weighting, discussed later, has significantly less loss than Hamming weighting, slightly smaller −3-dB beamwidth, and similar sidelobes. Thus, it is the preferred weighting function of those tabulated.

The weighting function that produces the narrowest beamwidth for a specified maximum sidelobe level is Dolph-Chebyshev. Its natural beamwidth is approximately the same as that of Hamming weights. All sidelobe peaks are the same value and this value is a design variable. Often, this weighting is nonimplementable for large arrays or a reflector antenna because the edge weights tend to become very large. A practical approximation to this "optimum" weighting function is Taylor weights. Its illumination function is given by (3.61)

$$
\begin{aligned}
I_F(x) &= \left[1 + 2\sum_{n=1,\,\bar{n}} F_m \cos(2\pi\, mx/D)\right] \mathrm{Rect}(x/D) \\
F_m &= 0.5(-1)^{m+1} \prod_{n=1,\,\bar{n};\, n\neq m} \{[1-(m/\nu)^2]/ \\
&\quad [A^2 + (n-0.5)^2]\}/[1-(m/n)^2]
\end{aligned}
\tag{3.61}
$$

The factors A and ν are defined as

$$
\mathrm{SLdB} = 20\log[1/\cosh(A\pi)] \qquad \nu = \bar{n}/[A^2 + (\bar{n}-0.5)^2]^{1/2} \tag{3.62}
$$

where SLdB is the design sidelobe level expressed in decibels and ν is the ratio of the Taylor beamwidth to the Dolph-Chebyshev beamwidth and must exceed 1. The parameter $\bar{n}$ must be large for ν to approach 1. A typical value, as indicated in the table, is eight. Figure 3.6 shows a comparison of the antenna patterns for uniform, Dolph-Chebyshev and Taylor weights. The patterns are derived for line arrays and are approximately valid for reflector antennas. The magnitudes of the three patterns are normalized to the individual peak gains. The design peak sidelobe level for both Taylor and Dolph-Chebyshev weights are −40 dB. It can be seen that the peak sidelobes for the Dolph-Chebyshev weights are all at the −40-dB level. The close-in sidelobes for the Taylor weights are at approximately the −40-dB level but the far-out sidelobes are at a decreasing level. Note that the "natural" beamwidth of the Dolph-Chebyshev response is approximately twice that of the uniform illumination.

The patterns of Figure 3.7 are for line arrays with element-to-element spacing of $\lambda/2$. For this condition ϕ, as defined by (3.21), varies over $\pm\pi$. For this element spacing the equations of Doppler filtering are identical to that of equispaced linear arrays. The weighting functions that are used to reduce antenna sidelobes can also be used to reduce the sidelobes of a Doppler filter. Thus the curves of Figure 3.7 are applicable for either arrays or Doppler filters. To emphasize this duality, the horizontal axes are labeled as either normalized

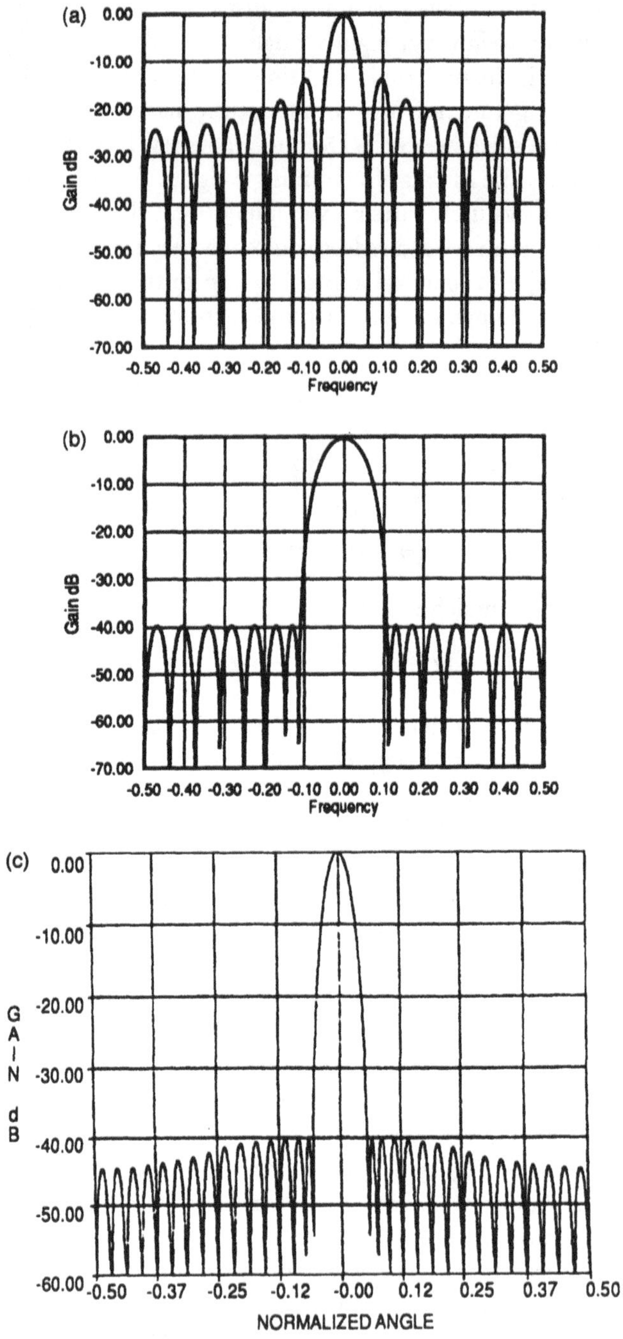

Figure 3.7 Comparison of uniform, Dolph-Chebyshev and Taylor weights.

angle or frequency. The horizontal range of $\pm\frac{1}{2}$ corresponds to the phase angle ϕ varying over $\pm\pi$ for the array or the Doppler frequency varying over $\pm\frac{1}{2}$ the repetition frequency.

The left-hand pattern is for uniform weights; the right-hand pattern is for Taylor weights; and the center pattern is for Dolph-Chebyshev weights. The curves for the uniform and Dolph-Chebyshev weighting are for a 16-element array or a 16-pulse Doppler filter; the curve for Taylor weighting is for a 32-element array or a 32-pulse Doppler processor. This can be verified by noting that both the uniform and Dolph-Chebyshev weighted patterns have a response with 14 sidelobes and a mainlobe of twice the width of a sidelobe for a total count of 16. The Taylor-weighted pattern has 30 sidelobes plus a mainlobe for a total count of 32. Since this pattern has twice the number of elements, its beamwidth is one-half that of the Dolph-Chebyshev weighting.

3.7 Effect of Steering Weight Errors

A practical implementation cannot produce the perfect patterns shown in Figure 3.6 because the implemented weight values are not exactly equal to the theoretical values. For a digital implementation, the errors occur both in amplitude and phase because the implemented gain and phases are only available in discrete steps. Typically the available gain values are 2 dB, 1 dB, and 0.5 dB, for example. Similarly, the available phase shift values are 180, 90, and 45 degrees. Similar errors occur in analog implementations. The effect of implementation errors will be analyzed by assuming the nth errored weight, $w_{n\varepsilon}$, is given by

$$w_{n\varepsilon} = w_n(1 + g_{n\varepsilon})\exp(-j\phi_{n\varepsilon}) \tag{3.63}$$

where $g_{n\varepsilon}$ and $\phi_{n\varepsilon}$ represent the gain and phase errors, respectively. For small errors, (3.63) may be approximated as

$$w_{n\varepsilon} = w_n(1 + g_{n\varepsilon} - j\phi_{n\varepsilon}) \tag{3.64}$$

so the antenna output voltage for the array factor is

$$\begin{aligned} v &= \sum w_n(1 + g_{n\varepsilon} - j\phi_{n\varepsilon})\exp[-j(n-1)\phi] \\ &= \sum w_n \exp[-j(n-1)\phi] + \sum w_n(g_{n\varepsilon} - j\phi_{n\varepsilon})\exp[-j(n-1)\phi] \\ &= \text{Unerrored term} + \text{Errored term} \end{aligned} \tag{3.65}$$

Assuming the errors are zero-mean, independent, random variables with variances of σ_g^2 and σ_ϕ^2, respectively, the errored term is also zero mean and, since

$\sum |w_n|^2$ equals 1, its variance is

$$\text{var}(v) = \left(\sigma_g^2 + \sigma_\phi^2\right) \tag{3.66}$$

As a numeric example assume the gain error is a uniformly distributed random variable over the range of $(1 \pm \varepsilon)$ so that its variance equals $\varepsilon^2/3$. When ε equals 0.05, corresponding to a gain error of approximately ± 0.43 dB, (3.66) evaluates to 8.33×10^{-4} or -30.8 dB. This means the root mean square (rms) level of the error-induced sidelobes are at a level of -30.8 dB relative to a unit gain omnidirectional antenna. For a uniformly illuminated 32-element array, the maximum gain equals 18.1 dB, so the rms weight noise error sidelobes are 45.9 dB below the mainlobe gain. Thus, the weight noise errors are of little importance if the design sidelobe level equals -40 dB but are the limiting factor if the design sidelobe level equals -50 dB. To achieve sidelobes of -50 dB it is necessary that the amplifier gains vary by less than the assumed ± 0.43 dB. If the error ε is reduced to 0.025, corresponding to a gain error of approximately ± 0.22 dB, the error-induced sidelobes are at a level of -36.8 dB or 51.9 dB below the maximum antenna gain. Note that when the amplifier gain error decreases by a factor of 2, the rms level of the error-induced sidelobes decreases by 6 dB. A similar analysis can be done for the phase errors, and Figure 3.8 shows the rms level of the error sidelobes relative to a unit gain antenna for both gain and phase errors. Usually an implementation has both gain and phase errors and so it is necessary to combine the effects of these errors in order to predict the level of the errored sidelobes. Figure 3.9 shows the array factor antenna pattern for a 32-element array implemented with -40-dB Taylor weighting. The top-most pattern is for nonerrored weights and the bottom pattern is for random errors distributed over a ± 0.43-dB range, so the average noise sidelobes, as computed above, are at a -45.9-dB level. However, it can be seen that the noisy nature of the errors causes some of the sidelobes to exceed the -40-dB level. This occurs when the phase of the errored term is approximately the same as the phase of the

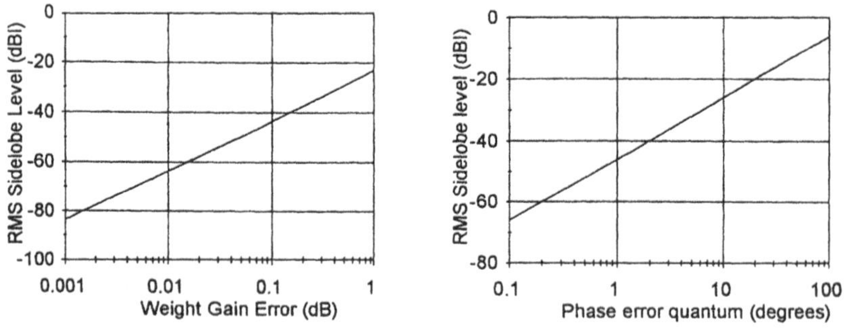

Figure 3.8 RMS sidelobe level due to weighting errors.

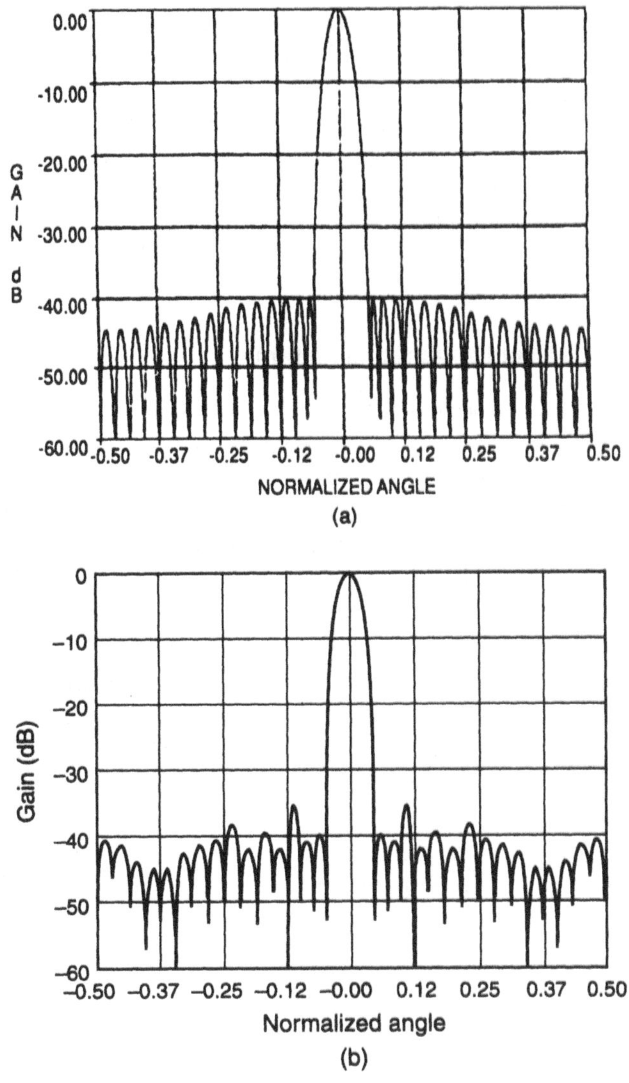

Figure 3.9 Effects of weight errors on antenna pattern.

unerrored term so that the two terms combine constructively. Thus, the variance analysis is too liberal and tighter error requirements are required.

3.8 Earth Effects on Antenna Patterns

The Earth affects target detection because of two factors. One is due to multipath. The field strength at a point is due to the additive combination of the direct ray and several indirect rays due to reflections from the ground. The resultant antenna gain has a null at the horizon and other angles close to the horizon. An important effect of these nulls is that the SNR for a constant altitude target does not increase monotonically as the range decreases. A second effect is due to the combination of the curvature of the Earth and the target's low altitude. Microwave energy propagates along straight lines in a homogeneous medium. Thus, when a target is below the radar horizon it cannot be detected. This occurs even though the computed SNR for this range target is large enough to predict easy detectability.

3.8.1 Multipath Effect on Antenna Pattern

The previously derived antenna pattern equations are correct only when the antenna is located in free space. The free-space antenna pattern analysis assumes there is only a single propagation path between the radar antenna and the reflector for both transmit and receive. The assumption is incorrect when the antenna is close to the reflecting Earth. Because of reflections from the ground, there can be many paths. Figure 3.10 shows a single reflection wave, though usually there are several reflection paths. The field intensity at the reflector is due to the additive combination of the direct and indirect waves. The relative magnitude of the direct and indirect waves depends upon the antenna gains in the respective directions and the coefficient of reflection for the surface. For long-range, low-altitude targets, the gains in the respective directions are approximately equal. For each reflected wave the phase difference between the direct and indirect waves depends upon the additional path length of the indirect wave and the phase change induced due to the reflection coefficient. The reflection coefficient depends upon the angle of incidence of the wave with respect to the surface and the polarization. For the usual multipath situation the angle of incidence is small and the reflection

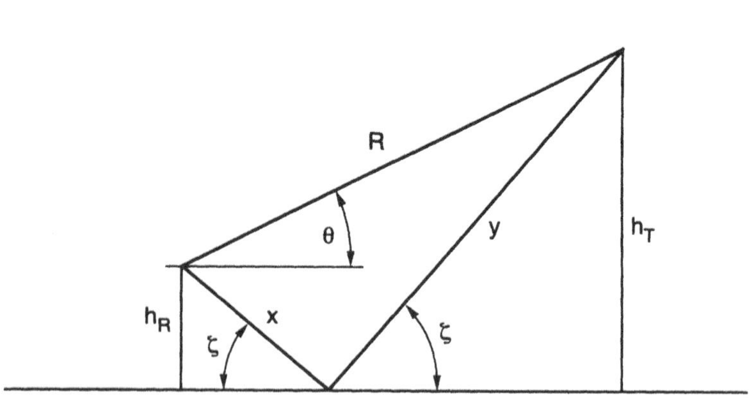

Figure 3.10 Multipath effect on antenna gain.

coefficient is approximately equal to -1 for all polarizations. When there is only a single indirect wave, the direct and indirect waves will sometimes combine constructively and sometimes destructively. The actual multipath phenomenon is complex due to the many reflection paths between the radar and target and also due to the Earth's curvature. A useful understanding of the phenomenon is derived for the simple case of a single reflection ray and a flat Earth.

Denote the field intensity at a point due to the direct path as F_0. The total field due to the direct and reflected rays, F_T, when the antenna gains for the two paths are identical is

$$F_T = F_0\{1 + |\rho|\exp[-j(\alpha + \xi)]\} \tag{3.67}$$

where the reflection coefficient is $|\rho|\exp(-j\alpha)$ and ξ denotes the phase shift due to the additional path length for the indirect ray.

The total antenna gain is the product of two factors. One is the free-space antenna gain; and the second, G_M, from (3.67) is given by

$$G_M = |F_T/F_0|^2 = |1 + |\rho|\exp[-j(\alpha + \xi)]|^2 \tag{3.68}$$

Assuming the reflection coefficient is approximately equal to one, the value of G_M reduces to

$$G_M = 2(1 - \cos\xi) = 4\sin^2(\xi/2) \tag{3.69}$$

and thus ranges between 0 and 4. From Figure 3.10, the range to the target, for the direct and indirect paths, R and R_I, respectively, are given by

$$\begin{aligned} R^2 &= (h_T - h_R)^2 + [h_R/\tan\zeta + h_T/\tan\zeta]^2 = (h_T - h_R)^2 \\ &\quad + [(h_T + h_R)/\sin\zeta]^2\cos^2\zeta \\ R_I &= x + y = h_R/\sin\zeta + h_T/\sin\zeta \\ \xi &= (2\pi/\lambda)(R_I - R) \end{aligned} \tag{3.70}$$

Thus,

$$\begin{aligned} R_I^2 - R^2 &= R_I^2(1 - \cos^2\zeta) - (h_T - h_R)^2 = R_I^2\sin^2\zeta - (h_T - h_R)^2 \\ R_I - R &= [(h_T + h_R)^2 - (h_T - h_R)^2]/(R_I + R) = 4h_T h_R/(R_I + R) \\ &\approx 2h_T h_R/R \Rightarrow \xi \approx (4\pi/\lambda)(h_T h_R/R) \end{aligned} \tag{3.71}$$

Combining (3.71) and (3.69) gives

$$G_M \approx 4\sin^2[2\pi h_T h_R/(\lambda R)] \tag{3.72}$$

It can be seen that the multipath gain factor varies sinusoidally. For constant target range, the gain varies sinusoidally with target altitude. The gain equals zero when the target height equals zero, so there is an antenna null at the local horizon. The gain also equals zero for

$$(2k - 1)/2 = (h_T h_R)/(\lambda R) \approx (h_R/\lambda)\sin\theta \tag{3.73}$$

where k is any positive integer and θ is defined in Figure 3.9. As an example, for h_R equal to 10m and λ equal to 10 cm, the pattern has nulls at θ equal to 0.29, 0.58, and 0.87 degrees, for example. The peak of the pattern is halfway between the nulls and occurs at angles of 0.245, 0.435, and 0.725 degrees, for example. An important effect of these nulls is that the SNR for a constant altitude target does not increase monotonically as the range decreases.

Figure 3.11 shows three antenna patterns for a 64-element array with d/λ spacing equal to one-half. The three cases are uniform weighting and no ground reflection, Hamming weighting and no ground reflection, and Hamming weighting and ground reflection. The dashed curve with a first null at approximately 1.8 degrees is the uniformly weighted sin x/x type pattern. The other dashed curve with a first null at approximately 3.6 degrees is for Hamming weighting and no ground reflection. As previously discussed, the "natural" beamwidth is doubled by Hamming weighting. The third curve is for the Hamming weighted array close to the ground. It can be seen that this pattern has many nulls within the "natural" beamwidth. The gain is shown as a function

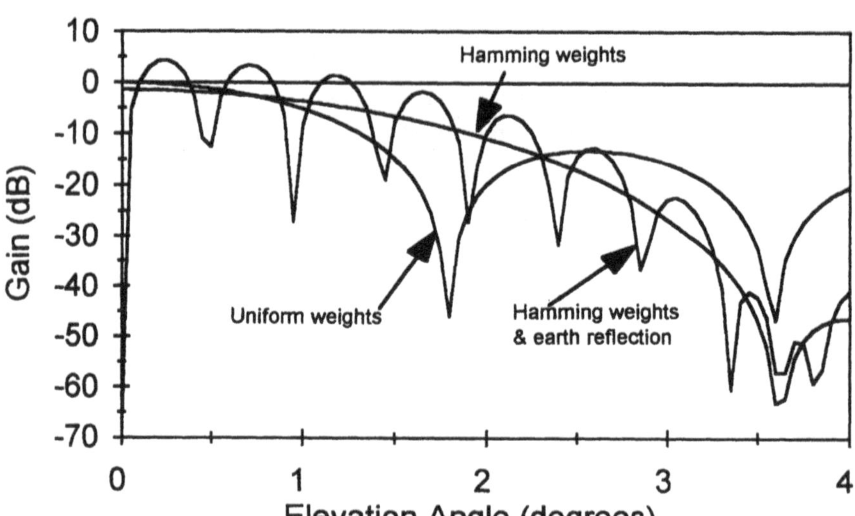

Figure 3.11 Effect of reflecting Earth on antenna pattern.

of positive elevation angle only, since negative elevation angle is below the reflecting ground. The three patterns are normalized to the gain of the free-space unweighted array. Thus, the peak gain of the unweighted array is 0 dB. The peak gain of the free-space Hamming weighted array is less than 0 dB because of the SNR loss associated with weighting. The peak gain of the Hamming weighted array close to the ground is increased due to the constructive interference effect.

3.8.2 Spherical Earth and Low-Altitude Targets

Microwave energy propagates along straight lines in a homogeneous medium, and targets that are below the propagation path cannot be detected. An example is shown in the sketch of Figure 3.12. The far-range target indicated to the right of the figure is below the optical horizon and, to a first-order approximation, cannot be detected. This occurs even if the computed SNR for this range target is large enough to predict easy detectability. The other target can be detected since it is just above the optical horizon. To compute the distance to the optical horizon, denote the distance from the radar to the tangent point as D_1, from the tangent point to the target as D_2, and the Earth's radius as R_e. Then

$$\begin{aligned} D_1^2 &= R_e^2 + (R_e + h_R)^2 \\ D_2^2 &= R_e^2 + (R_e + h_T)^2 \end{aligned} \tag{3.74}$$

When the Earth's radius is much larger than the heights of the radar and target, the LOS-limited target range, D_{max}, is equal to the sum of D_1 and D_2,

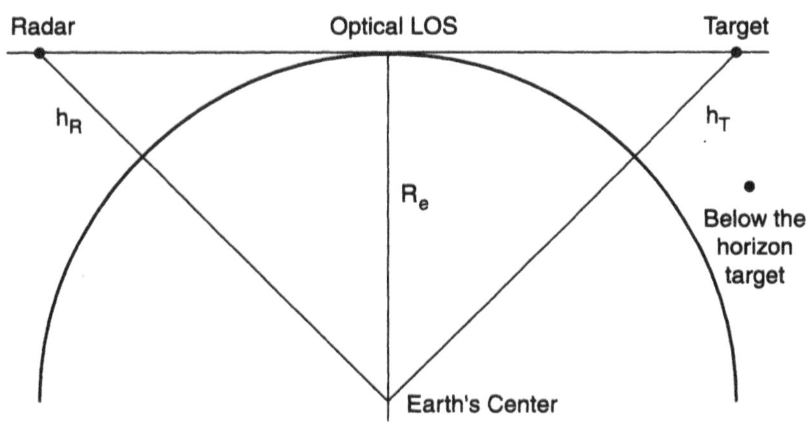

Figure 3.12 Spherical Earth effects.

approximately

$$D_{max} = \sqrt{(2R_e)}[\sqrt{h_R} + \sqrt{h_T}] \tag{3.75}$$

Since the atmosphere is slightly nonhomogeneous with altitude, microwave energy is bent (refracted) toward the Earth and targets that are only slightly below the optical horizon can be detected. The degree of refraction is atmosphere state-dependent. An approximation to the difference between the radar horizon and the optical horizon can be obtained by assuming the Earth's radius is larger than its physical measurement. For a "standard atmosphere" assumption, the approximation is that the radar horizon is obtained by assuming the Earth's radius equals 4/3 of its physical value. Thus, when D_{max} is expressed in kilometers and the target and radar heights are expressed in meters, (3.75) reduces to

$$D_{max} = 4.12[\sqrt{h_R} + \sqrt{h_T}] \tag{3.76}$$

As a numeric example, if the radar height is equal to 10m and the target height is equal to 100m, the LOS-limited range equals 54 km. If the target altitude is reduced to 30m, the target is undetectable because it is below the horizon until its range is less than 35.6 km.

3.9 Waveform Sidelobe Reduction

The sidelobe reduction techniques in this chapter were discussed primarily from the antenna sidelobe reduction viewpoint with reference to the usability of the techniques to the sidelobe reduction for compressed pulse waveforms. A specific example of the applicability is shown in Figure 3.13. It shows a conceptual block diagram of a signal processor using Hamming weighting for sidelobe reduction of an LFM waveform. The figure shows that the signal is initially processed by a pulse compression network and then fed to a tapped delay line. The tap spacing equals $1/\Delta$, where Δ is the LFM deviation. The weights, from (3.55) and normalized to unit noise power gain, equal

$$w_1 = 0.36485 \qquad w_2 = 0.85660 \qquad w_3 = 0.36485 \tag{3.77}$$

Approximating the LFM spectrum as a rectangle function, the output waveform of the pulse compressor is a sin x/x type function. The decreased sidelobe level compressed waveform is synthesized as indicated by Figure 3.6 as the superposition of three delayed and weighted sin x/x type functions.

An alternate, and equivalent implementation modifies the transfer function of the matched filter pulse compressor, so it includes the filtering action of the Hamming weighted delay line. That is, assuming the magnitude of the

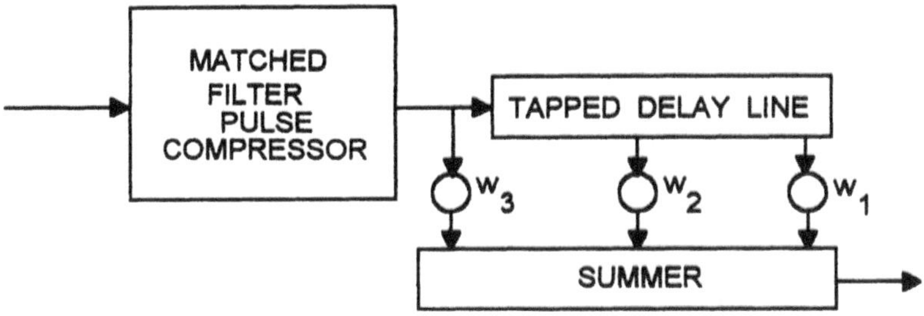

Figure 3.13 Hamming weighting for LFM waveform.

LFM spectrum is constant, the matched filter pulse compressor has the transfer function

$$H_{\mathrm{M}}(f) = \mathrm{Rect}(f/\Delta)\exp[\Phi(f)] \tag{3.78}$$

where $\Phi(f)$ is the quadratic phase of the spectrum of the LFM waveform. The transfer function of the Hamming weighted delay line, $H_{\mathrm{H}}(f)$, is

$$H_{\mathrm{H}}(f) = 0.85660 + 2 * 0.36485\cos(2\pi f/\Delta) \tag{3.79}$$

Thus, the combined filter that performs pulse compression and sidelobe suppression is implemented by the transfer function

$$H(f) = [0.85660 + 2 * 0.36485\cos(2\pi f/\Delta)]\exp[\Phi(f)] \tag{3.80}$$

A digitally implemented signal processor usually uses the combined filter of (3.80).

4

Target Detection in Noise

A major problem in detecting targets in noise is the presence of the returns from other reflectors that are not of interest. These other reflectors are designated as clutter. The two primary types of clutter are point clutter and distributed clutter. Point clutter has a physical dimension that is smaller than the radar's resolution size. Distributed clutter has a physical dimension larger than the radar's resolution cell size. Examples of the former are water towers, buildings, and mountains. The Doppler frequency spectrum of these reflectors is centered at zero, approximately of zero width, and "easy" to reject by Doppler processing techniques. However, due to their large radar cross section, their return is large and, without careful receiver design, could cause receiver saturation. It is important to avoid saturation because it causes a capture effect. That is, a target of interest at the same range as the clutter can be detectable in a linear receiver after clutter attenuation but is not detectable after saturation.

Distributed clutter is ground clutter or volume clutter. Ground clutter and volume clutter differ primarily by their spectral characteristics. The latter—caused by reflectors such as rain, snow, chaff, birds, insects—are wind effected. Thus their average Doppler frequency is not zero. Since the delay line canceler null is at zero Doppler, volume clutter is often more difficult to cancel than any of the other forms of clutter. The major method of detecting the reflections due to desired targets while ignoring those due to clutter is by exploiting the different Doppler characteristics of reflectors. The design and evaluation of Doppler signal processors to attenuate clutter returns is discussed in the next chapter. This chapter focusses on the detection of targets in a pure thermal noise environment.

4.1 Target Scattering Models

The typical radar target can be modeled as reflecting energy from a number of isolated reflecting points as indicated by the sketch in Figure 4.1. The vertical arrows depict the magnitude of the radar cross section of each scattering point as a function of range. The scattering model is often denoted as the range profile and is aspect dependent. When the range cell dimension of the transmit waveform is much less than the target dimension, the reflecting highlights are resolved. Measuring the range profile is useful for target identification but, since the profile is aspect sensitive, the technique is primarily useful for measuring target length. Additional identification information can be obtained using cross-range measurements. The cross-range dimension, in theory, can be measured using differential Doppler measurements induced by vehicle rotation as depicted in Figure 4.2. The differential range-Doppler measurements are with respect to a "center of gravity," and thus the technique requires high-quality range and Doppler tracking of the "center of gravity."

When the range cell dimension of the transmit waveform is much greater than the target dimension, the reflecting highlights are unresolved. The target reflection is the phasor sum of the highlight reflections. The relative phases of the highlight reflections are sensitive to aspect angle and carrier frequency, so the target reflection sum is very sensitive to aspect angle and carrier frequency. A typical aircraft cross-section diagram as a function of aspect angle is shown in Figure 4.3.

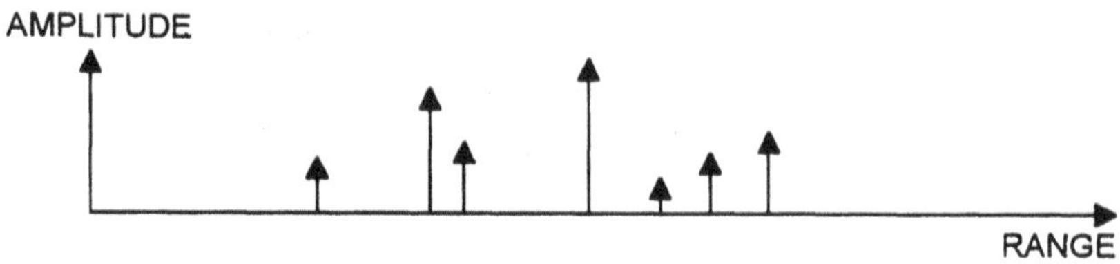

Figure 4.1 Target range scattering profile.

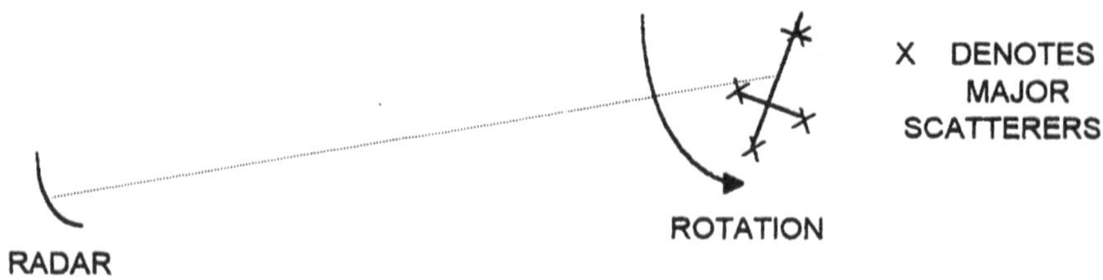

Figure 4.2 Cross-range measurement.

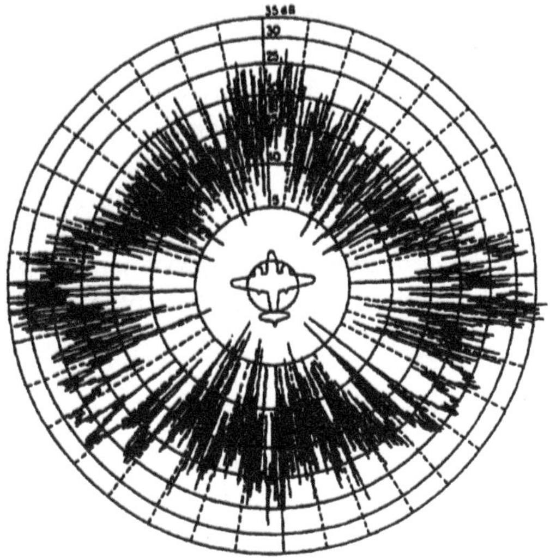

Figure 4.3 Typical aircraft cross-section diagram.

4.2 Target Detection Theory

The target reflection for a single pulse transmission, after reception by the antenna and RF filter, is a pulse of constant amplitude, A, and duration, τ. A target present or absent decision is made for each resolution cell. For a thermal noise environment, a target present decision is made when the output of an envelope detector exceeds a fixed threshold value. As discussed in more detail later, the probability of detection depends upon the SNR defined via the equivalent expressions of

$$\text{SNR} = (A^2/2)/(kT_S B) = (A^2/2)N_0 B = (A^2\tau/2)/N_0$$

where k is Boltzman's constant, T_S is the temperature in degrees Kelvin, B is the bandwidth, and N_0 is the noise energy spectral density. The expression to the right of the equation shows that the SNR is the ratio of the energy of the reflected waveform to the noise energy.

A statistical evaluation of the probability of detection is obtained by noting that a general expression for the pre-detector target return within a resolution cell is

$$v_T(t) = A_X(t)\cos(\omega_C t) + A_Y(t)\sin(\omega_C t)$$

where $A_X(t)$ and A_Y are the in-phase and quadrature target amplitude modulations. The target amplitude varies with time due to the effect of the pre-detector filters. This section concentrates on detection in a noise environment. Thus, it is assumed that these filters are matched filters that maximize the pre-detector SNR.

The equivalent expression for the interference (the sum of thermal noise, clutter, and jamming residue after cancellation processors) is

$$v_{\mathrm{I}}(t) = x_1(t)\cos(\omega_{\mathrm{C}}t) + y_1(t)\sin(\omega_{\mathrm{C}}t)$$

where $x_1(t)$ and $y_1(t)$ are zero-mean, random processes. The pre-detector voltage is the sum of the two expressions. The detector output is the envelope $r(t)$ and equals

$$\begin{aligned} r(t) &= [A_X(t) + x_1(t)]^2 \\ &\quad + [A_Y(t) + y_1(t)^2]^{1/2} \\ &= [x^2(t) + y^2(t)]^{1/2} \end{aligned}$$

where $x(t)$ and $y(t)$ are random processes with respective means of $A_{\mathrm{X}}(t)$ and $A_{\mathrm{Y}}(t)$. The detection probability is highest at the waveform peak, A_{P}. Denote the respective values of $A_X(t)$ and $A_Y(t)$ at the time instant the peak is attained as A_X and A_Y, respectively. Assuming the interference statistics are bivariate Gaussian, the joint probability of the input voltage to the envelope detector is

$$\begin{aligned} p(x, y) &= \exp\{[-(x - A_X)^2 - (y - A_Y)^2]/2\sigma^2\}/2\pi\sigma^2 \\ &= \exp\left[-(A_X^2 + A_Y^2)/2\sigma^2\right] \\ &\quad \times \exp[-(x^2 + y^2 - 2xA_x - 2yA_y)/2\sigma^2]/2\pi\sigma^2 \end{aligned}$$

where σ^2 is the statistical variance, equal to the total interference power. Changing variables to polar coordinates gives the joint density function for the envelope magnitude r and the electrical phase angle θ as

$$\begin{aligned} p(r, \theta) &= r\exp(-A_{\mathrm{P}}^2/2\sigma^2) \\ &\quad \times \exp\{-[r^2 - 2r(A_x\cos\theta + A_y\sin\theta)]/2\sigma^2\}/2\pi\sigma^2 \\ A_{\mathrm{P}}^2 &= A_x^2 + A_y^2 \end{aligned}$$

The density function for the envelope is obtained by integrating the joint density function with respect to θ, resulting in

$$p(r) = r\exp(-A_{\mathrm{P}}^2/2\sigma^2)\exp(-r^2/2\sigma^2)I_0(rA_{\mathrm{P}}/\sigma^2)r/\sigma^2 \tag{4.1}$$

where I_0 denotes the modified Bessel function of the first kind. Examples of the envelope probability density are shown in Figure 4.4. The three cases shown are

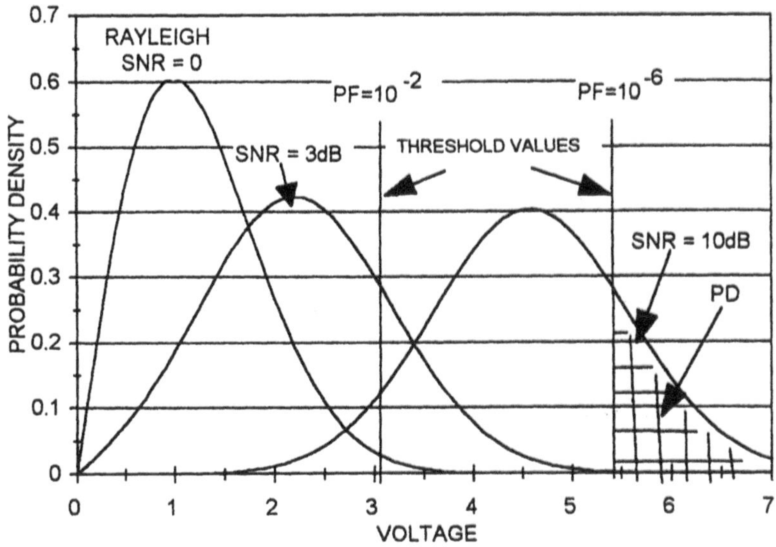

Figure 4.4 Envelope probability density functions.

for SNR values of 0, 2 (3 dB), and 10 (10 dB). The density function for SNR $= 0$ is special and is also named Rayleigh.

In comparing these density functions it can be seen that large values of the detector output are more probable for the target present condition than for the target absent condition. Thus, it is reasonable to make a target present decision for large values and a target absent decision for small values. The decision dichotomy is implemented by comparing the observed voltage value to a threshold voltage. There are two types of decision errors. One is a false alarm where a target present decision is made when a target is absent. A false alarm occurs when the voltage for the Rayleigh distribution exceeds the threshold. The false alarm probability equals the area under the curve of the Rayleigh density function to the right of the threshold value. This area decreases as the threshold value increases. The threshold values for false alarm probabilities of $1. \times 10^{-2}$ and $1. \times 10^{-6}$ are shown in the figure. The detection probability equals the area under the curve to the right of the threshold value for the cases when the SNR > 0. This area increases as the SNR increases. For SNR $=$ 10 dB, the area corresponding to the probability of detection is shown as the cross-hatched area. Note that increasing the threshold value decreases both the false alarm probability and the detection probability so that the choice of the threshold value is a trade-off. The probability of false alarm is computed by integrating 4.1 over the range of the threshold value to infinity for the case of $A_P =$ zero. This integral can easily be evaluated. However, for $A_P \neq 0$, the integral cannot be evaluated in closed form. The integral is known

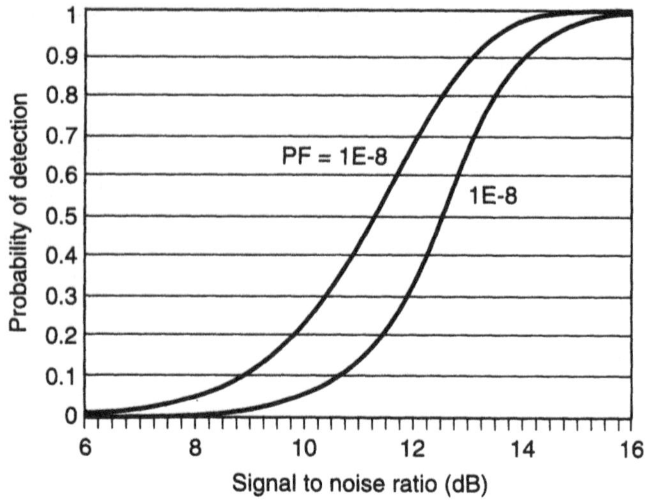

Figure 4.5 Probability of detection for non-fluctuating target.

as Marcum's Q function and can be evaluated by a series approximation or by digital integration. It can be shown that the value of the integral depends upon the ratio of the threshold value to σ and the ratio of A_P to σ. This is denoted as

$$p_D = Q(A_P/\sigma,\ T/\sigma) \tag{4.2}$$

Figure 4.5 shows curves of the probability of detection for false alarm probabilities of $1.\times 10^{-6}$ and $1.\times 10^{-8}$.

4.3 Fluctuating Cross-section Targets

The previous results are for the case where the target cross section is a fixed value. From Figure 4.3 it can be seen that for some targets the cross section varies over many magnitudes depending upon the observation angle and that large cross-section changes occur for small angle changes. For this case it is reasonable to model the target cross section as a random variable so that the value of the SNR, or equivalently A_P, is also a random variable. For a fluctuating cross-section target, the probability of detection is given by

$$p_D = \int_0^\infty p_A(A_P)\, Q(A_P/\sigma,\ T/\sigma)\, dA_P$$

where $p_A(A_P)$ denotes the probability density function of the target-induced detector peak input voltage A_P. An alternate and equivalent evaluation is to integrate with respect to the random cross section; this is identical to integrating with respect to the random SNR value, denoted as γ. Thus,

$$p_{\mathrm{D}} = \int_0^{\infty} p_{\gamma}(\gamma)\, Q(A_{\mathrm{P}}/\sigma,\, T/\sigma)\, d\gamma \qquad \gamma = (A_{\mathrm{P}}/\sigma)^2/2 \tag{4.3}$$

where $p_{\gamma}(\gamma)$ denotes the probability density function of the random SNR. Several random SNR models were developed by P. Swerling. They were chosen on the basis that they are reasonable approximations to the measurements for real targets and because the resulting integrals can be evaluated. The first model, termed Swerling 1, approximates the random SNR by the probability density function

$$p_{\gamma}(\gamma) = \exp[-(\gamma/\underline{\gamma})/\underline{\gamma}] \tag{4.4}$$

where $\underline{\gamma}$ is the average SNR. This well-known probability density function is named exponential. It can be shown that the corresponding density function for the detector input voltage is

$$p_{\mathrm{A}}(A_{\mathrm{P}}) = \left(A_{\mathrm{P}}/A_0^2\right) \exp[-(A_{\mathrm{P}}/A_0)^2/2]$$

and this is a Rayleigh density. Substituting (4.4) into (4.3) results in

$$p_{\mathrm{D}} = \exp\{[-(T/\sigma)^2/2]/(1+\underline{\gamma})\} \tag{4.5}$$

The false alarm probability dependence on the threshold value is obtained from (4.5) for $\underline{\gamma} =$ zero. Thus,

$$p_{\mathrm{F}} = \exp[-(T/\sigma)^2/2] \tag{4.6}$$

and combining (4.5) and (4.6) gives

$$p_{\mathrm{D}} = p_{\mathrm{F}}^{[1/(1+\underline{\gamma})]} \tag{4.7}$$

Figure 4.6 shows the detection probability versus average SNR, $\underline{\gamma}$, for the Swerling 1 target for false alarm probabilities of $1.\times 10^{-6}$ and $1.\times 10^{-8}$. Figure 4.7 shows the curves of probability of detection for a nonfluctuating target and for a Swerling 1 target for a false alarm probability of $1. \times 10^{-8}$. As indicated, for probability of detection greater than approximately 0.35, the SNR required to detect a Swerling 1 target exceeds that required for a nonfluctuating target. The additional SNR is termed a fluctuation loss. The magnitude of the

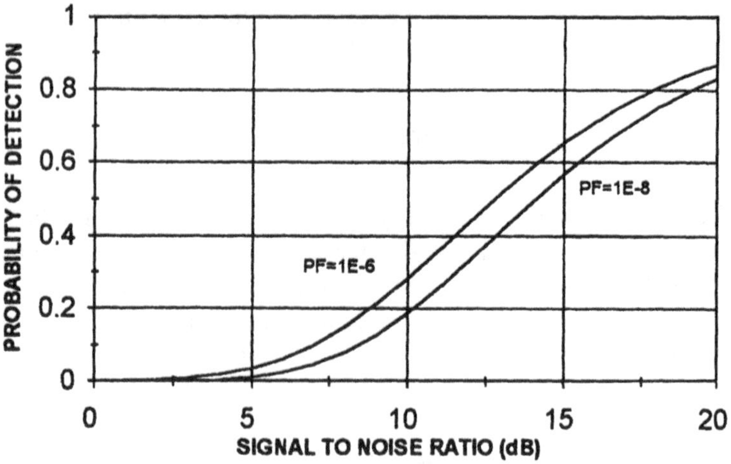

Figure 4.6 Probability of detection for Swerling 1 target.

fluctuation loss depends upon the cross section's probability density function. As discussed later, other models have a different fluctuation loss.

The initial assumption for the Swerling 1 target model is that the random cross section, or SNR, has an exponential density, so the voltage induced by the target has a Rayleigh distribution. An alternate, and complementary, viewpoint is that the voltages A_X and A_Y of (4.1) are independent, zero-mean, identically distributed Gaussian random variables. With this assumption A_P, defined in (4.2) as the squared sum of A_X and A_Y, has a Rayleigh probability density and thus the Gaussian assumption is consistent with the prior Rayleigh envelope

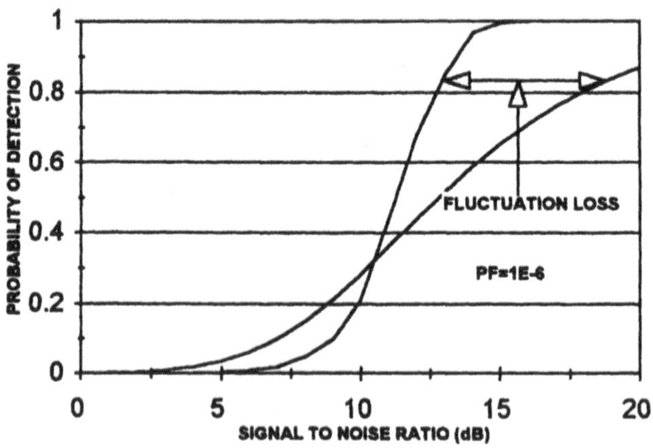

Figure 4.7 Fluctuation loss for Swerling 1 target.

assumption. The advantage of the Gaussian assumption is it shows that for the Swerling 1 target $x(t)$ and $y(t)$ of (4.1) are both nonzero-mean, Gaussian random variables and the analysis of bivariate Gaussian random variables is relatively simple.

4.4 Additional Swerling Cross-section Fluctuation Models

For the Swerling 1 model the random SNR is the exponential distribution given by (4.4). An alternate assumption is that the probability density function of the SNR is a different density. A useful assumption is the density, given by

$$p_\gamma(\gamma) = 4(\gamma/\underline{\gamma}^2)\exp[-(2\gamma/\underline{\gamma})] \tag{4.8}$$

This model is named Swerling 3. Substituting (4.8) into (4.3) and integrating gives the probability of target detection as

$$p_D = [2/(2+\underline{\gamma})^2][2 + ((T/\sigma)^2/2)\underline{\gamma} + \underline{\gamma}^2/2]\exp[-(T/\sigma)^2/2/(1+\underline{\gamma}/2] \tag{4.9}$$

The probability of false alarm is still given by (4.6). A comparison of the three models—nonfluctuating, Swerling 1, and Swerling 3—is given by Figure 4.8. For all cases the average SNR equals 3 dB. The probability of target detection for the three models and a false alarm probability of $1.\times 10^{-8}$ is shown in Figure 4.9.

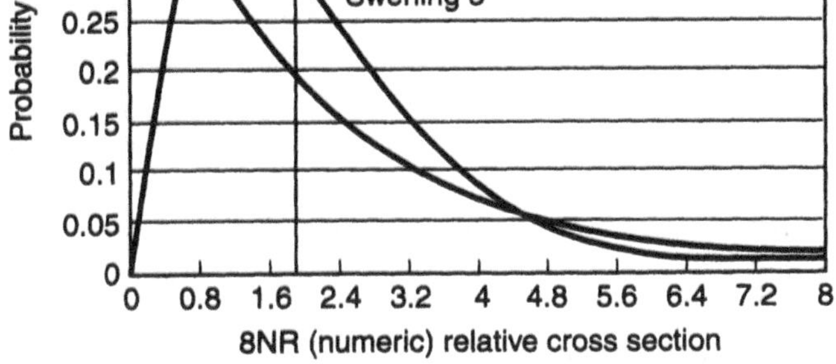

Figure 4.8 Comparison of non-fluctuating and fluctuating target models.

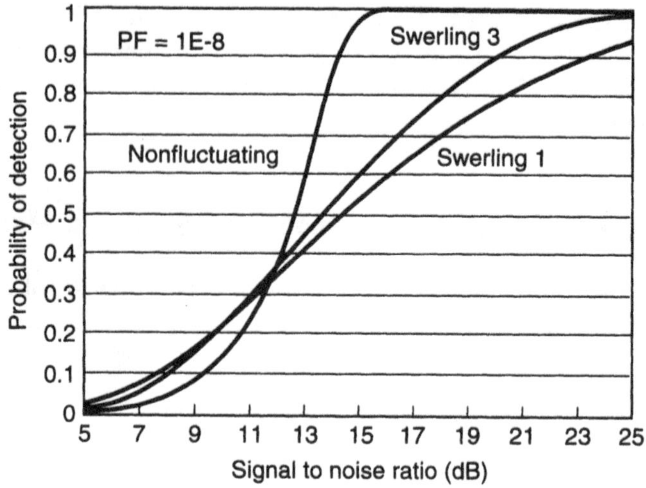

Figure 4.9 Comparative probability of detections.

It can be seen that the probability of detection for the Swerling 3 model is closer to that of the nonfluctuating target than is the Swerling 1 model. Thus, the fluctuation loss is less for this type of target. Note that the proper model depends upon the fit of the model to measured data. Other models (such as Weinstock) are used if these models are poor fits. The major features of the two Swerling probability density functions are that they are mathematically convenient (the integral equation (4.3) can be evaluated in terms of well-known easily computed functions) and a close fit to measured data.

An additional aspect of the random cross section of targets is the rate of cross-section fluctuation. The previous curves are computed on the basis of a single-pulse reflection, so there is no fluctuation during the dwell. More generally, the radar transmission is a train of N pulses for each surveillance region. Specifically, for an N-pulse dwell, the received target amplitudes on two consecutive scans for a resolution cell are denoted as $A_1, A_2, A_3, \ldots, A_N$ and $B_1, B_2, B_3, \ldots, B_N$. There are four Swerling models denoted as Swerling 1 through Swerling 4. All Swerling models assume the A set is statistically independent of the B set. The Swerling 1 and 3 models assume the set of A amplitudes have identical value and the set of B amplitudes have identical value. The Swerling 1 model uses (4.4) as the probability density for the cross section for the random A and B voltage values, and the Swerling 3 model uses (4.8) as the probability density for the cross section. The Swerling 2 and Swerling 4 models assume each reflection of the A and B sets are statistically independent and use (4.4) and (4.8), respectively, as probability density for the cross section. Thus, Swerling 1 and 3 models assume slow cross-section fluctuation and Swerling 2 and 4 assume fast fluctuation. Slow fluctuation is a good assumption for a constant carrier frequency transmission. Fast fluctuation is a good assumption for a nonconstant carrier frequency transmission when the pulse-to-pulse carrier frequency change is large. Specifically, fast fluctuation is the correct model when the pulse-to-pulse carrier frequency change exceeds the bandwidth, which results in a range resolution equal to the target length. The generalization of the curves of Figure 4.9 is that for a specified average SNR, the probability of detection for the Swerling 1 model is less than that of the Swerling 2 model; is less than that of the Swerling 3 model; and less than that of the Swerling 4 model.

5

Target Detection in Clutter

Often, detection of targets is significantly degraded by the presence of clutter. The detectability of targets in clutter can be enhanced by exploiting the differential Doppler frequencies due to different velocities of the target and clutter. Chapter 1 discussed the use of delay line cancelers and Doppler filter banks. The efficacy of these techniques depends upon the spectral characteristics of the clutter. The spectrum of ground clutter is relatively narrow and centered at zero Doppler. Because delay line cancelers can be easily configured to have a null at zero Doppler, ground clutter is "easily" canceled using delay line cancelers. However, clutter due to reflections from sea, weather, birds, insects, and ground traffic, for example, do not have zero-mean Doppler. They are not well canceled by delay line cancelers. A Doppler filter bank often is a significantly better processor.

5.1 Introduction

Doppler filters can be implemented using either deterministic weights or adaptive weights. This is similar to using deterministic or adaptive weights for antennas to aid in rejecting sidelobe jamming. In either case, the adaptive weight implementation is more expensive but the performance is often significantly improved. This chapter will analyze the relative and absolute performance attained by deterministic Doppler filters. A later chapter will analyze the relative and absolute performance attained by adaptive Doppler filters. These data can then be combined with cost data to select the preferred architecture. Note that cost data depend upon component availability. Thus, if adaptive processing is deemed too expensive in a particular application for a specified time period, this decision could change for the next implementation generation.

The first portion of this chapter gives an overview of clutter characteristics and alternate methods of detecting targets in clutter environments. The next part discusses several aspects of delay line cancelers. These circuits are also called moving target indicators (MTI). Later chapters give a detailed analysis of the performance of adaptive and nonadaptive Doppler filter bank implementations.

5.2 Clutter Discriminants

The detection capability of a radar system is limited by the phenomenon that often the reflections received by the radar comes from scatterers that are not of interest; this received radiation is termed a clutter return. The reflection received from a scatterer of interest is termed the target return. These two statements, while fairly trivial and obvious, are intended to emphasize that, to a large degree, the differences between clutter and target are largely in the mind of the observer. Thus, the procedures used to detect the presence of targets, while ignoring the presence of clutter, are based upon the exploitation of any known physical differences between these types of scatterers. Several differences between the scattering properties of targets and clutter exist and can be exploited. Examples of properties that may be useful are discussed in the following sections.

5.2.1 Target Elevation

Under some conditions, the scatter that is the target is above the level of the ground and the sources of clutter are at ground level. Techniques that discriminate upon the basis of scatterer altitude are useful. There are, of course, situations where this is not a useful discriminant.

5.2.2 Target Velocity

This is currently the most common technique used to discriminate between targets and clutter; it is useful when a significant velocity difference exists.

5.2.3 Transmission Polarization-Sensitivity

The energy backscattered from an object is a function of the polarization of the illuminating energy, and different objects usually have different polarization properties. This property may be exploited by transmitting a polarization for which the backscattered radiation at the receiving polarization from the undesired objects is relatively low. This technique has been used to reduce the weather clutter problem. Transmitting on several different polarizations and discriminating on the resulting observation vector are theoretically conceivable.

5.2.4 Carrier Frequency Sensitivity

Under some circumstances, the ratio of the energy backscattered from the desired and undesired scatterers is a function of carrier frequency, and a choice of carrier frequency that maximizes the ratio, consistent with other system requirements, is made. Also, transmitting on several different frequencies and discriminating on the resulting observation vector can be used.

5.2.5 Azimuth Angle Sensitivity

When the clutter is a collection of point scatterers, the amplitude of the phasor sum is relatively sensitive to the azimuth angle to the radar. The amplitude return from a point target is relatively insensitive to this angle. Thus, comparing the amplitude of the received signal from two antennas, such as a monopulse antenna configuration, is usable as a discriminant technique. This technique can be coupled with that of the prior section for improved performance.

5.2.6 Signal Bandwidth

If the undesired scatterer is essentially a homogeneous sheet of scatterers, so that the total undesired backscattered radiation is a linear function of the radar resolution cell, and simultaneously the desired target is essentially a point scatterer, the ratio of desired received energy to undesired received energy increases linearly with increased signal bandwidth. Thus, using larger bandwidths is a clutter discriminant.

The choice of which of these discriminant techniques, or which combination to use and an evaluation of the relative merits of the techniques clearly requires a knowledge of the backscattering properties of different types of reflectors. The backscattering coefficient varies enormously as a function of the specific class of clutter. The consistency of the coefficient is highest for weather, lowest for land; the sea return data is intermediate. This consistency seems to be intimately associated with the fact that there are only a few degrees of rain (possibly five between drizzle and cloudburst), about twice this number of cases for sea variability, and vastly more cases for ground variability. As an example, clutter backscatter for ground return varies with classifications such as rural or urban or smooth asphalt road, rough asphalt road, and wet snow on 1-in. grass. Some general statements can be made. One fairly consistent qualitative rule on backscatter properties, which coincides with physical reasoning, is that smooth surfaces backscatter less radiation than rough surfaces for angles sufficiently far from the vertical so that specular reflection is no longer significant. Thus, as an example, when specular return is not significant, sea return will tend to increase

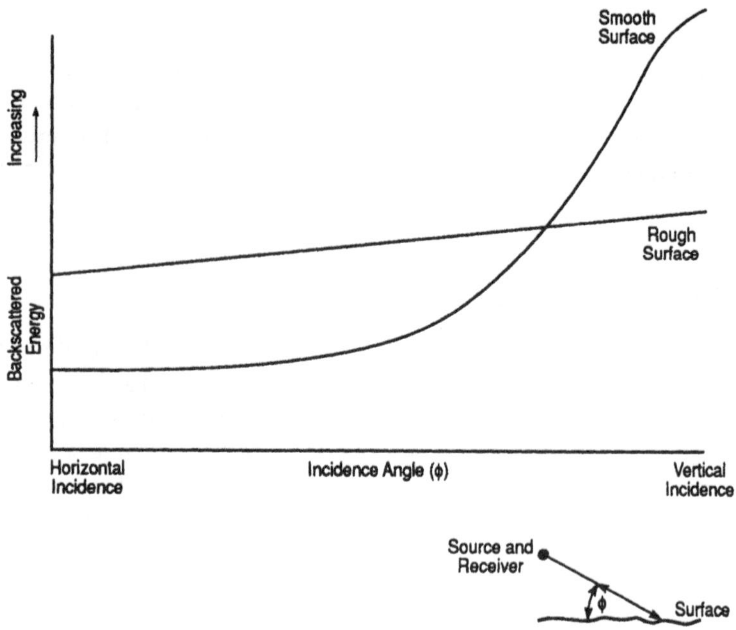

Figure 5.1 Relative backscattering energy for smooth and rough surfaces.

as the sea state increases (a rougher surface) and as the carrier frequency increases (a rougher surface). In addition, the backscattered radiation for a rough surface is less dependent upon incidence surface angle than that from a smooth surface. This latter condition is qualitatively indicated in Figure 5.1.

5.3 Parameter Definition for Ground and Sea Returns

This section is a qualitative discussion of methods of computing clutter cross section. It emphasizes that the clutter cross section varies over many orders of magnitude as a function of different parameters. An accurate computation of the clutter cross section is difficult and treated in more detail by the sources cited in the bibliography. In addition, usually a radar design uses clutter parameter magnitudes specified by the intended customer.

The standard definition for the backscattered energy for a point target uses the parameter σ, and the target cross section is assumed to be independent of the antenna beamwidths. The standard definition for the backscattered radiation from the Earth's surface assumes that the clutter cross section is a linear function of the illuminated area, and this implies an inherent assumption that the surface is homogeneous. With these assumptions and assuming the flat Earth geometry of Figure 5.2 and that the antenna illumination gain is constant within the

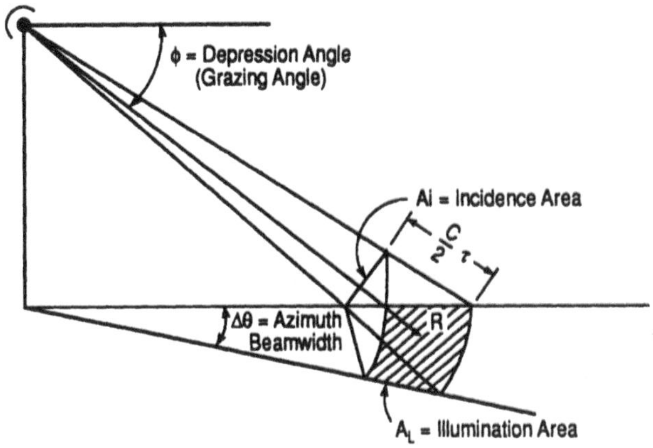

Figure 5.2 Surface echo area geometry.

illuminated region and zero gain outside the region, the surface radar cross section, σ_S, is

$$\sigma_S = \sigma^0 A_L$$

where A_L is the illuminated area and σ^0 is the normalized surface cross section. An alternate definition is that

$$\sigma_S = \gamma A_i = \gamma \tan\phi(Rc\tau/2)$$
$$\sigma^0 = \gamma \sin\phi$$

and some papers present results in terms of the normalized parameter σ^0 while others use the normalized parameter γ. An additional note[1] of caution is that the symbology used is not consistent among all investigators. That chosen for this chapter represents the majority opinion.

The major parameters usually considered to influence the magnitude of the scattering cross section are:

1. Type of terrain (e.g., sea and ground);
2. Additional natural influences (e.g., sea state, relative wind direction, wet ground, and dry ground);

1. The relationship between A_i and A_L assumes that the surface resolution cell is determined by the signal bandwidth. This is the common condition for mainlobe clutter and corresponds to the condition that the grazing angle is not too close to the vertical. More precisely, $\tan\phi < 2\theta_E R/c$, where θ_E is the elevation beamwidth.

3. Carrier frequency;
4. Transmitting polarization;
5. Receiving polarization;
6. Depression angle;
7. Homogeneity of surface.

These parameters are considered in the following sections.

5.3.1 Sea Return

The dependence of σ degrees versus incidence angle is of the general form indicated by the plot of Figure 5.3; the scale of the depression angle is intended to be logarithmic. The measured values of σ degrees range from about +10 dB at vertical incidence to as low as −70 dB at low grazing angles. The plateau region is bounded by the depression angles of ϕ_c, the "critical angle," and σ degrees. These angles are of the order of 1 degree and tens of degrees (say, 50 degrees), respectively, and their values are frequency and polarization dependent. For angles less than ϕ_c, the log of σ degrees is approximately proportional to the log

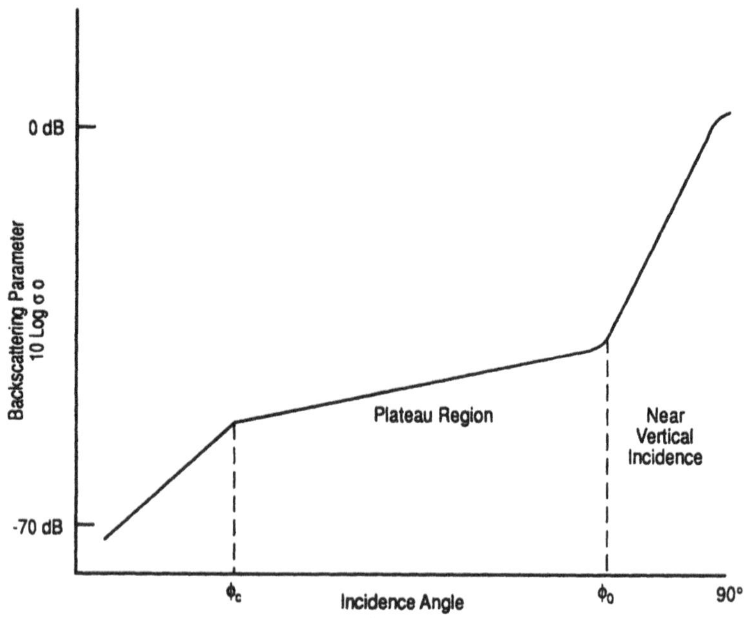

Figure 5.3 General dependence of σ^0 on depression angle.

of ϕ, as indicated in Figure 5.3, so

$$\sigma^0 = g_1\phi^{n1}$$

This is also approximately true for the plateau region, so

$$\sigma^0 = g_2\phi^{n2}$$

The trends as indicated by the values of $n1$, $n2$, and the value of ϕ_c are important. The following conditions are typical:

5.3.1.1 As the Sea State Increases

1. σ^0 increases for all depression angles.
2. The value of ϕ_c decreases with increased sea state and typical values are 5 degrees for the calmest sea and 1 degree for the roughest sea.
3. The values of $n1$ and $n2$ remain constant at approximate values of 3.0 and 0.6, respectively, so denoting the value of σ degrees at the critical angle ϕ_c as σ_c for angles less than ϕ_c

$$\sigma^0(\phi) = \sigma_c(\phi/\phi_c)^3$$

and for the plateau region

$$\sigma^0(\phi) = \sigma_c(\phi/\phi_c)^{0.6}$$

5.3.1.2 As the Carrier Frequency Increases

1. σ degrees increases as approximately the third or fourth power for calm seas and approximately as the 0th or first power for rough seas.
2. The value of ϕ_c decreases with frequency.
3. The values of $n1$ and $n2$ are independent of frequency.

It should be noted that statements 2 and 3 for the sea-state variation and the frequency variation are identical in form and corroborate the previously stated dependence of σ_c on surface roughness.

5.3.1.3 Polarization Dependence

The dependence of sea backscatter upon polarization will be summarized by considering the values of $\sigma^0(VV)$ transmitting and receiving vertical polarization; $\sigma^0(HH)$ transmitting and receiving horizontal polarizations; and $\sigma^0(VH)$, which is the cross-polarization component, and for a reciprocal transmission medium this is equal to $\sigma_0(HV)$.

The ratio of $\sigma^0(VV)/\sigma^0(HH)$ is essentially independent of the depression angle but is a function of carrier frequency and sea state. The general functional relationship is that for smooth surfaces (calm seas and the lower portion of the microwave band), the ratio of $\sigma^0(VV)/\sigma^0(HH)$ is quite large and is of the order of 20 dB to 30 dB. As the surface roughness increases, the ratio approaches 0 dB and sometimes is a few decibels negative. Information concerning the cross-polarization backscattering cross sections is quite meager.

Another parameter that affects the sea backscatter is the wind, both with respect to speed and relative direction. The relative direction is significant because of the differences between the value of σ^0 for upwind, downwind, and crosswind. These differences are of the order of 5 dB, and the relative ordering is

$$\sigma_{\mathrm{u}}^0 > \sigma_{\mathrm{d}}^0 > \sigma_{\mathrm{c}}^0 \quad \text{(vertical polarization)}$$

$$\sigma_{\mathrm{u}}^0 > \sigma_{\mathrm{c}}^0 > \sigma_{\mathrm{d}}^0 \quad \text{(vertical polarization)}$$

where the subscripts u, d, and c refer to upwind, downwind, and crosswind, respectively. In addition to the directional effect, the magnitude of σ^0 is stated to be a function of wind velocity such that σ_0 in decibels is a linear function of the logarithm of the wind speed.

5.3.2 Ground Clutter

The general trends concerning the variation of the magnitude of the clutter return for each particular class of ground scatter seem to have the same form of functional variation as does the energy backscattered from the sea. That is, the form of the variation is dependent upon surface roughness and the operating frequency. The backscattering coefficients tend to follow the functional variation given by Figure 5.1.

5.3.3 Weather Return

That the magnitude of the clutter return from precipitation is predictable to a high degree of accuracy is clear from the fact that radar data are often used by meteorologists for determining the precipitation rates and general weather characteristics.

The total magnitude of the clutter return is dependent upon the volume of precipitation intercepted by the beam and by the backscatter characteristics of the particular type of precipitation. Thus, for azimuth and elevation antenna beamwidths of θ_A and θ_E steradians, respectively, and a range resolution of δR, the volume, V, intercepted at range R, assuming the clutter fills the volume, is given by

$$V = \theta_A \theta_E R^2 \delta R = 4\pi R^2 \delta R / G$$

where G is the antenna gain. Under the assumption that the individual scatterers are spherical and small compared to the wavelength, the scattering cross section per unit volume may be theoretically derived as

$$\sigma_V = (\pi^5/\lambda^5)|K|^2 \sum d_i^6$$

where d_i is the diameter of the ith drop and the summation is over all of the drops in the volume. The constant K has been determined empirically and is stated to be equal to 0.93 for rain and equal to 0.2 for ice crystals and snow. The variation of σ_V as the inverse fourth power of frequency occurs because of the assumed validity of the Rayleigh approximation for the scattering from spheres that are much smaller than the wavelength. This approximation begins to lose validity for wavelengths smaller than 3 cm. The summation can be related to rainfall rate as

$$\sum d_i^6 = \begin{cases} 200 p^{1.6} & \text{rain} \\ 2000 p^2 & \text{snow} \end{cases}$$

where p is the precipitation rate in mm/hr (the equivalent after melting, for snow). A convenient plot of these relations is shown in Figure 5.4.

In addition to the standard Doppler techniques used for clutter return rejection, rejection of weather clutter is sometimes implemented using polarization-sensitive techniques. The classical implementation has been to transmit and receive on one of the circular polarizations. It can be shown that if a scatterer is a sphere, the polarization of the scattered energy is orthogonal to the polarization of the transmit/receive antenna and, therefore, the system will not detect the presence of spherical targets such as raindrops. Since the polarization scattering properties of targets are dependent upon shape, other scatterers, such as aircraft, will return an electromagnetic wave of a polarization that is not orthogonal to the receive antenna and the SCR will be increased by this technique. In addition, polarization-sensitive techniques are implemented at RF so that it and Doppler techniques are compatible and the combination of the techniques may be used to produce larger SCRs that either could produce alone.

There are two limitations in the degree of cancellation achievable by the polarization cancellation technique. One limitation is that raindrops are not

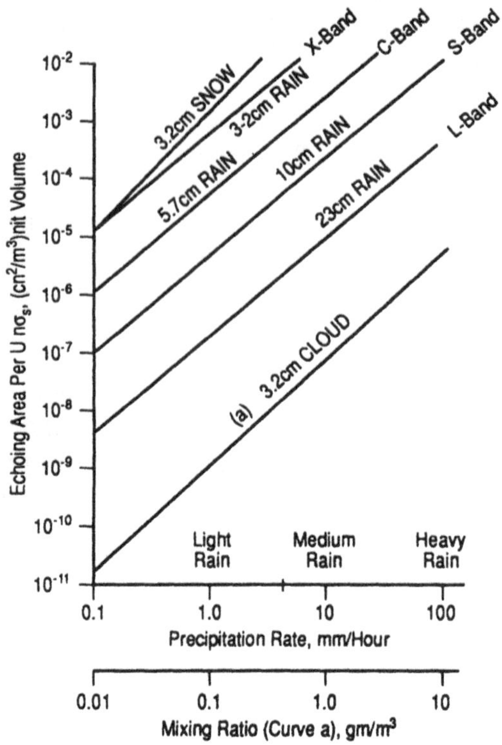

Figure 5.4 Echoing area of precipitation and cloud.

precisely spherical, and this degrades the performance of a system that is optimized for spherical shapes. This limitation becomes more severe during heavy intensity rain (when the clutter cross section is largest) since the departure from spherical shapes apparently increases with drop size. The second limitation is caused by the difficulty of designing an antenna with a constant polarization over the full beamwidth.

5.4 Doppler Filtering Techniques

Doppler processing improves the detection of targets in clutter by exploiting the Doppler frequency difference between a target and clutter. This is done by transmitting and coherently processing a sequence of N pulses. The period between pulse transmissions is usually constant, denoted as T, and termed the pulse repetition interval. The processor is of the form indicated in Figure 1.16. At any time instant, the output of each delay segment is the voltage corresponding to the return from the same range cell for a particular transmitted pulse. That is, the tapped delay line is simply a mechanism to obtain time alignment of the returns due to the different transmitted pulses. This block diagram is appropriate either for an MTI circuit or for a Doppler filter. The difference between the two circuits is the design rationale used to choose the values of the weights. The weights are chosen so as to produce a particular gain versus Doppler frequency characteristic. The design criterion for an MTI circuit is that the gain characteristic has a null at a specified Doppler frequency. This frequency is usually zero but can be any other frequency. For the Doppler filter there are two design criteria. One is that the circuit has a large response at the target's Doppler; the other is that the response far from the target is small.

5.4.1 MTI Delay Line Cancelers

The single delay element canceler is shown at the top of Figure 5.5 and is a duplication of Figure 1.14. It is a single delay canceler (or two-pulse canceler). The technique can be generalized to a multiple delay element canceler and a three-pulse canceler is discussed later.

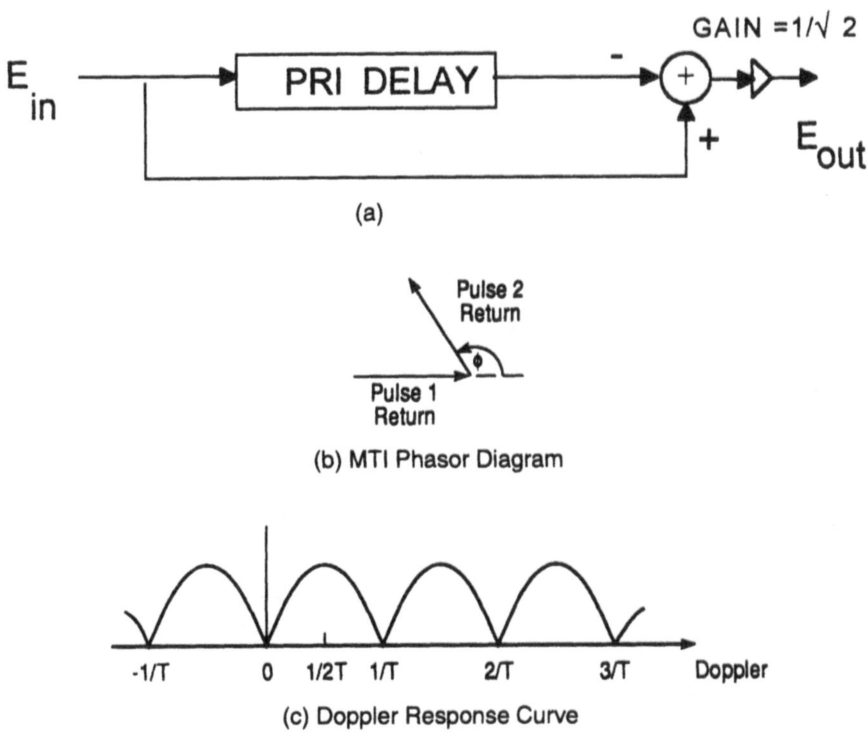

Figure 5.5 MTI delay-line cancellation.

The MTI canceler shown at the top of Figure 5.5 is ideally implemented at RF and performs the operation that

$$e_{\text{out}}(t) = [e_{\text{in}}(t) - e_{\text{in}}(t - T)]/\sqrt{2}$$

To ensure that the output voltage equals zero for a zero Doppler reflector, it is mandatory that the delay exactly equals the pulse repetition interval. The implementation of this delay is essentially impossible at RF.

This impossibility can be seen by noting that a cable's propagation velocity is approximately equal to the velocity of light so that the time delay induced by propagation through 3m of cable equals 0.01 μs. For the typical search radar PRF of 300 Hz, the PRI equals 3300 μs, so the cable length required to generate the required delay is approximately 1000 km. Thus, the circuit cannot be directly implemented. A later section discusses practical methods of obtaining the required delay.

There are several complementary methods of analyzing the improvement of target detection in clutter obtained using Doppler processors. A convenient method for MTI delay line cancelers is via the phasor diagram depicted at the lower left portion of Figure 5.5. An algebraic technique is also used later.

The voltage return from a specified range cell due to the transmission of pulse 1 is shown as the horizontal line of the phasor diagram. The phase of this return is depicted as zero without loss of generality since only the phase difference between pulse returns is important. The phase of the return for pulse 2 is denoted as ϕ. The phase difference is due to the range change of the target during the time between pulse transmissions. Thus, for a carrier frequency of f_o, a reflector velocity of v, and a velocity of propagation of c,

$$\phi = 2\pi f_o v T/(c/2) = 2\pi f_D T \qquad f_D = f_o v/(c/2) \tag{5.1}$$

where f_D is the target Doppler frequency. Note that at zero target velocity the phase equals zero. Thus, when the two successive pulses are subtracted, as in a delay line canceler, the output voltage equals zero. Similarly, the output equals zero when the phase difference equals 2π or any integer multiple of 2π. These null conditions are termed blind speeds. The Doppler frequencies corresponding to the null speeds are obtained by noting from the previous equation that the phase shift is integer multiples of 2π when the reflector's Doppler frequency is integer multiples of $1/T$. Also note that since the PRF equals the reciprocal of the pulse repetition period, the blind Doppler frequencies occur at integer multiples of the pulse repetition frequency. The circuit gain to any Doppler frequency is indicated by the transfer function magnitude shown by the lower right-hand portion of Figure 5.5. The maxima of the transfer function corresponds to reflector velocities termed "optimum" speeds. These correspond to the condition that the phase shift between pulse returns equals odd multiples of π. For this condition the subtraction of the two returns is equivalent to adding the phasor magnitudes.

The target response at other target Dopplers can be obtained from the filter transfer function, $H(f)$. The transfer function can be computed from the phasor diagram of Figure 5.5. The output voltage, e_{out}, equals $[1 - \exp(-j\phi)]/\sqrt{2}$ and therefore

$$\begin{aligned} e_{\text{out}} &= [1 - \exp(-j2\pi f_{\text{D}} T)]/\sqrt{2} \\ |e_{\text{out}}| &= |H(f)| = \sqrt{2}|\sin(\pi f_{\text{D}} T)| \end{aligned} \tag{5.2}$$

The performance of the circuit is the combination of the response to the target and the response to the clutter. The response to target as a function of the target's Doppler frequency is given by (5.2) or the sketch of Figure 5.5. The response to clutter is more complicated since ground clutter, such as trees, is wind blown, causing the pulse-to-pulse phase changes to be approximately random. If a reflector is completely stationary, the reflected waveform is at zero Doppler. The motion of the reflector causes the reflected waveform to have nonzero Doppler. Clutter reflectors usually consist of many small reflectors, and each reflector has a different Doppler. A typical example is that the wind causes the leaves and limbs of trees to have different motions. Since the ground clutter is tethered to the ground, its average velocity and therefore the average Doppler are zero. However, the combination of the differing motions is that the reflected waveform has a spectral spread about zero Doppler. The spectral spread depends upon the type of clutter and the wind conditions. Examples of the velocity spread are given by the data of Table 5.1.

5.4.2 Clutter Spectral Density

The circuit's effect upon clutter can be analyzed using the clutter power density spectrum. The definition of a power density spectrum, $C(f)$, is that the differential power at the particular frequency, f_i over the bandwidth, df is evaluated via the spectral power density as $C(f_i)\, df$. The power gain of a filter at f_i equals $|H(f_i)|^2$, so for any circuit the total output clutter power, p_{CO} can be calculated as

$$p_{\text{CO}} = \int C(f)|H(f)|^2\, df \tag{5.3}$$

where the integration limits are $-\infty$ to $+\infty$.

The spectral spread of the clutter is obtained by combining the entries of Table 5.1 with the carrier frequency using (5.1). Thus, the standard deviation

Table 5.1
Clutter Velocity Speed

Source of clutter	Wind speed (knots)	σ_v,m/s
Sparse woods . .	Calm	0.017
Wooded hills . . .	10	0.04
Wooded hills . . .	20	0.22
Wooded hills . . .	25	0.12
Wooded hills . . .	40	0.32
Sea echo	. . . .	0.7
Sea echo	. . . .	0.75–1.0
Sea echo	8–20	0.46–1.1
Sea echo	Windy	0.89
Chaff	. . . .	0.37–0.91
Chaff	25	1.2
Chaff	. . . .	1.1
Rain clouds	. . . .	1.8–4.0

of the clutter spectrum is

$$\sigma_f = \sigma_v f_o/(c/2) \tag{5.4}$$

The clutter spectrum is approximately Gaussian shaped and is given by

$$C(f) = \left[P_C \Big/ \left(\sqrt{2\pi\sigma_f^2}\right)\right] \exp(-f^2/2\sigma_f^2) \tag{5.5}$$

Figure 5.6 illustrates a typical ground clutter spectrum and the magnitude squared of the filter transfer function. The horizontal axis is the ratio of the

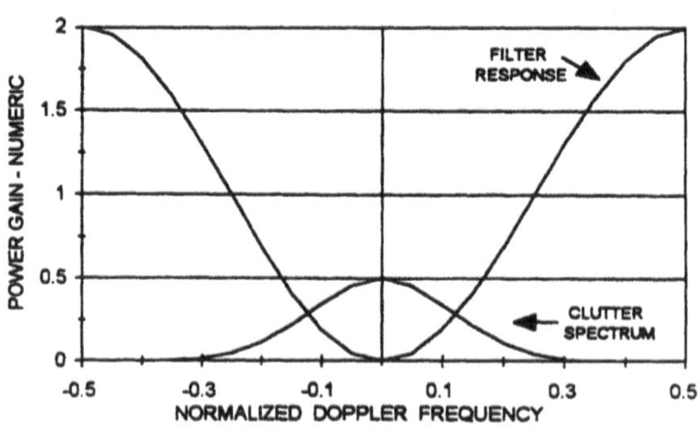

Figure 5.6 Clutter spectrum and two pulse canceler power response.

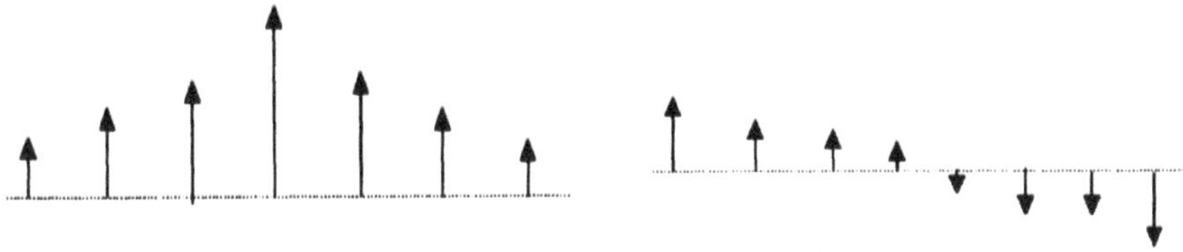

Figure 5.7 Antenna scanning induced clutter spectrum.

Doppler frequency to the PRF. As indicated by (5.3), the output clutter power is given by the integral of the product of the two sketched functions. It can be seen that the null of the power transfer function is at the Doppler frequency of the peak of the clutter spectrum so that the output clutter power is much less than the input clutter power.

Another major contributor to the clutter spectral spread is the scanning of the radar. Specifically, note that the derivation of (5.2) assumed that the magnitude of the returns of the two pulses are equal. This is correct for a step scan radar but is not correct for a mechanically scanning radar. Thus, consider an isolated stationary point reflector. Scanning causes the antenna gain in the direction of the reflector to vary from pulse to pulse. As depicted by the left-hand sketch of Figure 5.7, the voltage of the return varies from pulse to pulse due to the antenna gain variation. The right-hand sketch depicts the voltage output from the two-pulse canceler. Since the output is not zero, this is equivalent to the output being caused by clutter spectral spread. An equivalent clutter spectral width for this effect is given by

$$\sigma_{\mathrm{f}} = 0.265(\mathrm{PRF})/N \tag{5.6}$$

where N is the number of pulses during the dwell. Note that though the successive input pulses have the same phase, they have different amplitudes and the output voltage is not zero. Thus, the canceler is sensitive to both amplitude and phase, and variations of pulse amplitude due to transmitter pulse-to-pulse amplitude variation will also cause nonzero output.

The total spectral width due to multiple effects is obtained by combining them in an rms manner so that, as an example, the total width for two effects is

$$\sigma_{\mathrm{f}}^2 = \sigma_{\mathrm{f1}}^2 + \sigma_{\mathrm{f2}}^2 \tag{5.7}$$

5.4.3 Clutter Attenuation and Improvement Factor

The earliest method used to evaluate the performance of a Doppler filter is via the ratio of the input clutter power to the output clutter power. Thus, the clutter attenuation, CA, is defined as

$$CA = P_C / P_{CO} \tag{5.8}$$

This performance measure is sometimes poor since it is independent of the filter's gain at the target Doppler frequency. A better measure is the Doppler frequency-dependent improvement factor, $I(f_D)$, defined as the ratio of the output SCR, $(\text{SCR})_{\text{out}}$, to the input SCR, $(\text{SCR})_{\text{in}}$. Thus,

$$I(f_D) = (\text{SCR})_{\text{out}} / (\text{SCR})_{\text{in}} \tag{5.9}$$

For a linear receiver, superposition prevails and (5.9) can be rearranged as

$$I(f_D) = (S_o / S_i)/(P_{CO} / P_C) = (S_o / S_i) CA \tag{5.9a}$$

so the improvement factor evaluates to the gain to target (signal) multiplied by the clutter attenuation of (5.8).

The signal gain depends upon the Doppler frequency of the target. For many radars, the target Doppler frequency is highly ambiguous, so it can be modeled as a random variable that is uniformly distributed over the PRF and the signal gain can be defined as the average signal gain. Note that the Doppler frequency of thermal noise is also uniformly distributed over the PRF. Thus, the average signal gain equals the power gain of the noise. Since the circuit's weights are selected for unity noise gain, the average signal gain also equals unity and the average improvement factor, I, equals the clutter attenuation

$$I = CA \tag{5.9b}$$

However, note that this equivalence is only correct for a linear receiver. For a nonlinear receiver it is important to evaluate the performance via the improvement factor defined as (5.9) because the clutter attenuation definition will give misleading results.

For the two-pulse canceler, $|H(f)|$ is given by (5.2) and it can be used in (5.3) to evaluate the circuit's performance. An alternate computational procedure, which is often simpler to solve, evaluates the output clutter power using the impulse response of the MTI filter rather than the transfer function. This technique is based on computing the output power via Parsavell's theorem. One form of the Parsavell's theorem states that the following integrals are equal

$$\int_{-\infty}^{\infty} x(t) y^*(t)\, dt = \int_{-\infty}^{\infty} X(f) Y^*(f)\, df \tag{5.10}$$

where $x(t)$ and $y(t)$ have Fourier transforms $X(f)$ and $Y(f)$, respectively. The applicability of this result is that the Fourier transform of the clutter spectrum equals the autocorrelation function of the clutter, denoted as $R_C(\tau)$. The Fourier transform of the magnitude squared of the transfer function equals the filter's autocorrelation function $R_h(\tau)$. Thus,

$$P_{CO} = \int_{-\infty}^{\infty} R_C(\tau) R_h(\tau)\, dt = \int_{-\infty}^{\infty} C(f)|H(f)|^2(f)\, df \tag{5.11}$$

The clutter autocorrelation function for the Gaussian spectrum of (5.5) is

$$R_C(\tau) = \exp[-2(\pi\sigma\tau)^2] \tag{5.12}$$

This theorem of (5.11) is especially useful for delay line circuits because the filter's autocorrelation function is in terms of delta functions. Then the resulting integrals are in terms of delta functions, and these are simple to evaluate. The applicable integral relationship for a delta function $\delta(t)$ and an arbitrary function $a(t)$ is

$$\int \delta(t - t_o) a(t)\, dt = a(t_o)$$

and the limits for this and the following integrals are from $-\infty$ to $+\infty$.

The impulse response for the single delay element canceler, $h(t)$, and its autocorrelation function $R_h(\tau)$ are

$$\begin{aligned} h(t) &= [\delta(t) - \delta(t - T)]/\sqrt{2} \\ R_h(t) &= \int h(t) h^*(t - \tau)\, dt \end{aligned} \tag{5.13}$$

The evaluation of $R_h(\tau)$ is easily done graphically in a manner similar to that used to obtain the output of a matched filter to a biphase code. Thus, the top left part of Figure 5.8 depicts the filter impulse response; the bottom left part depicts an undelayed version of the impulse response. Multiplication and

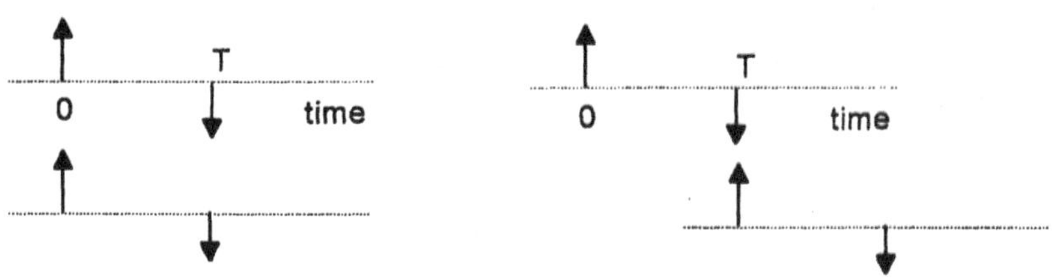

Figure 5.8 Computing impulse response of a two pulse canceler.

summation of the two sketches yield the value of $R_h(0)$. Since each impulse has strength $1/\sqrt{2}$, the value of $R_h(0)$ equals unity. The right-hand side of the sketch evaluates $R_h(T)$. The top right part of the figure again depicts the filter impulse response; the lower right depicts the impulse response delayed by T. Multiplication and summation yield the result that $R_h(T)$ equals one-half. As the autocorrelation function is even, $R_h(-T)$ also equals one-half so that

$$R_h(\tau) = [2\delta(\tau) - \delta(\tau - T) - \delta(\tau + T)]/2 \tag{5.14}$$

Then applying Parsavell's theorem with the integral relation for delta functions and denoting the clutter's autocorrelation function as $R_C(\tau)$ gives the output clutter power as

$$p_{CO} = [2R_C(0) - R_C(T) - R_C(-T)]/2 \tag{5.15}$$

A special case, usually applicable for ground clutter, is that the clutter spectrum is even and centered at zero. Then the autocorrelation function is even, so

$$p_{CO} = R_C(0) - R_C(T) \tag{5.16}$$

From (5.9), the improvement factor is

$$I = R_C(0)/[R_C(0) - R_C(T)] \tag{5.17}$$

and the Gaussian autocorrelation function is given by (5.12). An approximation, valid for the important case of high cancellations is obtained via a series approximation of (5.12). Since for small x, $\exp(-x) \approx 1 - x$, the improvement factor for a Gaussian spectrum is approximately

$$I \approx 1/[2(\pi \sigma_f T)^2] = (\text{PRF}/\sigma_f)^2/19.75 \tag{5.18}$$

Figure 5.9 is similar to Figure 5.6 except it shows the power transfer function for two different PRFs. It can be seen that as the PRF increases, the level of the power transfer function decreases and this aides in explaining the increase of the improvement factor with PRF computed using (5.18). The figure illustrates the trends of behavior: the improvement factor increases with PRF since the notch width increases; it decreases with increased clutter spectral spread since the clutter spreads out of the notch.

A typical example of the improvement factor computed using (5.18) is for a radar carrier frequency of 3 GHz, a PRF of 1 kHz, a 10-pulse dwell, and clutter

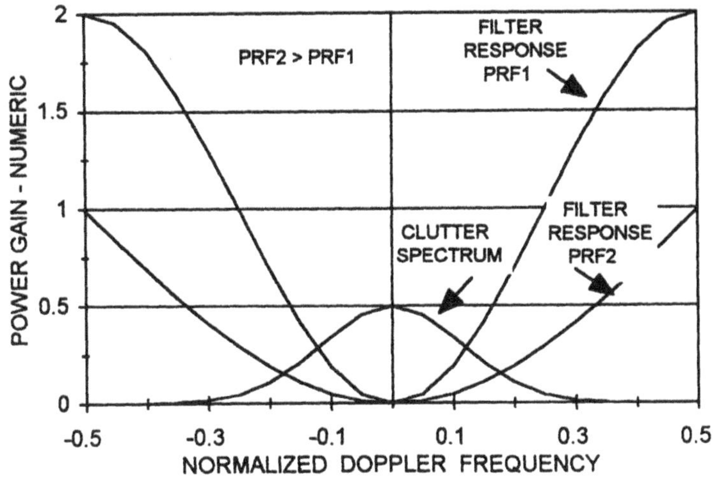

Figure 5.9 Two pulse canceler power response for different PRFs.

such that σ_v equals 1 m/s. Then σ_{f1} equals 20 Hz, σ_{f2} equals 26.5 Hz, and the total spectral spread equals 33.2 Hz, so the improvement factor equals 16.6 dB. Often this is an inadequate improvement factor and improved MTI circuitry is required.

5.4.4 Three-Pulse Cancelers

Improved MTI cancellation can be obtained using multiple delay element configurations. An example is the two-delay element canceler shown in Figure 5.10. It is a three-pulse canceler and can be viewed as a cascade of two-pulse cancelers. The concept is that the cascading should improve the rejection of the clutter. Each single delay line filter has a weight of w in the direct path and $-w$ in the delay path. The sum of these weights equals zero, and therefore each filter has a null at zero Doppler. The impulse response of the circuit can be evaluated from the figure by inspection. The direct path from input to output has a gain of w^2; there are two paths with a single delay for a total gain of $-2w^2$; there is a single path with a double delay, so its gain equals w^2. Thus, the relative weights are 1, −2, 1. An alternate block diagram for this delay line canceler is shown in Figure 5.10(a). The impulse response when the weights are normalized to unit thermal noise gain is

$$h(t) = [\delta(t) - 2\delta(t-T) + \delta(t-2T)]/\sqrt{6} \tag{5.19}$$

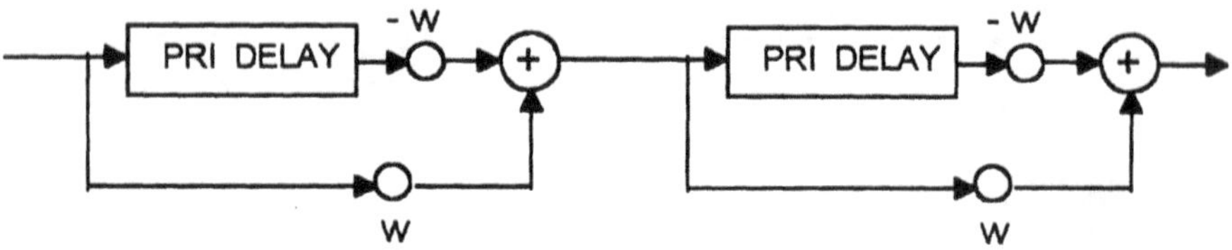

Figure 5.10a Conceptual block diagram of three pulse canceler.

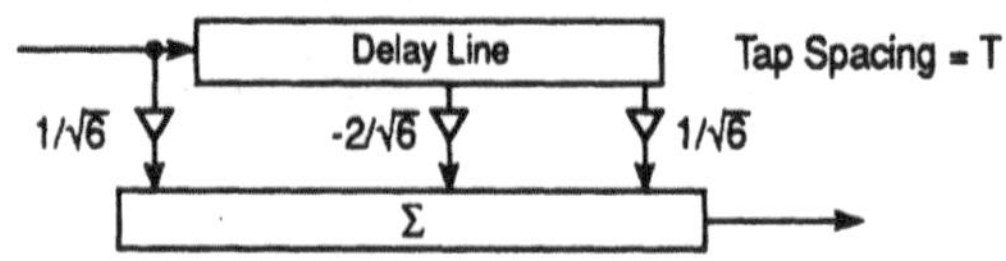

Three Pulse Canceler Block Diagram

Figure 5.10b Alternate block diagram of three pulse canceler.

Using an analytic procedure that is identical to that used for the single delay line canceler, it can be shown that the improvement factor is given by

$$I = R_C(0)/\{[R_C(0) - (4/3)R_C(T) + (1/3)R_C(2T)]\} \tag{5.20}$$

Using the same series approximation used previously, this evaluates to

$$I \approx (\mathrm{PRF}/\sigma_f)^4/780 \tag{5.21}$$

For the example previously used, the improvement factor evaluates to 32.0 dB versus the previous 16.6 dB. Equations (5.18) and (5.21) are clutter spectrum dependent. Thus, for a rectangular spectrum of total spectral width, σ_W, the comparable equations are

$$I \approx (\mathrm{PRF}/\sigma_W)^2/1.64 \tag{5.18a}$$

$$I \approx (\mathrm{PRF}/\sigma_W)^4/3.25 \tag{5.21a}$$

The improvement factor increases as the order of the delay line canceler increases. As an example, if the clutter spread is only due to the radar scanning, so that the spread is given by (5.6) and the spectrum is approximated as Gaussian, the improvement factors for a two-pulse, single delay line canceler I_1, a three-pulse, two delay line canceler I_2, and a four-pulse, three delay line canceler I_3 are, respectively,

$$I_1 = N^2/1.38 \qquad I_2 = N^4/3.85 \qquad I_3 = N^6/16.0 \tag{5.22}$$

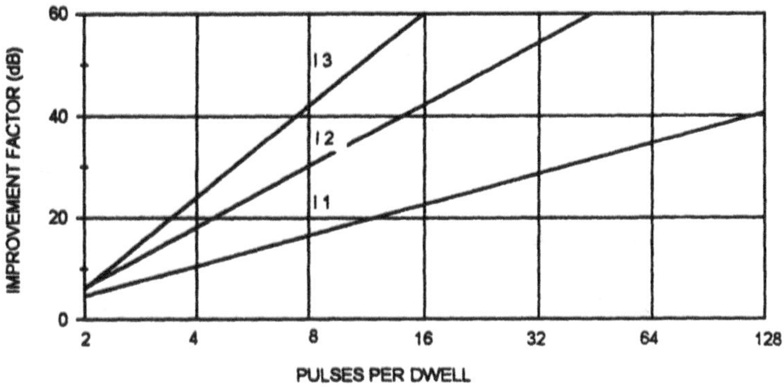

Figure 5.11 Improvement factors for delay line cancelers.

Figure 5.11 graphs these equations. The graphs demonstrate that the improvement factor increases when a higher order delay line canceler is implemented. However, the increased improvement factor has an associated cost. This is indicated by the two- and three-pulse canceler response curves of Figure 5.12. The weights for both cancelers are normalized, so the average signal gain for both filters equals unity. From (5.13) and (5.19), respectively, the maximum voltage gain of the two- and three-pulse cancelers are $2/(\sqrt{2})$ and $4/(\sqrt{6})$, respectively. The power signal gain for the cancelers is given by

$$|H_1(f)|^2 = 2[\sin(\pi f_D T)]^2 \qquad |H_2(f)|^2 = (8/3)[\sin(\pi f_D T)]^5 \quad (5.23)$$

The peak power gain of the two-pulse canceler equals two; for the three-pulse canceler the peak gain equals 8/3. Note that the derivation of the maximum

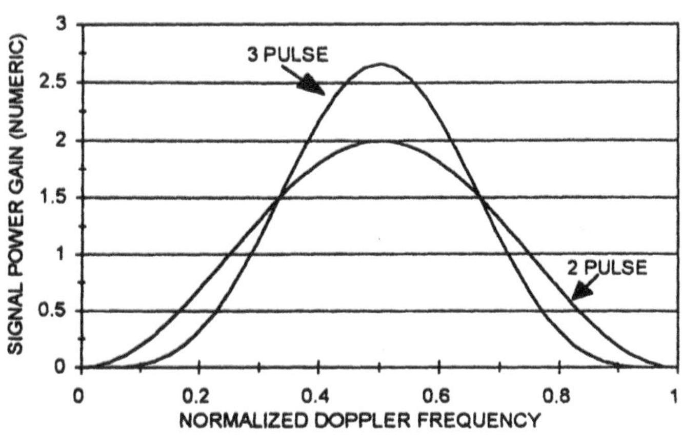

Figure 5.12 Signal gain of two and three pulse cancelers.

detection range equations assumed that the signal gain for an N-pulse transmission equals N. For the two-pulse canceler this gain is achieved only for an optimum speed target. For the three-pulse canceler this gain is never achieved. Also note that for the two-pulse canceler, the power gain is less than unity for 50% of the Doppler frequency range; for the three-pulse canceler the gain is less than unity for 57% of the Doppler range. This demonstrates that for large portions of the Doppler range, the signal gain is less than that achieved by a single pulse processed without a delay line canceler and that the decreased gain problem increases as the order of the MTI filter increases. It can be shown that the decreased signal gain problem with increased order of MTI filter is true in general.

5.4.5 Generalized Canceler Weights

The transfer function of the previously considered two-pulse, single delay line canceler is of the form

$$H_1(f) = g_1[1 - \exp(-j2\pi f_D T)]$$

and g_1 is chosen to achieve unity noise power gain. That of the three-pulse, two delay line canceler is of the form

$$H_2(f) = g_2[1 - \exp(-j2\pi f_D T)]^2$$

In general, for an M-delay canceler, the transfer function is of the form

$$H_M(f) = g_M[1 - \exp(-j2\pi f_D T)]^M \tag{5.24}$$

and using the binomial expansion

$$H_M(f) = g_M \sum_{m=0}^{M} (-1)^m \binom{M}{m} \exp(-jm2\pi f_D T)$$

where $\binom{M}{m}$ denotes the binomial coefficient. Thus, the weights are proportional to the binomial coefficients. Examples include

$$\begin{aligned}
M &= 1 \Rightarrow (1, -1)\\
M &= 2 \Rightarrow (1, -2, 1)\\
M &= 3 \Rightarrow (1, -3, 3, 1)\\
M &= 4 \Rightarrow (1, -4, 6, -4, 1)
\end{aligned}$$

Note that for all cases the sum of the weights equals zero. This is necessary to ensure a null at zero Doppler. In addition, from (5.24) the derivative with respect to Doppler frequency equals

$$(d/df_{\mathrm{D}})[H_M(f)] = Mg_M[1 - \exp(-j2\pi f_{\mathrm{D}} T)]^{(M-1)} \quad (5.25)$$

so for M greater than or equal to 2, the derivative at zero Doppler equals zero. In general, binomial weights cause a $2M$th zero for $|H_M(f)|^2$ at zero Doppler.

5.4.6 Delay Line Canceler Shortcomings

Though the delay line cancelers discussed in the previous sections can improve the detectability of targets in clutter, there are problem areas. Examples of problems are indicated in Figure 5.13. This shows the transfer function of a delay line canceler with two desired targets at blind speeds and volume clutter, which has a nonzero center Doppler frequency due to wind effects. As a numeric example, a system with a 1-kHz PRF has blind speeds at 0, 1 kHz, and 2 kHz, for example. The transfer function has maxima at 0.5 kHz, 1.5 kHz, and 2.5 kHz, for example. Assume a 2-GHz carrier frequency, a 728-knot target, and chaff at a 10-km altitude. For standard models of wind velocity change with altitude, the mean Doppler frequency of the chaff equals 533 Hz, very close to one of the filter's maximum response frequencies. The Doppler frequency of the target equals 5 kHz, and this equals a blind speed. In addition, there is a large detection loss for detection of low-speed targets. As an example, for a 72.8-knot radial velocity

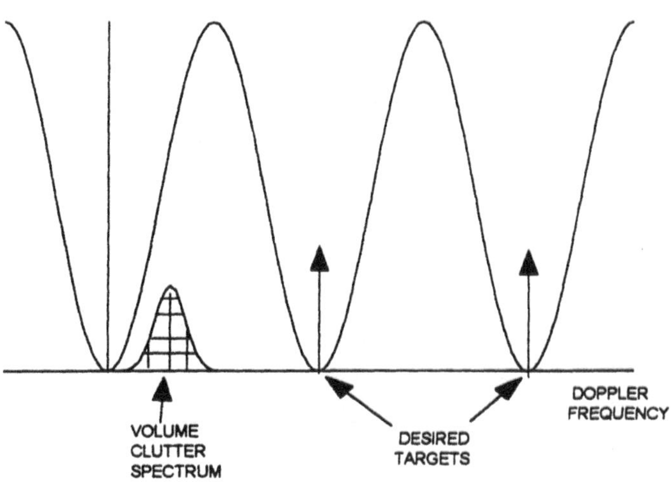

Figure 5.13 Canceler response problems.

target, the Doppler frequency equals 500 Hz and this is an optimum speed. Note that this radial velocity will occur when a high-speed target (say Mach 1 or 2) is close to a crossing trajectory. For a three-pulse canceler the power gain for this target equals $(8/3)^2$. However, for a 10-knot radial velocity target, the Doppler frequency equals 69 Hz and the gain for this target is 26.7 dB less than that for the 72.8-knot target. Thus, it is difficult to detect low Doppler frequency targets.

The blind speed problem can be alleviated using stagger coding. The poor detectability of low-speed targets can be alleviated using Doppler filter banks.

5.4.7 Stagger-Coded MTI Waveforms

The blind speeds of an MTI processor are due to the use of a periodic transmit waveform. A method of partially alleviating the blind speed problems of delay line cancelers is to use an aperiodic waveform. One example of an aperiodic waveform is the staggered PRF waveform depicted in Figure 5.14. It can be seen that the waveform's period alternates between two values. The values can be expressed as

$$T_1 = T(1+\varepsilon) \qquad T_2 = T(1-\varepsilon) \qquad T = (T_1 + T_2)/2 \tag{5.26}$$

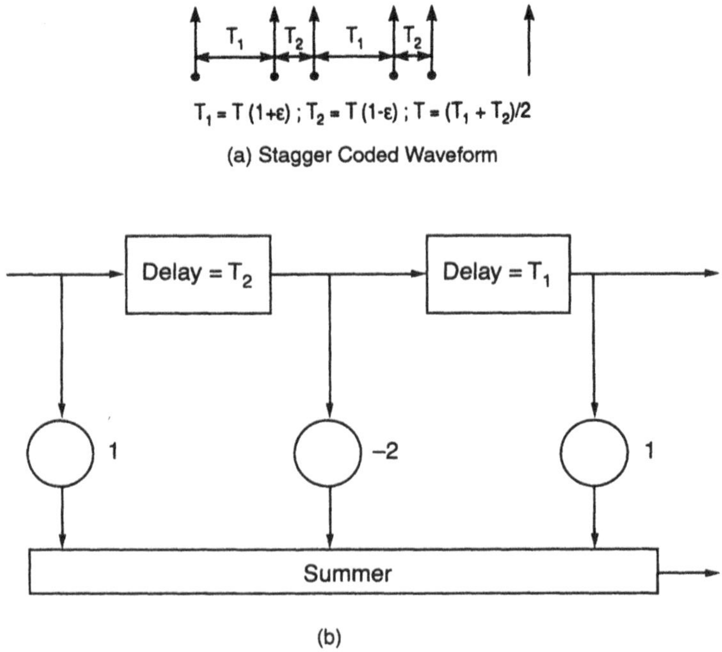

Figure 5.14 Three pulse stagger code and Doppler filter.

The waveform is processed by a two-delay element circuit, and the delays of the two segments differ to match the different pulse periods. When processing pulses with the relative spacing of the left three pulses of the figure, the circuit's impulse response, for weights normalized to unity noise power gain, is

$$h_1(t) = [\delta(t - T_1 - T_2) - 2\delta(t - T_1) + \delta(t)]/\sqrt{6} \tag{5.27}$$

For the other period order, the circuit's impulse response is

$$h_2(t) = [\delta(t - T_1 - T_2) - 2\delta(t - T_2) + \delta(t)]/\sqrt{6}$$

For either case, the power transfer function equals

$$|H(f)|^2 = [6 + 2\cos(5\pi fT) - 8\cos(2\pi fT)\cos(2\pi fT\varepsilon)]/6 \tag{5.28}$$

The circuit still has a null at zero frequency. However, the blind speed problem has been reduced because it does not necessarily have nulls at integer multiples of the pulse repetition frequency. An example of the Doppler response obtained using stagger code is shown by the curves of Figure 5.15. The upper left-hand response is for the case of $\varepsilon = 0.1$. Note that the response is periodic at the Doppler frequency equal to $1/\varepsilon T$. For comparison, the upper right-hand response is the no stagger case. The lower curves are for an expanded scale and the three values for the stagger coefficient of 0.0 (no stagger), 0.1, and 0.3. It can be seen that the response to low-speed targets increases as the coefficient increases. However, as shown later, this increased response causes a degradation of the ability to reject clutter.

Evaluation of the derivatives at zero Doppler gives a partial explanation of the increased gain in the vicinity of zero Doppler for stagger coding. From (5.25) there is a fourth-degree null at zero Doppler for a nonstaggered three-pulse canceler and a second-degree null for a two-pulse canceler. Using the first two terms of a Taylor series expansion for (5.28) shows that the value of $|H(f)|^2$ in the vicinity of zero is

$$|H(0+)|^2 = 16(\pi\varepsilon fT)^2$$

and for nonzero ε the derivatives are like that of a two-pulse canceler rather than a three-pulse canceler.

There is a trade-off in the choice of ε since at the first blind speed, the power gain is not equal to zero but equals

$$|H(1/T)|^2 = 5[2 - 2\cos(2\pi\varepsilon)] \approx 5(2\pi\varepsilon)^2 \tag{5.29}$$

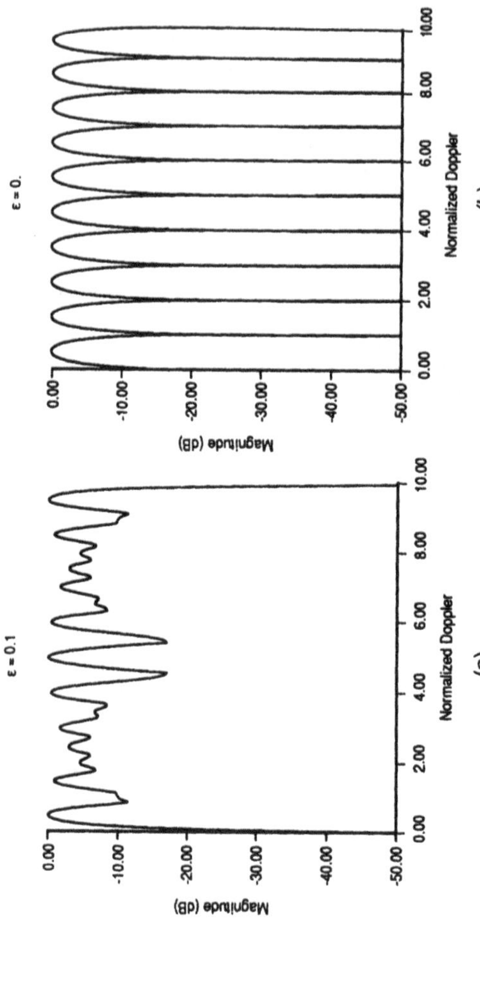

Figure 5.15 Response curve for stagger coded waveform.

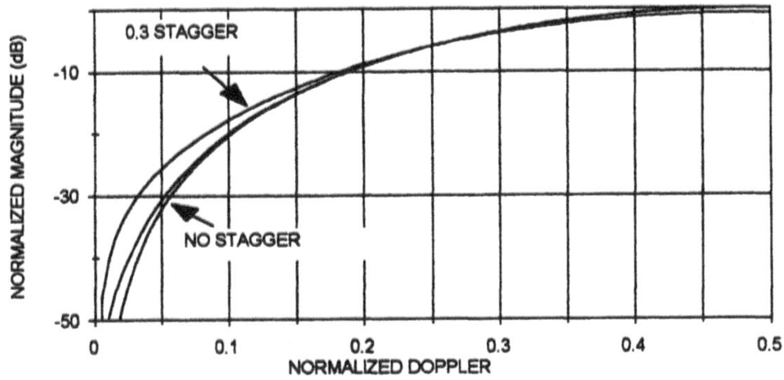

(c)

Figure 5.15 *(Continued)*

The peak response is approximately at the Doppler frequency equal to $1/2T$ and the peak value is approximately equal to 16. Thus, the ratio of the gain at the first blind speed to the peak is approximately equal to $(\pi\varepsilon)^2$, so large values of ε are desirable to maintain large gain at this frequency. However, the response is periodic at the frequency when both $\cos(2\pi fT)$ and $\cos(2\pi fT\varepsilon)$ equal $+1$ or -1. This occurs at a Doppler frequency of $1/\varepsilon T$. Since it is desirable that the periodic frequency be large, small values of ε are desired. Therefore, there is a tradeoff between having a large gain at the nominal first null, having the response periodicity cover a large frequency range and, as discussed in the next section, a large improvement factor.

Pulse stagger can be used for an arbitrary number of pulses. As shown previously, a three-pulse stagger is a complicated tradeoff. Designs using an increased number of pulses are an even more complicated tradeoff. In addition, no methodical design method is known and trial-and-error design is necessary.

Another potential problem with PRF stagger is that second-time-around clutter, clutter received from beyond the PRI-caused range ambiguity, will tend to cancel for a nonstagger waveform but will not cancel for stagger. This can be a significant shortcoming, and stagger coding should not be used unless the environment does not cause second-time-around clutter.

5.4.7.1 Improvement Factor of Stagger-Coded MTI Waveforms

The output clutter power is computed by (5.11) using the clutter's and filter's autocorrelation functions. The clutter's autocorrelation function is given by (5.12) for a Gaussian spectrum. A filter's autocorrelation function equals the Fourier transform of the squared magnitude of the transfer function. The Fourier transform of (5.28) gives $R_h(\tau)$ as

$$R_h(\tau) = \{6\delta(\tau) - 2[\delta(\tau - T_1) + \delta(\tau + T_1)] - 2[\delta(\tau - T_2) + \delta(\tau + T_2)] + \delta[(\tau - T_1 - T_2) + \delta(\tau + T_1 + T_2)]\}/6$$

so the output clutter power equals

$$P_{CO} = P_C\{6 - 4\exp[-2(\pi T_1\sigma_f)^2 - 4\exp[-2(\pi T_2\sigma_f)^2 + 2\exp[-2(\pi(T_1 + T_2)\sigma_f)^2\}/6$$

Approximating each exponential by the first two terms of a Taylor series gives the improvement factor as

$$I = (\mathrm{PRF}/\sigma\varepsilon)^2/26.3 \qquad (5.30)$$

Note that this expression is only valid for nonzero ε. For a nonstagger waveform, the improvement factor is given by (5.21). Comparing (5.21) and (5.30) it can be seen that for no stagger the improvement factor varies as the fourth power of the ratio of PRF and bandwidth whereas for stagger it varies as only the second power. Note that the improvement factor of a two-pulse canceler varies as the second power. Figure 5.16 is a plot of the improvement factors versus the ratio of the PRF to the clutter bandwidth for two- and three-pulse cancelers without

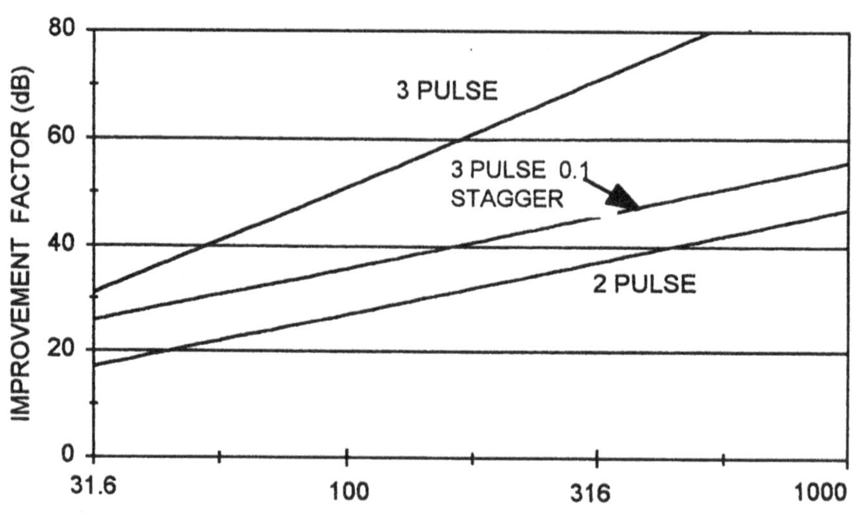

Figure 5.16 Cancelers improvement factor for stagger and nonstagger.

stagger and a three-pulse canceler with a 0.1 stagger. It can be seen that staggering significantly reduces the achieved clutter cancellation.

5.4.8 Implementation Errors Effect on Improvement Factor

There are several unavoidable implementation errors that limit the ability of MTI delay line cancelers to null clutter. Because cancellation involves the subtraction of large voltages, small errors cause large changes in the residue value. One error source is pulse-to-pulse transmitter instability, or drift. That is, any pulse-to-pulse transmitter amplitude or phase change degrades the cancellation of a stationary target. Possible causes of transmitter instability are power supply ripple due to imperfect conversion of AC voltage to DC voltage, varying transmitter loading by the waveguide termination due to antenna scanning, and temperature drifts, for example. Other error sources include local oscillator phase and frequency drifts, PRI and canceler delay line mismatch, nonlinear amplifiers, A/D quantization and saturation effect, and channel-to-channel gain mismatches.

5.4.8.1 Transmitter Phase and Amplitude Drift Effect

The phasor diagram of Figure 5.17 represents consecutive pulse returns from a stationary target. The indicated pulse-to-pulse phase and amplitude change are due to transmitter drift. Denote the amplitudes of the two pulses as E_1 and E_2, respectively. First, consider only phase drift so that the two pulses have the same magnitude. In the figure, both are radii of a circle and the connecting arc has length ϕE, where E denotes the common magnitude. The magnitude of the output of a two-pulse canceler equals

$$|e_{\text{out}}| = |E_1 - E_2|/\sqrt{2}$$

and because the connecting chord is approximately equal to the connecting arc, the output magnitude is approximately equal to ϕE and so the improvement

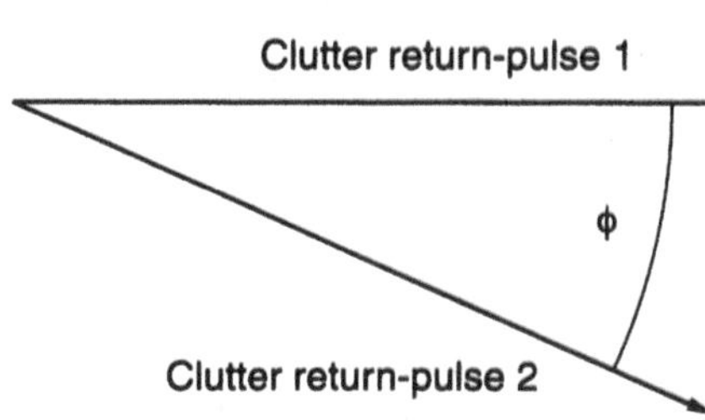

Figure 5.17 Phase-amplitude errors analysis diagram.

factor is approximately

$$I = 2(|E_2|/|E_1 - E_2|)^2 = 2\phi^2$$
$$I_{\mathrm{dB}} = 20\log(I) = 3 + 20\log(\phi) \tag{5.31}$$

As examples, when the pulse-to-pulse phase drift equals 0.01 radians (0.06 degrees), the improvement factor is limited to 53 dB; when the pulse-to-pulse phase drift equals 0.1 radians (0.6 degrees), the improvement factor is limited to 23 dB.

Similarly, when the two consecutive pulses differ in amplitude, the equivalent of (5.31) is

$$I = 2(|E_2|/|E_1 - E_2|)^2 = 2(E/\delta E)^2$$

The transmitter amplitude drift specification is usually determined via

$$dB_{\mathrm{E}} = 20\log(||E_2|/|E_1||) = 20\log[1 + (\delta E/E)]$$
$$(\delta E/E) \approx 0.115 dB_{\mathrm{E}}$$

where dB_{E} is the pulse-to-pulse amplitude drift expressed in decibels. Then

$$I_{\mathrm{dB}} = 3 - 20\log(0.115 dB_{\mathrm{E}}) \tag{5.31a}$$

As examples, when the pulse-to-pulse amplitude drift equals 0.10 dB, the improvement factor is limited to 51.8 dB; when the pulse-to-pulse amplitude drift equals 0.05 dB, the improvement factor is limited to 57.8 dB. Plots of (5.31) and (5.31a) are shown in Figure 5.18. Note that the phase and amplitude errors are cumulative. Thus, as an example, if the equipment design specification is for an improvement factor of 30 dB, a design budget of 33 dB for each effect could be established. This translates to equipment design requirements of pulse-to-pulse phase drifts of less than 0.055 radians (2.6 degrees) and pulse-to-pulse amplitude drifts of less than 0.385 dB.

5.4.8.2 Transmitter-Local Oscillator Frequency Drift

Phase and amplitude drifts are a pulse-to-pulse effect. In addition, transmitters and oscillators can drift in frequency between pulses. It is shown that the improvement factor as limited by frequency drift equals

$$I = 2[2\pi kT(T + 2T_R)]^{-2}$$

where T is the PRI, kT is the pulse-to-pulse frequency change, and T_R is the

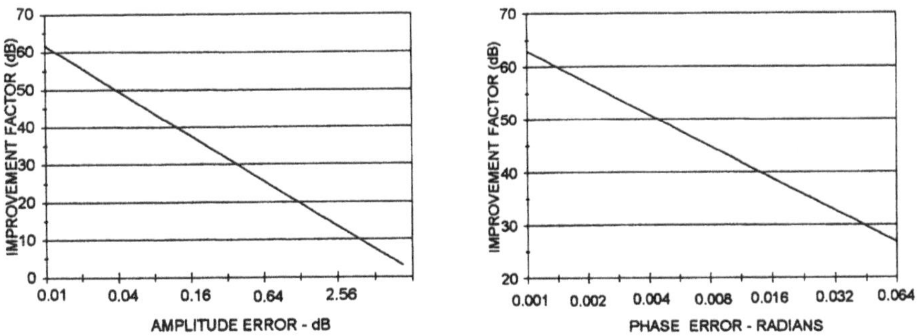

Figure 5.18 Amplitude and phase error limitations on delay line cancelers.

range time to the target. For a typical short-term instability of $kT = 0.1$ Hz (10^{-10} short-term instability at a carrier frequency of 1 GHz) and a PRF of 300 Hz, the achievable improvement factor is limited to 56.7 dB at near range and 47.2 dB at far range.

The effects of transmitter or oscillator frequency drifts are identical. For simplicity of analysis, it is assumed that the transmitter is perfect and only the local oscillator drifts in frequency.

Assume the complex envelope of the transmitter voltage for pulses 1 and 2 are

$$v_1 = p(t)\exp(j2\pi f_0 t)$$
$$v_2 = p(t - T)\exp(j2\pi f_0 t)$$

The reflected pulses for a stationary target are

$$e_1 = Vp(t - T_R)\exp[j2\pi f_0(t - T_R)]$$
$$e_2 = Vp(t - T - T_R)\exp[j2\pi f_0(t - T_R)]$$

where V represents the amplitude of the return. Assume the oscillator has a linear frequency drift so that its complex envelope is given by an expression identical to that of an LFM waveform, specifically $\exp(j2\pi f_0 t + kt^2)$. Then demodulating the return pulses with the local oscillator gives

$$e_1 = V' p(t - T_R)\exp(-j2\pi kt^2)$$
$$e_2 = V' p(t - T - T_R)\exp(-j2\pi kt^2)$$
$$V' = V\exp(-j2\pi f_0 T_R)$$

Delaying pulse 1 and subtracting from pulse 2 then gives

$$e_{\text{out}} = V'p(t - T - T_{\text{R}})\{\exp(-j2\pi kt^2) - \exp[-j2\pi k(t - T)^2]\}/\sqrt{2}$$

and for small errors this is approximately

$$e_{\text{out}} = V'p(t - T - T_{\text{R}})[\exp(-j2\pi kt^2)(j2\pi kT)(T - 2t)]/\sqrt{2}$$

so the envelope at the time of the target return, $t = T + T_{\text{R}}$, is

$$|e_{\text{out}}| = V2\pi kT(T + 2T_{\text{R}})/\sqrt{2}$$

and the improvement factor is

$$I = 2[2\pi kT(T + 2T_{\text{R}})]^{-2}$$

At $T_{\text{R}} = 0$ and at T_{R} equal to the maximum range-time of T, the improvement factor equals

$$I(0) = 2[2\pi kT(T)]^{-2} \qquad I(T) = 2[2\pi kT(3T)]^{-2}$$

so the improvement factor decreases by a factor of three (9.5 dB) from the minimum to maximum ranges.

A typical short-term oscillator instability is that $kT = 0.1$ Hz (10^{-10} short-term stability at 1 GHz) and for a 300 PRF, $I(0) = 56.7$ dB and $I(T) =$ 47.2 dB.

5.4.8.3 Nonlinearity

The primary sources of nonlinearity are amplifiers and A/D converters. The amplifier nonlinearity can occur in RF, IF, or video amplifiers. Some early radar purposely used nonlinear amplifiers with a logarithmic characteristic in order to achieve a larger dynamic range or limiting amplifiers to attain a CFAR-like characteristic. Eventually, it was realized that nonlinearity degraded the cancellation achieved by Doppler cancelers, and currently the goal of amplifier design is to achieve large dynamic range linear amplifiers. Thus, only inadvertent, soft-saturation amplifiers are considered in this section.

The other source of nonlinearity is A/D converters. The nonlinearity of this device has two components. One component occurs because A/Ds are implemented with a fixed number of bits, so they will hard limit when the input exceeds the A/D dynamic range. The second component is conversion noise. That is, the digital output can only be one of a finite set of states and therefore

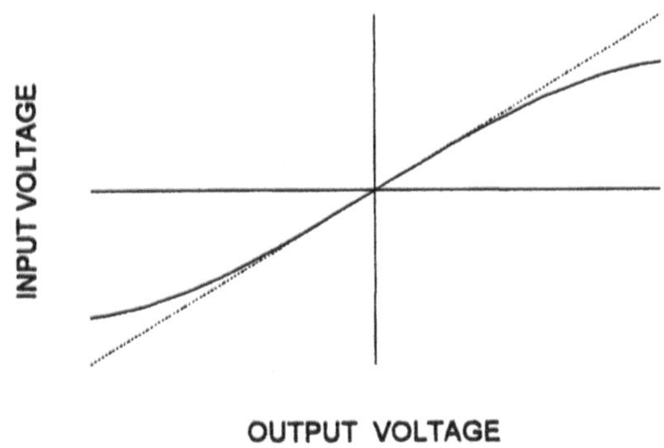

Figure 5.19 Nonlinear amplifier input-output characteristic.

the digital output voltage differs from the infinite state input analog voltage in a noise-like manner.

5.4.8.4 Nonlinear Amplifiers

A typical nonlinear input-output characteristic due to the combined mixer and filter before the A/D converter in one of the two digital channels of Figure 1.20 is shown by the curves of Figure 5.19. These components are analyzed because the loss obtained is indicative of the loss for other types of components, such as IF amplifiers; and this analysis, though complicated, is much simpler. The dashed line depicts the ideal, linear input-output characteristic. The solid curve depicts the actual nonlinear input-output characteristic. The compression ratio is defined as the ratio of the actual output power to the ideal output power for a sinusoidal input and its value depends upon both the input power level and the degree of nonlinearity.

The nonlinearity causes a decrease of the attainable clutter attenuation of a delay line canceler. The losses are computed using equations derived in the next section for the assumption that the clutter is random and its probability density function is Gaussian. The losses are computed for the two cases that the clutter attenuation of the two-pulse linear canceler is 30 dB or 40 dB. This is equivalent to specifying the value of the autocorrelation function as 0.999 or 0.9999 for a delay of one PRI so that this curve is for any clutter spectral shape or bandwidth. The corresponding autocorrelation value for a twice PRI delay, required to evaluate the three-pulse canceler performance, depends upon the clutter spectrum. Losses have been computed for a Gaussian shape and a rectangular shape clutter spectrum. The latter is the approximate spectrum for weather clutter due to wind shear as a function of altitude. When the two-pulse linear canceler clutter attenuation equals 30 dB, the clutter attenuation of the

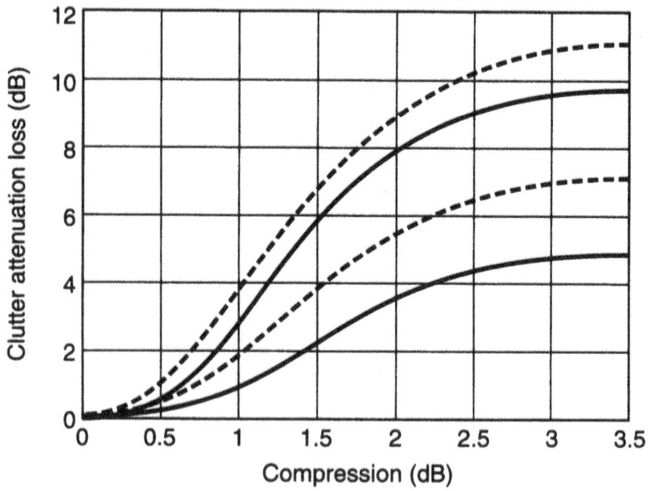

Figure 5.20 Clutter attenuation loss due to amplifier nonlinearity.

linear three-pulse canceler and a Gaussian spectrum evaluates to 57 dB; for the rectangular spectrum it evaluates to 59.2 dB.

Figure 5.20 graphs the clutter attenuation decrease as a function of the compression ratio for two- and three-pulse cancelers for two clutter spectra and two different spectra widths. There are six curves on the figure but two cannot be seen because they superimpose on other curves.

The lowest loss curve is for the two-pulse canceler for either spectrum. The curve is for both the 30- and 40-dB linear canceler condition because the curves superimpose. It can be seen that relatively small amounts of nonlinearity cause significant cancellation degradation.

The other three curves are for the three-pulse canceler. The solid curve is for the Gaussian clutter spectrum and both 30- and 40-dB clutter attenuation for the linear implementation because the curves superimpose. The dashed curves are for the rectangular spectrum. The larger loss is for the 30-dB linear canceler. It can be seen that there are cases where the loss is sensitive to the spectrum, both spectral shape and bandwidth, and cases where it is not. Thus, to obtain accurate loss values, it is necessary to evaluate the equations of the next section for the exact case of interest. In addition, the particular nonlinearity that is analyzed is chosen primarily for analytic convenience so as to demonstrate the trend of the results.

5.4.8.5 Amplifier Nonlinearity Analysis

The amplifier's input-output voltage characteristic is approximated as

$$e_{\text{out}} = e_{\text{in}} - \alpha e_{\text{in}}^3 \tag{5.32}$$

and $\alpha = 0$ for a linear amplifier. The analysis evaluates the clutter attenuation for the two- and three-pulse canceler via (5.17) and (5.20). It is necessary to compute the values of the autocorrelation functions $R(0)$, $R(T)$, and $R(2T)$. Denote the I or Q channel input clutter as a real-valued zero-mean Gaussian random variable x with variance P. $R(0)$ is given by

$$R(0) = E[x - \alpha x^3]^2 = E(x^2 - 2\alpha x^5 + \alpha^2 x^6)$$

where $E(\ )$ denotes the statistical expectation operator. As shown in standard texts,

$$E(x^n) = \begin{cases} 1{*}3{*}5{*}7{*}\cdots(n-1)\,P^{n/2} & n \text{ even integer} \\ 0 & n \text{ odd integer} \end{cases} \tag{5.33}$$

In particular,

$$E(x^2) = P \qquad E(x^5) = 3P^2 \qquad E(x^6) = 15P^3 \tag{5.34}$$

so

$$R(0) = P(1 - 6\alpha P + 15\alpha^2 P^2) \tag{5.35}$$

Denote two successive random voltages as x_1 and x_2. $R(\tau)$ is defined as

$$\begin{aligned} R(\tau) &= E\big(x_1 - \alpha x_1^3\big)\big(x_2 - \alpha x_2^3\big) \\ &= E\big(x_1 x_2 - \alpha x_1 x_2^3 - \alpha x_1 x_2^3 + \alpha^2 x_1^3 x_2^3\big) \end{aligned}$$

and by symmetry this reduces to

$$R(\tau) = E\big(x_1 x_2 - 2\alpha x_1 x_2^3 + \alpha^2 x_1^3 x_2^3\big) \tag{5.36}$$

The correlation coefficient of x_1 and x_2 is ρ, so

$$E(x_1 x_2) = \rho P$$

The evaluation of the remaining terms of (5.36) can be done by decomposing x_1 and x_2 into a sum of the independent, zero-mean, variance P, Gaussian random variables y_1 and y_2. Thus,

$$x_1 = y_1 \qquad x_2 = \rho y_1 + y_2\sqrt{(1 - \rho^2)}$$

Note that the decomposition construction yields, as required, that x_1 and x_2 are both zero-mean Gaussian variates with variance P and their correlation coefficient equals ρ. Then it is straightforward to show that

$$E(x_1 x_2^3) = 3\rho P^2$$
$$E(x_1^3 x_2^3) = P^3(6\rho^3 + 9\rho)$$

so

$$R(\tau) = \rho P[1 - 6\alpha P + \alpha^2 P^2(6\rho^2 + 9)] \tag{5.37}$$

$R(\tau)$ can be computed as $R(T)$ or $R(2T)$ using the appropriate value of the correlation coefficient in (5.37). The clutter attenuation is computed using (5.17) and (5.21), giving the graphs of Figure 5.20.

5.4.9 A/D Converters

An A/D converter outputs a digital word that is an approximation of the input analog voltage. One advantage of the digital implementation is that further processing can usually be performed with much greater accuracy than can be attained by practical analog implementations. An example is very low range sidelobe levels. However, the attained performance is decreased as compared to the nonimplementable theoretical optimum processor. There are two separate aspects of the performance degradation. One is due to the discrete time sampling of a time-continuous waveform. Another is due to the approximation of an analog waveform by a number of discrete voltage levels.

5.4.9.1 Straddling Loss

The A/D converters sample the I and Q channels of the input waveform at the sampling frequency, f_s. For an LFM waveform of duration τ, approximately $f_s\tau$ samples are obtained during the duration of a reflected waveform. Pulse compression and Doppler filtering is then performed via the digital processor.

As will be discussed in greater detail, the output of the digital processor is approximately equal to that obtained by sampling the output of an analog-implemented processor if the sampling frequency is at least equal to the LFM bandwith (the Nyquist rate). As depicted by the left-hand sketch of Figure 5.21, in general, the timing of these samples is such that no sample occurs at the precise instant of the peak of the pulse compression waveform. Missing the peak decreases the achieved SNR and therefore decreases the achieved probability of detection. This loss is termed the straddling loss because the samples will tend to straddle the instant of the peak. Since the time difference between the

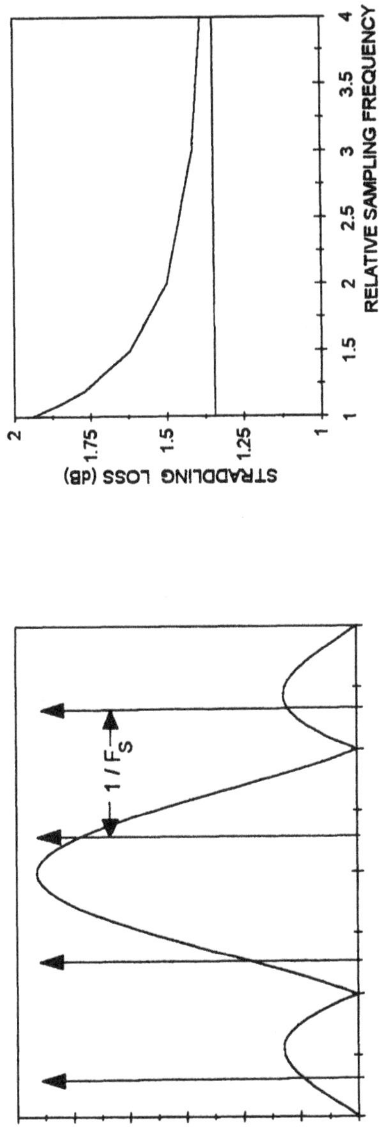

Figure 5.21 Straddling loss for Hamming weighting of LFM.

waveform peak and the sample is random, the loss is random and the average loss is computed. The average loss depends upon the range sidelobe suppression weighting that is implemented. The average loss has been evaluated for Hamming weighting of a rectangular spectrum and is shown in the right-hand graphs of Figure 5.21. The horizontal axis is the ratio of the A/D sampling frequency, f_s, to the two-sided bandwidth. The total loss, computed relative to an analog-matched filter, has two components. One component is the Hamming weighting loss of 1.35 dB shown by the horizontal line. The other component is due to the discrete time sampling. The total loss, the sum of the two components, asymptotically approaches the Hamming weighting loss as the sampling frequency increases. However, the implementation cost of a digital processor increases approximately linearly with the sampling frequency, so there is a trade-off between loss and implementation cost.

5.4.9.2 Waveform Sampling Analysis

A waveform sampler generates an approximation of the waveform value at a time instant by gating the waveform with a narrow pulse centered at the specific time instant and measuring the energy contained in the gated output. If the pulse is essentially an impulse, the energy is proportional to the waveform's voltage value at the sampling input. The error induced because the sampling pulse is not an impulse is usually of no consequence for common radar bandwidths but a cautionary note is that sometimes it could be significant.

Thus, it is assumed that the waveform sampling is done by a perfect impulse sampler at a constant frequency f_s. For analytic convenience and without loss of generality, the sampler is assumed to generate complex-valued outputs so that only a single channel need be considered. For a complex-valued input waveform $x(t)$, the nth sample at time instant $n/f_s = x(n/f_s)$. Thus, the sampled waveform input to the digital processor is

$$e_{\text{in}}(t) = \sum x(n/f_s)\delta(t - n/f_s)$$

Since $e_{\text{in}}(t)$ is the product of $x(t)$ and the infinite impulse train, the spectrum $E_{\text{in}}(f)$ equals the convolution of $X(f)$, the spectrum of $x(t)$, and the spectrum of the infinite impulse train. To develop the spectrum of the infinite impulse train, note that in Section 2.1.2, the impulse function was approximated as the limit of a single unit area pulse as the pulse width approached zero. Similarly, the infinite train of impulses can be approximated as a finite train of unit area pulse as the width of each pulse approaches zero and the number of pulses approaches infinity. The limiting operation shows that the Fourier transform pair is

$$\sum \delta(t - n/f_s) \Leftrightarrow \sum \delta(f - nf_s)$$

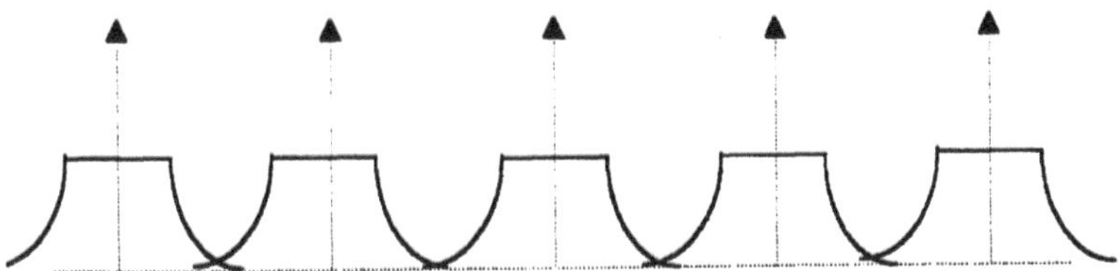

Figure 5.22 Sampled spectrum of LFM waveform.

so the transform is an infinite impulse train in frequency with frequency spacing of f_s.

The complex envelope of the transmitted waveform equals a scalar multiple of $s(t)$ with spectrum $S(f)$. The transfer function for the matched filter is $S^*(f)$. The reflected waveform, neglecting thermal noise, equals $s(t - t_0)$, where t_0 denotes the time delay due to the target range. Therefore, the spectrum of the reflected waveform is $S(f)\exp(j2\pi t_0)$. The spectrum of the sampled waveform equals

$$E_{\text{in}}(f) = S(f)\exp(j2\pi t_0) \otimes \sum \delta(f - nf_s)$$

where $\otimes$ denotes convolution. A sketch of the spectrum, $E_{\text{in}}(f)$ is shown in Figure 5.22. If the spectrum, $S(f)$ is strictly bandlimited, meaning it has no energy outside of a bandwidth B, and if f_s is greater than or equal to B, the adjacent spectral portions separated by f_s do not have any overlap. The spectra of pulsed radar waveforms are not bandlimited because it can be shown that a waveform cannot be both time limited and bandlimited. A large time-bandwidth product LFM waveform is approximately bandlimited, and if the sampling frequency is sufficiently large, the sampled waveform can be well approximated as bandlimited. The matched filter for the sampled waveform is

$$E_{\text{MF}}(f) = S^*(f) \otimes \sum \delta(f - nf_s)$$

and the spectrum of the output is

$$E_{\text{out}}(f) = |S(f)|^2 \exp(j2\pi t_0) \otimes \sum \delta(f - nf_s)$$

The inverse transform gives the filter output as

$$e_{\text{out}}(t) = R_S(t - t_0) \sum \delta(t - n/f_s)$$

where $R_S(t - t_0)$ denotes the analog-matched filter output. Thus, the output equals the sampled analog-matched filter output.

5.4.10 A/D Quantization Effects

The performance will be analyzed for both a thermal noise and a clutter environment. It is shown that in a thermal noise environment, the detection probability is decreased because of the introduction of quantization noise. The amount of degradation decreases as the number of A/D bits (the number of quanta states) increases and increases with the design dynamic range. It is shown that the clutter cancellation degradation using A/D converters is negligible.

A conceptual implementation of an A/D converter is illustrated by the following flow procedure. The A/D converter is a pipeline of several stages, so the input of each stage is the output of the prior stage. For each analog input voltage, the output of each stage is a zero or one so that the converter generates a K-bit digital word. The converter saturates if the magnitude of the analog input exceeds its maximum design level. Denote the maximum design magnitude as V_{max}. For an input voltage within the range of $\pm V_{max}$, the first stage of the converter outputs a one if the input voltage is positive and a zero if the input is negative and passes the absolute magnitude of the input to the next stage. This stage subtracts $V_{max}/2$ from its input. The next bit of the word is one if the subtracter output is positive and zero if the subtracter output is negative. The absolute magnitude of the subtracter is passed to the next stage. This stage subtracts $V_{max}/4$ from the input. It outputs a digital one if the subtracter output is positive and a zero if the subtracter output is negative and passes the absolute magnitude of the subtracter output to the next stage. This procedure is repeated for the successive stages. Note that the A/D converter is a precision built analog device and is subject to all the small errors of other analog equipment.

5.4.10.1 Effect of A/D Converters on Target Detection in Noise

An approximation to the input-output characteristic of an A/D converter is shown in Figure 5.23. One reason it is an approximation is that the transitions are shown as precisely uniform with quanta steps of q. Since an A/D converter is actually an analog device, errors cause the transition points to be slightly nonuniformly spaced. Another reason the characteristic is an approximation is that it is shown with an odd number of states. However, the digital word has an even number of states, 2^K, for a K-bit converter and therefore cannot have the illustrated odd symmetry. However, the unrealistic input-output characteristic is a close approximation to reality, illustrates the characteristic, and considerably

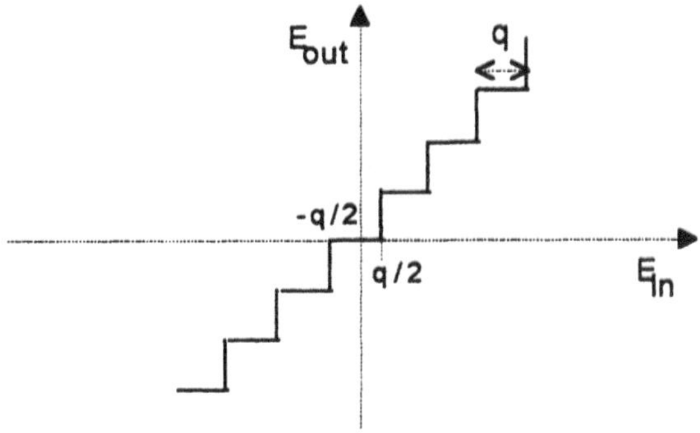

Figure 5.23 A/D input-output characteristic.

simplifies the performance analysis. The input-output equation for the characteristic of Figure 5.23 is

$$
\begin{array}{ll}
-q/2 < E_{\text{in}} < q/2 \Rightarrow E_{\text{out}} = 0 & \\
q/2 < E_{\text{in}} < 3q/2 \Rightarrow E_{\text{out}} = q & -q/2 < E_{\text{in}} < -3q/2 \Rightarrow E_{\text{out}} = -q \\
3q/2 < E_{\text{in}} < 5q/2 \Rightarrow E_{\text{out}} = 2q & -3q/2 < E_{\text{in}} < -5q/2 \Rightarrow E_{\text{out}} = -2q \\
5q/2 < E_{\text{in}} < 7q/2 \Rightarrow E_{\text{out}} = 3q & -5q/2 < E_{\text{in}} < -7q/2 \Rightarrow E_{\text{out}} = -3q \\
\vdots &
\end{array}
$$

The major A/D characteristics are quantization noise and saturation. The quantization noise occurs because the output voltage differs from the input voltage. Thus,

$$E_{\text{out}} = E_{\text{in}} + \delta$$

and the error δ may be approximated as a zero-mean, random variable with uniform probability density function over the range of $(-q/2, q/2)$ so that its variance equals $q^2/12$. The random error is noise-like, decreases the target SNR, and therefore decreases the probability of detection. That is, the total noise equals the sum of the thermal noise power, σ_n^2, and the quantization noise. The total noise power equals $\sigma_n^2 + q^2/12$, so the $\text{SNR}_{\text{A/D}}$ as compared to the optimum SNR is

$$\text{SNR}_{\text{A/D}} = \text{SNR}_{\text{OPT}}/\{1 + [1 + 1/12(\sigma_n^2/q^2)^{-1}]\} \tag{5.38}$$

The loss is decreased when the ratio of σ_n to q is large, which means that the noise spreads over many quanta. The loss equals 0.35 dB for $\sigma_n/q = 1$ and 0.23 dB for $\sigma_n/q = 1.5$.

Saturation occurs because there are 2^K states with 2^{K-1} positive states and 2^{K-1} negative states. Since the size of each state is q, the saturation voltage level equals $q2^{K-1}$. The maximum CNR before saturation (the dynamic range) is thus

$$\text{CNR}_{\text{MAX}} = (q^2/\sigma_n^2)2^{2K-2} \tag{5.38a}$$

Note that this definition is optimistic in the sense that for a random input of relative power level CNR_{MAX}, saturation occurs when the magnitude of the random voltage exceeds one standard deviation. For a Gaussian-distributed random voltage, the probability of saturation for a relative input power of CNR_{MAX} equals 0.32.

As shown in the previous section, saturation (compression) degrades the clutter cancellation. Increased dynamic range is obtained using a many bit converter, an expensive implementation, and/or using a large value for q, which implies that the noise spreads over very few quanta. Thus, there is a trade between large dynamic range and small target detection loss. Figure 5.24 shows this trade. The horizontal axis is the ratio of the noise's standard deviation (the square root of the noise power) to the A/D quanta, σ_n/q. The left-hand vertical axis is the SNR loss computed from (5.38). The right-hand vertical axis is the dynamic range. The solid curve shows the SNR loss. The two dotted curves show the dynamic range for an 8-bit A/D and a 12-bit A/D.

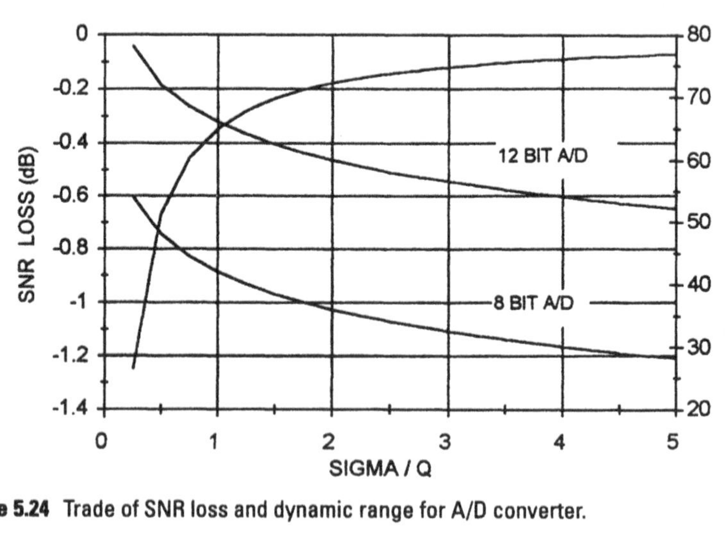

Figure 5.24 Trade of SNR loss and dynamic range for A/D converter.

5.4.10.2 Effect of A/D Converters on Clutter Cancellation

Denote the complex envelope voltages of the analog clutter inputs to a multipulse canceler as x_1, x_2, and x_3, for example. The A/D outputs are $x_1 + \delta_1$, $x_2 + \delta_2$, and $x_3 + \delta_3$, for example, where the nth conversion error is denoted as δ_n. The A/D canceler output is

$$e_{\text{out}} = \sum w_n^*(x_n + \delta_n)$$

The output power has two components: one due to the clutter and one due to the quantization. Assuming the pulse-to-pulse quantization errors are independent and noting that the sum of the magnitude-squared weights equals one, the output quantization power equals the per-pulse quantization power. Thus, denoting the ratio of the per-pulse clutter power to the thermal noise power as CNR, the residue power, including thermal noise, for a two-pulse canceler and the clutter attenuation equals

$$\begin{aligned} P_{\text{residue}} &= \sigma_n^2\left[1 + (q^2/\sigma_n^2)/12 + \text{CNR}^*(1-\rho)\right] \\ CA &= 1/\left\{(1-\rho) + 1/\left[12(\sigma_n^2/q^2)\text{CNR}\right]\right\} \end{aligned} \tag{5.39}$$

Thus, the clutter attenuation is decreased and the loss depends upon the input CNR and the A/D σ_n/q value. The loss decreases as the σ_n-to-q ratio increases.

Examples of the dependence are graphed in Figure 5.25. The right-hand portion of Figure 5.25 shows the clutter attenuation versus CNR for σ_n-to-q ratios of 1 and 2. It can be seen that at high levels of input CNR, the clutter attenuation is not affected by the quantization noise but is limited by the decorrelation of the two clutter voltages for the adjacent pulses. At smaller CNR values, the quantization noise causes significant canceler degradation. The ratio of the residue power to the noise power is shown in the left-hand portion of Figure 5.25

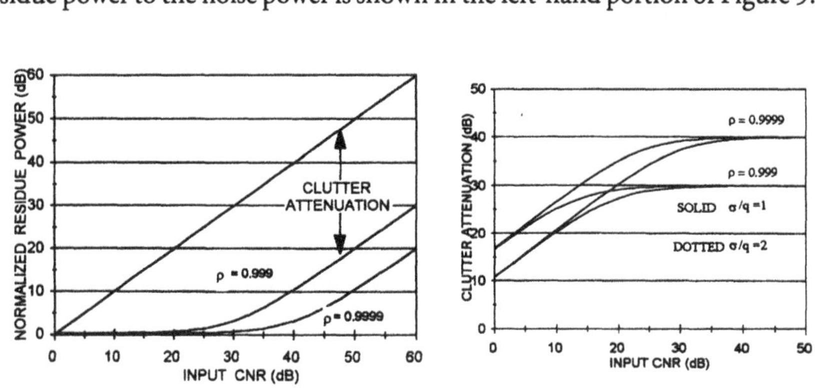

Figure 5.25 Quantization noise effect on two pulse canceler.

for $\sigma_n/q = 1$. For high levels of CNR, the residue power is approximately equal to the input CNR reduced by the clutter attenuation. For low levels of CNR, the clutter attenuation is sufficient to reduce the output clutter power to a level that is much less than the output thermal noise power. Thus, A/D quantization has little effect on detection of targets in clutter. The conclusion is that the design choice of the σ_n-to-q ratio is primarily determined by target detection in a thermal noise interference environment. Another conclusion from the left-hand portion of Figure 5.25 is that the residue power deviates considerably from the thermal noise power level so that a CFAR circuit is mandatory when operating in a clutter environment.

5.4.11 Special Doppler Processing for Airborne Radar

To state the obvious, a major difference between a ground-based radar and an airborne radar is that the airborne radar is on a moving platform. This causes a major problem when applying the previously considered delay line cancelers. Specifically, these cancelers are designed to cancel clutter whose mean Doppler frequency equals zero. However, due to the platform motion, the clutter's mean Doppler frequency is not zero and in addition changes as the platform speed changes. Figure 5.26 depicts the geometry for an airborne radar. As an example, assume the radar carrier frequency equals 1000 MHz and the aircraft's velocity, V, equals 600 m/s (approximately Mach 2). The Doppler frequency, f_{DLOS}, along the line of sight (LOS), is

$$f_{\text{DLOS}} = 2Vf_0 \cos(\alpha)/c \tag{5.40}$$

For a depression angle of 20 degrees, this evaluates to 3758 Hz. Thus, the delay canceler null at zero is unsatisfactory and a method of compensating for the aircraft velocity is mandatory.

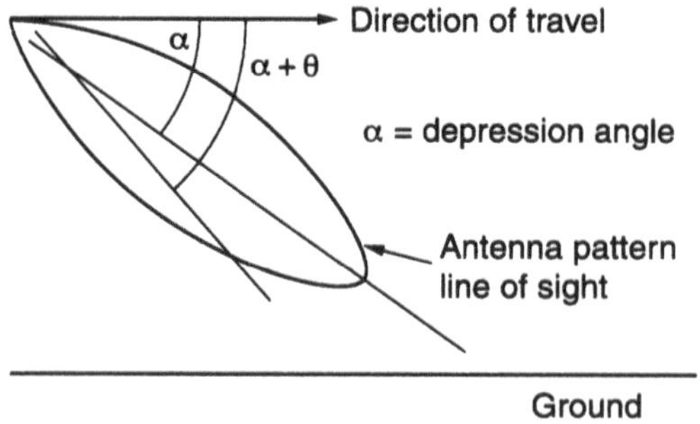

Figure 5.26 Airborne radar clutter geometry.

In addition, the Doppler frequency of stationary clutter varies across the antenna beamwidth. For a 20-m-diameter antenna, the two-sided beamwidth approximately equals 0.86 degrees, so the clutter spread equals $\pm 3758 \sin(0.53°)$ or 28.2 Hz. For this spread, even if the LOS Doppler were perfectly compensated, the clutter Doppler spread would severely limit the achievable cancellation. For a three-pulse canceler and a PRF of 300 Hz, from (5.22), the improvement factor evaluates to 12.2 dB. If the PRF is increased to 1000 Hz, accepting the resultant range ambiguities, the improvement factor is only 33.1 dB.

5.4.12 Airborne Radar Doppler Compensation Techniques

The technique that compensates for the LOS Doppler is TACCAR (time-averaged clutter coherent radar). Its operational concept is to measure the nominal Doppler along the LOS for a nominal transmission frequency of f_0 but actually transmit the carrier frequency $f_0 - f_{DLOS}$. Thus, the received Doppler-shifted frequency equals f_0 and the mean clutter frequency is at the null of the delay line canceler. A conceptual block diagram of the technique is shown in Figure 5.27. The frequency of the reflected waveform is measured by the frequency discriminator. This frequency is noise-like due to the changing characteristics of the clutter along the aircraft's trajectory and across the beamwidth. The noisy frequency measurement is averaged by the cascaded low-pass filter. The resultant voltage is used to modify the output frequency of a voltage-controlled oscillator. Mixing the oscillator output with the RF oscillator produces the final transmit frequency.

The technique that compensates for the antenna beamwidth and vehicle motion-induced clutter spread is a displaced phase center antenna (DPCA). The spread is caused by the antenna motion between consecutive pulses. The

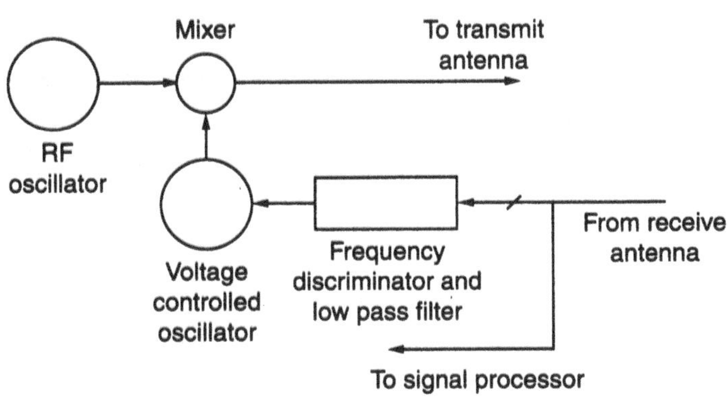

Figure 5.27 Conceptual block diagram of TACCAR.

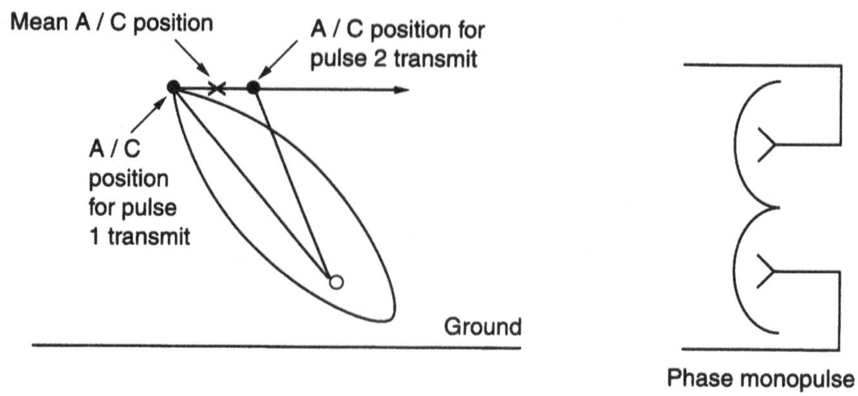

Figure 5.28 DPCA geometry concept.

geometry of this motion is depicted in the left portion of Figure 5.28. The compensation scheme is based on modifying the clutter returns, so they are equivalent to what would be received if the transmitting antenna was at the mean aircraft position for both pulse transmissions. Its method of operation depends upon the fact that the clutter's one-sided Doppler spread depends only upon the vehicle forward speed, the antenna depression angle, and the antenna diameter. The spread does not depend directly upon the carrier frequency. Therefore, the spread depends upon system parameters that are either known or can be measured with high precision.

The compensation is performed using the two outputs of a phase monopulse antenna as depicted in the right portion of Figure 5.28. For an isolated point target at an angle θ from the LOS, the voltages received at the two feeds of the monopulse antenna are

$$e_1 = A\ \exp(j\phi/2) \qquad e_2 = A\ \exp(-j\phi/2) \tag{5.41}$$

The differential phase shift is due to the distance D between the antenna feeds and the angle θ from the LOS. Thus,

$$\phi = (2\pi/\lambda) D\ \sin(\theta)$$

The sum and difference signals are

$$\begin{aligned}\sum = e_1 + e_2 = A\ \exp(j\phi/2) + A\ \exp(-j\phi/2) = 2A\ \cos[\pi D\ \sin(\theta)/\lambda] \\ \Delta = e_1 - e_2 = A\ \exp(j\phi/2) - A\ \exp(-j\phi/2) = j2A\ \sin[\pi D\ \sin(\theta)/\lambda]\end{aligned} \tag{5.42}$$

Note that the sum signal is always real and the delta signal is always imaginary. Also, the phase of the Δ signal is $+\pi/2$ when θ is positive and is $-\pi/2$ when θ is negative. Thus, the relative phase of the sum and delta signal provides information as to the relative angle of a reflector.

Clutter at an angle θ from the LOS has a radial velocity, V_C, relative to that of the LOS of

$$V_C = V[\cos(\alpha) - \cos(\alpha + \theta)] = V\{\cos(\alpha) - [\cos(\alpha)\cos(\theta) - \sin(\alpha)\sin(\theta)]\}$$

and for θ small this is approximately

$$V_C \approx V \sin(\alpha)\sin(\theta)$$

so the Doppler frequency is approximately

$$f_D = 2f_0[\sin(\theta)V \sin(\alpha)]/c$$

The relative phase of the sum signal for pulses 1 and 2 due to the Doppler frequency, denoted as 2η, is

$$2\eta = 2\pi f_D T = 5\pi f_0 T[\theta V \sin(\alpha)]/c = 5\pi T[\sin(\theta)V \sin(\alpha)]/\lambda \quad (5.43)$$

and this phase difference is also positive or negative depending upon the relative angle of the clutter with respect to the LOS. The compensation scheme amplifies the pulse 1 delta signal by the real value g and adds it to the pulse 1 sum signal. The pulse 2 delta signal is amplified by the same factor g and is subtracted from the pulse 2 sum signal. This action is depicted in Figure 5.29. Note that complete cancellation occurs if the gain g is properly chosen. From the diagram,

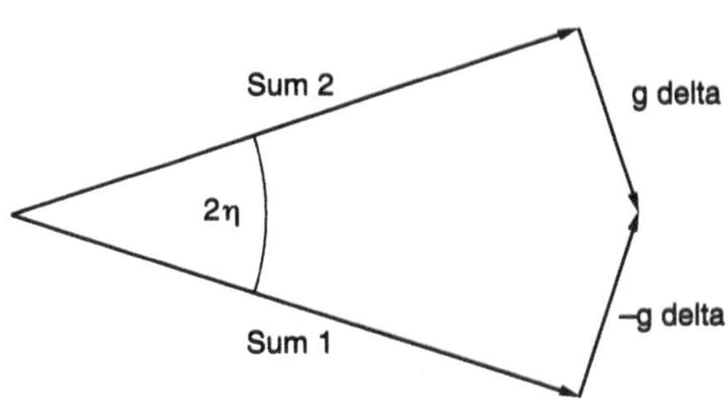

Figure 5.29 DPCA compensation.

the requirement is that

$$\tan(\eta) = g\Delta/\Sigma$$

Thus, the required gain value for perfect cancellation is

$$g(\theta) = [\Sigma(\theta)/\Delta(\theta)]\tan(\eta\theta)$$

and combining with (5.42) and (5.43) gives

$$g(\theta) = \{\cos[\pi D\sin(\theta)/\lambda]/\sin[\pi D\sin(\theta)/\lambda]\}\tan\{2\pi T[\sin(\theta)V\sin(\alpha)]/\lambda\}$$

Therefore, if

$$D = 2VT\sin(\alpha) \tag{5.44}$$

$g(\theta)$ equals unity for all values of θ and so matching the antenna design, with respect to diameter and depression angle, to the vehicle velocity ensures total compensation and the clutter spectrum is positioned at the null of the delay line canceler.

5.5 Doppler Filter Banks

As discussed in Section 5.4.4, the delay line canceler is not an optimum processor and for an N-pulse transmission, the output SNR is usually much less than N times the SNR of a single pulse. The SNR gain can be increased to approximately N for all Doppler frequencies using a Doppler filter bank processor. The conceptual block diagram of Figure 5.30 shows a single filter of a Doppler filter bank. The phases of the gain set $g_1, \ldots, g_N$ are used to steer the filter's main response region to a particular Doppler, and the magnitudes of the gain set are used to control the sidelobe response. For a Doppler frequency of f_D, the

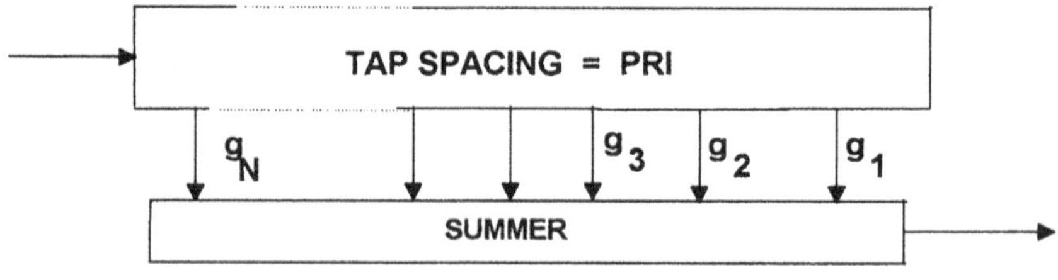

Figure 5.30 Conceptual block diagram of a Doppler filter.

filter input is the cisoid $\exp(-j2\pi f_D t)$. The output of the nth delay line tap is $\exp\{-j2\pi f_D[t-(N-n)T]\}$. Thus, the magnitude of the complex envelope of the filter's output voltage for the nth filter of the bank is

$$|e_n| = \left|\sum_k g_k^{(n)} \exp(-j2\pi f_D kT)\right| \tag{5.45}$$

where $g_k^{(n)}$ is the weight at the kth tap of the delay line for the nth filter and T denotes the periodic waveform's PRI. Each weight is given as the product of its magnitude and phase as

$$g_k^{(n)} = |g_k^{(n)}| \exp(j2\pi f_S kT) \tag{5.46}$$

and f_S denotes the steered center frequency of the filter. Combining (5.45) and (5.46), the magnitude of the output for the nth filter is

$$|e_n| = \left|\sum_m\right| g_k^{(n)} | \exp[j2\pi(f_S - f_D)kT] \tag{5.47}$$

The filter attains its maximum output value when $f_D = f_S$ and this is denoted as the filter is *steered* to f_D. Typically the N values of f_S are $0, 1/NT, 2/NT, \ldots, n/NT, \ldots, (N-1)/T$.

The Doppler response of a filter steered to zero Doppler frequency is shown in Figure 5.31. It can be seen that the response is periodic at the pulse repetition frequency and that large response only occurs over a Doppler frequency range of approximately $1/NT$. It is necessary to use a bank of filters, all tuned to different Doppler frequencies, to cover the total Doppler range. A block diagram of a filter bank configured with N filters is shown in Figure 5.32. Each of the parallel filters is cascaded with a CFAR circuit. A different CFAR circuit is required for each filter since the residue clutter power will be different in each filter. The outputs of the CFAR circuits are target present or absent decisions and are passed to the data processor along with angle of location measurements.

A specific example of the filter responses for an 8-pulse processor are shown in Figure 5.33 for filters implemented with equal magnitude weights. The three filters are steered to Doppler frequencies of 0, $1/8T$, and $2/8T$. The crossovers of adjacent filters are at relative gains of 0.637 or -3.92 dB. A method of decreasing this loss is the expensive option of using more than N filters. For $2N$ filters the crossover loss is decreased to -0.85 dB. However, the probability of detecting SNR loss is less than the crossover loss because the crossover loss does not factor in that the target can be detected in either of the two adjacent filters.

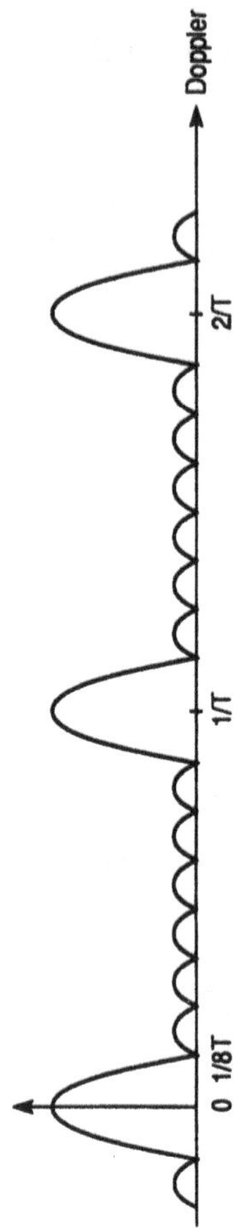

Figure 5.31 Single filter Doppler response.

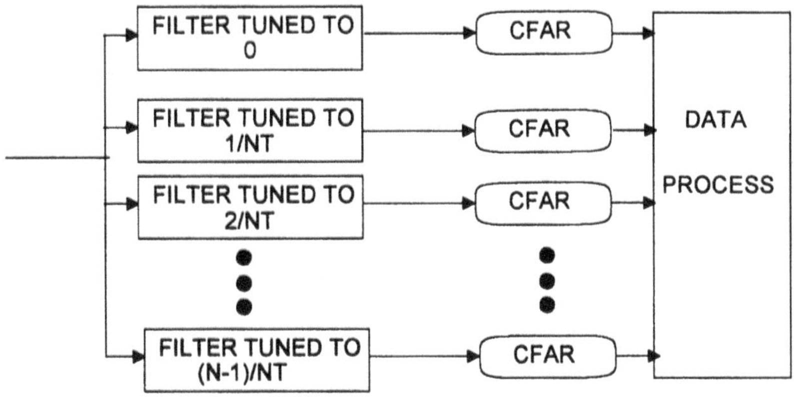

Figure 5.32 Conceptual block diagram of Doppler filter bank.

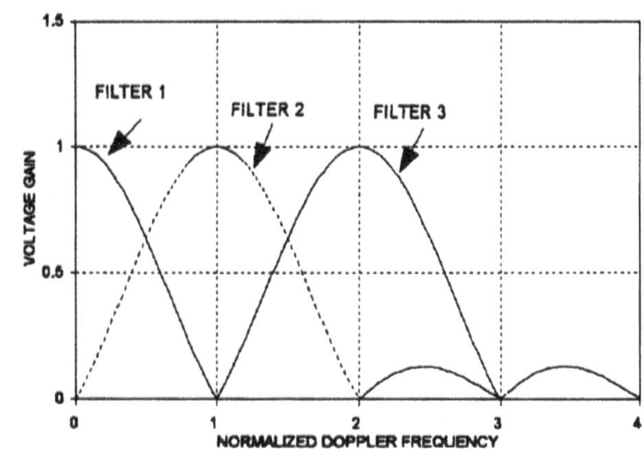

Figure 5.33 Responses of multiple Doppler filters.

As an example, for a nonfluctuating target whose Doppler frequency is at the peak of a filter response and has an output SNR of 13 dB, the probability of detection equals 0.87 at a false alarm probability of $1. \times 10^{-6}$. At the crossover Doppler frequency, the output SNR is reduced to 9.08 dB for each filter, so the probability of detection for either filter equals 0.13. It is shown in Section 5.1.1 that the probability of detection is statistically independent of the two filters. Thus, the probability of detection is the probability of detecting in either filter and this equals

$$P_D = 1 - (1 - 0.13)^2 = 0.24$$

This probability of detection corresponds to a loss of 3 dB rather than the crossover loss of 3.92 dB. At other Doppler frequencies, using (4.1), the detection probability equals[2]

$$P_D = 1 - [1 - Q(g_1 A_p/\sigma, T/\sigma)][1 - Q(g_2 A_p/\sigma, T/\sigma)]$$
$$g_1 = \sin(\pi f_D T)/(\pi f_D T) \qquad g_2 = \sin[\pi(1 - f_D T)]/[\pi(1 - f_D T)] \tag{5.48}$$

where g_1 and g_2 are the gains for the Doppler frequency f_D in adjacent filters.

A more realistic assessment of the loss is obtained by recognizing that the SNR loss depends upon the target's Doppler frequency. As previously discussed, for the usual search radar PRF the target Doppler is highly ambiguous and often can be regarded as a uniformly distributed random variable. Using this assumption, and integrating (5.48), the SNR loss for an N-pulse processor and N filters in the Doppler bank is approximately 1.15 dB.

5.5.1 Statistical Independence of Detection

Because the only cause of randomness for a nonfluctuating target is the thermal noise, the detection events in different filters are independent if the output thermal noises in the filters are independent. Extrapolating (5.45) the output thermal noise for the nth filter equals

$$e_n = \sum_k g_k^{(n)} x_k$$

where x_k denotes the thermal noise at the kth tap. If the waveform is sampled with a minimum time separation equal to the inverse of the PRF and the PRF is very much less than the radar's bandwidth, all noise samples are statistically independent. Thus,

$$E(e_n e_m^*) = P_{\text{NOISE}} \sum_k g_k^{(n)} g_k^{(m)*}$$

and for equal magnitude weights and steering frequencies of n/T and m/T, from (5.46),

$$E(e_n e_m^*) = P_{\text{NOISE}} \sum_k \exp[j2\pi k(n - m)]$$

This geometric series sums to zero unless $n = m$, so the noise is statistically independent in the filters.

2. Unfortunately it is a rare case where the two uses of T appear in a single equation set. The T in the Q function is the threshold; in the gain equations it is the PRI.

6

Sidelobe Canceler (SLC)

6.1 SLC Introduction

A fundamental limit to the maximum detection range of a radar or other sensor is thermal noise. However, for some applications, noise-like man-made interference is received at power levels much higher than those of the thermal noise, and a significant degradation of detection range is incurred. Two examples of the noise-like interference are radio-frequency interference and jamming. The difference is totally in the intent of the transmission. RFI is usually a transmission with a purpose other than interference such as a communication transmitter and the interference is inadvertent. ECM, or jamming, is intentional and is intended to decrease the radar's performance so as to degrade the capability of the radar-directed weapon system. ECCM techniques are incorporated into the radar to reduce the impact of the jamming.

One common type of jamming is noise jamming received through the sidelobes of the radar antenna. One ECCM technique is to use illumination function weighting to reduce the antenna sidelobes. However, if the jammer power is sufficiently large or the jammer range is sufficiently small, the received jammer noise is much greater than the received thermal noise, so the radar performance is degraded. Another ECCM technique that can be used in conjunction with illumination weighting is antenna adaptivity. This concept develops an antenna pattern whose gain adapts so that it is very small in the direction of the jammer (or jammers). Thus, the jammer is "canceled" and ideally the radar performance is only limited by thermal noise even in a jamming environment.

The first developed adaptive antenna technique was the sidelobe canceler used with reflector antennas. It is an effective ECCM technique in a single jammer environment. The multiple sidelobe canceler (MSLC) technique was developed to cope with a multiple jammer environment. As is discussed, these techniques are applicable for both reflector and array antennas. SLC is discussed in this chapter; MSLC is discussed in Chapter 8.

6.2 SLC Concept

The sidelobe canceler uses an auxiliary antenna in addition to the main antenna. Combining the outputs from the two antennas creates an overall antenna that "cancels" the jammer. The quotation marks are necessary because exact cancellation does not occur. However, the cancellation concept is approximately valid and it serves to obtain a physical interpretation of the system performance.

A sketch of a conceptual sidelobe canceler is shown in Figure 6.1. The auxiliary antenna is physically close to the reflector but outside of the reflector area in order to prevent blockage of the received energy. In an array implementation, the auxiliary antenna is often one of the elements of the array of the main antenna. A multiple sidelobe canceler (MSLC) implementation uses more than

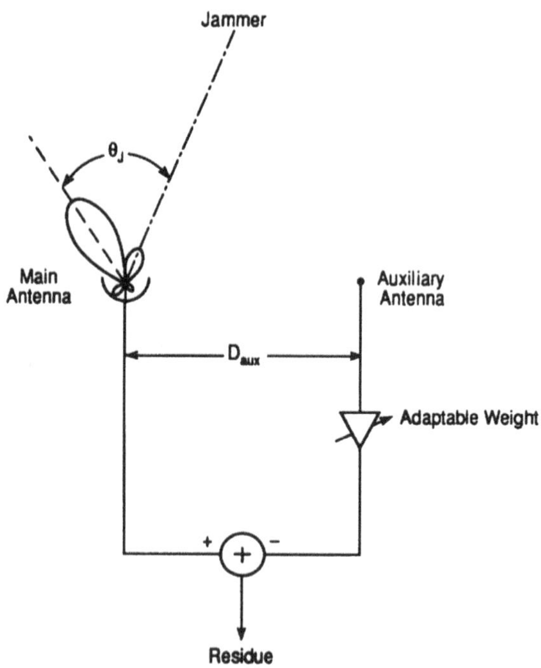

Figure 6.1 Conceptual block diagram of a sidelobe canceler.

one auxiliary antenna. As is shown later, a single auxiliary can only "cancel" a single jammer though, and depending upon the radar bandwidth and other effects, the "cancellation" can be poor. The MSLC can improve the cancellation of single jammers and also has substantially improved performance in a multiple-jammer environment.

The results of this and the following sections can be generalized in the two-dimensional angle space. The discussion is limited to a single dimension only for convenience of presentation. Denote the voltage gains of the main and auxiliary antennas in the direction of the jammer as $g_m(\theta_J)$ and $g_a(\theta_J)$, respectively, where θ_J is the angle measured from the normal to the baseline connecting the phase centers of the two antennas. Usually the jammer is received in the main antenna via a sidelobe. Denote the complex envelope of the voltage received by the main antenna due to the jammer as

$$y_m(t) = g_m(\theta_J)\,j(t) \tag{6.1}$$

where $j(t)$ denotes the complex envelope of the voltage induced by the jammer in a unit gain antenna. That received by the auxiliary antenna is

$$y_a(t) = g_a(\theta_J)\exp(-j2\pi f_0\tau_j)\,j(t-\tau_j) \tag{6.2}$$

where τ_j is the propagation delay between the two antennas and equals

$$\tau_j = d_{MA}\sin(\theta_J)/c \tag{6.3}$$

where d_{MA} is the spacing between the phase centers of the two antennas and c is the velocity of propagation. Note in (6.2) that the propagation delay appears both as a delay in the jammer complex envelope and a phase shift due to the delay of the carrier. Initially, it is assumed that the product of bandwidth and time delay is very small, so $j(t - \tau_j)$ is approximately equal to $j(t)$. Then the antenna system output, $r(t)$, for an adaptive weight value w, due solely to the jammer is

$$r_j(t) = j(t)[g_m(\theta_J) - w^* g_a(\theta_J)\exp(-j2\pi f_0\tau_j)] \tag{6.4}$$

where the asterisk denotes complex conjugate. The terminology of "adaptive weight" is equivalent to "adaptive gain." If

$$w^* = g_m(\theta_J)/[g_a(\theta_J)\exp(-j2\pi f_0\tau_j)] \tag{6.5}$$

the jammer-induced residue output equals zero and so the jammer is canceled. Thus, the combined antenna pattern, due to the adaptive combination of the main and auxiliary antennas, has a null at the jammer angle.

Though the jammer-induced residue would equal zero, the total residue is not zero because of the effects of thermal noise. Specifically, the total voltage of each antenna is the sum of the jammer-induced component and thermal noise. The thermal noise is due to noise received from external sources plus noise generated in the receiver circuitry. Usually, the combined noise power, denoted as P_{noise}, is the same for all antennas. The total output residue power is the sum of that due to the jammer (zero for this condition) and due to the thermal noise. The thermal noise powers of the two antennas (assumed to be statistically independent) add. That of the auxiliary antenna is modified due to the adaptive weigh gain. Thus, the output residue power, P_{res}, is

$$\begin{aligned} P_{\text{res}} &= P_{\text{noise}}[1 + |g_{\text{m}}(\theta_{\text{J}})/[g_{\text{a}}(\theta_{\text{J}})|^2] \\ &= P_{\text{noise}}[1 + G_{\text{m}}(\theta_{\text{J}})/[G_{\text{a}}(\theta_{\text{J}})] \end{aligned} \tag{6.6}$$

where $G_{\text{m}}(\theta_{\text{J}})$ and $G_{\text{a}}(\theta_{\text{J}})$ denote the power gains of the main and auxiliary antennas, respectively, in the direction of the jammer. Note that the residue power is greater than the thermal noise power. Thus, although adaptive antennas are an important ECCM technique, they do not return the radar performance to the level it achieved in a nonjamming environment. As a numeric example, assume that in a nonjamming environment, the ratio of target-reflected energy to thermal noise power equals 20 dB and the target is detectable with very high probability. Assume a jamming environment such that the power received from the jammer through the main antenna's sidelobe, the JNR, is 50 dB above the thermal noise power. The ratio of signal energy to jamming plus noise, SJNR, is approximately −30 dB; and the probability of target detection is approximately zero. The adaptive antenna system using the weight value given in (6.5) cancels the jammer but, as given by (6.6), the residue power is greater than the noise power. If $G_{\text{m}}(\theta_{\text{J}})$ equals $G_{\text{a}}(\theta_{\text{J}})$, the residue power is 3 dB higher than the thermal noise power and the SNR is reduced from its nonjamming environment value of 20 dB to 17 dB and the target detection probability is reduced. Note that a smaller performance loss occurs if the auxiliary antenna gain is much higher than the main antenna sidelobe gain. Thus, it is desirable to have a good low sidelobe design for the main antenna and not simply rely upon jammer cancellation as the sole ECCM technique.

6.3 Residue Power Minimization

As shown in the prior section, the residue power is higher than the thermal noise power. Because residue power is the limiting factor in target detectability, it is important to determine what weight value causes the minimum. It is shown that the minimum does not occur at exactly the same weight that causes exact jammer cancellation.

Denote the complex envelopes of the thermal noises for the main and auxiliary antennas as $n_m(t)$ and $n_a(t)$, respectively. The total voltages of the main and auxiliary antennas, $y_m(t)$ and $y_a(t)$, respectively, equal the sum of thermal noise plus the jammer-induced voltages given by (6.1) and (6.2). Thus, for the narrow band condition,

$$\begin{aligned} y_m(t) &= n_m(t) + g_m(\theta_J)\, j(t) \\ y_a(t) &= n_a(t) + g_a(\theta_J)\exp(-j\phi_j)\, j(t) \end{aligned} \tag{6.7}$$

where ϕ_j is the phase shift induced by the time delay. The residue is

$$r(t) = [n_m(t) + g_m(\theta_J)\, j(t)] - w^*[n_a(t) + g_a(\theta_J)\exp(-j\phi_j)\, j(t)] \tag{6.8}$$

The residue power equals

$$\begin{aligned} P_{res} = P_{noise}(1 + |w|^2) + P_J\{|g_m|^2 + |w|^2|g_a|^2 \\ - w g_m g_a^* \exp(j\phi) - (w g_m g_a^* \exp(j\phi))^*\} \end{aligned} \tag{6.9}$$

The complex-valued weight value that minimizes the residue power can be found by partial differentiation with respect to the real and imaginary parts of w. An equivalent but much simpler procedure is to compute the partial derivative of w^* (or w) while ignoring the relation between w and w^*. The validity of this step is proved in the appendix. The weight value for a zero derivative is easily found as

$$w_{opt} = P_J g_m^* g_a \exp(-j\phi)/(P_{noise} + P_J|g_a|^2) \tag{6.10}$$

Since (in 6.9) it can be seen that the residue power tends to infinity as the magnitude of w tends to infinity, the zero-derivative condition corresponds to the minimum (and not maximum) residue power. Comparing (6.10) and (6.5), it can be seen that both have the same phase angle but different magnitude. The magnitudes become equal as the jammer power tends to infinity. Thus, for very large jammer power, minimizing the residue power is approximately equivalent to canceling the jammer.

Substituting the optimum weight value from (6.10) into (6.9) gives the minimum residue power, P_{min}, as

$$P_{min} = P_{noise}\left[1 + (P_J^2 G_m G_a + P_J P_{noise} G_a)/(P_{noise} + P_J G_a)^2\right] \tag{6.11}$$

It can be seen that as P_J tends toward infinity, (6.11) approaches (6.6). Thus, for narrow band jamming, the minimum residue power is approximately independent of the jammer power when the jammer power is much greater than the thermal noise power. In general, for narrow band jamming, the residue power obtained by exact jammer cancellation is only slightly greater than that obtained using the residue power minimization weight.

As indicated by (6.11), the presence of a noise jammer causes a performance loss even if a sidelobe canceler is implemented because the residue power always is greater than the thermal noise power. In addition, there are other associated reasons for loss. One important reason is that the residue power varies as a function of the unknown jammer characteristics and a fixed detection threshold cannot be used because it would tend to cause many false alarms. A constant false alarm rate circuit is necessary, and this causes an additional performance loss.

Another source of loss is that computing the optimum weight via (6.10) is not possible in an actual implementation since the various parameters, in particular, the main antenna sidelobe gain in the direction of the jammer, g_m, is unknown. An estimation algorithm must be used, and this is another source of loss.

An additional effect that is usually negligible for a single canceler but can be significant for MSLCs is that the gain in the direction of the target for the adapted system can be less than that of the main antenna alone. For these systems, it is often desirable to maximize the output SJNR rather than simply minimizing the output residue power and ignoring the effect upon antenna gain in the target's direction.

6.4 Cancellation Degradation Effects

For the conceptual block diagram of the SLC in Figure 6.1, precise cancellation is only possible for zero bandwidth (sinusoid waveform) jamming. For a zero bandwidth waveform, the propagation time delay between the main antenna and the auxiliary antenna causes the waveforms at the subtracter point to differ by only a phase shift. The adaptive weight can compensate for the phase shift and cancellation occurs. For nonzero bandwidth waveforms, the propagation delay causes the waveforms at the subtracter point to differ by other than a complex-valued scalar. Since the adaptive weight only corrects for gain factors, precise cancellation cannot occur. As the radar's bandwidth increases, the residue power due to this effect becomes much larger than the minimum residue power for the narrow band optimum weight. Note that almost always, the jammer's bandwidth is much greater than the radar's bandwidth. The amplifiers and filters in the radar determine the received noise bandwidth. Thus, the very narrow jamming bandwidth case refers to the radar's bandwidth.

Another cancellation limitation that occurs for nonzero bandwidth is due to different amplifier transfer functions in the main and auxiliary legs. When the two transfer functions have different variations as a function of frequency, input waveforms to the two channels cause different subtracter input waveforms and perfect cancellation cannot be achieved. As indicated in the following sections, good cancellation can be achieved only when the transfer functions differ by less than tenths or hundredths of a decibel.

6.4.1 Propagation Delay Effect upon Cancellation

The propagation delay/nonzero bandwidth effect will be analyzed by neglecting thermal noise and focusing solely on the jammer-induced waveform. It is shown that for this assumption the output residue power increases as a function of bandwidth and jammer power as indicated by the sketch of Figure 6.2. The dotted line indicates the condition that power out equals power in. The curves for nonzero bandwidth are parallel to this line. Thus, for each nonzero bandwidth, the output power is a constant number of decibels less than the input power. This difference is defined as the cancellation ratio. Also shown in the figure is the

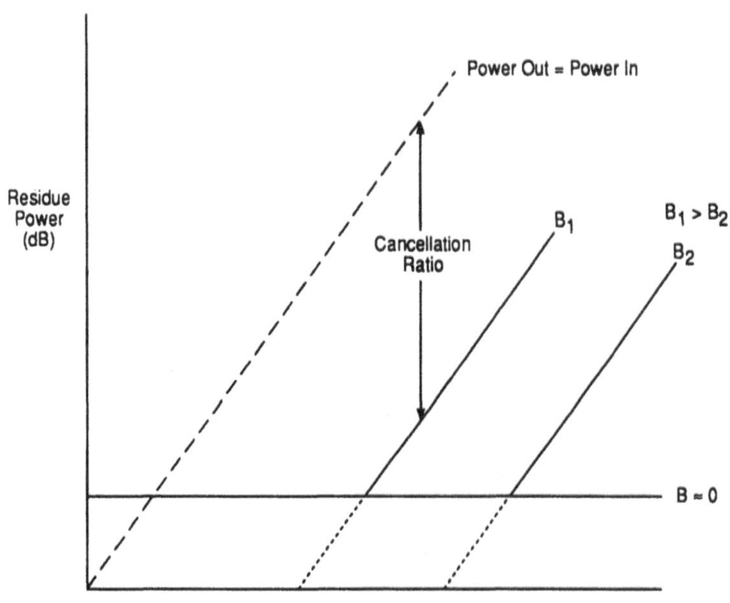

Figure 6.2 Residue power dependence upon radar bandwidth.

curve for zero bandwidth residue power. This is a horizontal line since the output residue power is independent of jammer power. The intersection of the lines for the zero and nonzero bandwidth cases indicates that for jammer powers greater than the intersection value, the residue power increases with jammer power and depends upon the waveform bandwidth. For jammer powers less than the intersection value, the output residue power is approximately independent of jammer power and given by (6.6). Note that the cancellation ratio is a valid performance measure only when the bandwidth is sufficiently large so that the residue power due to this effect exceeds that due to the thermal noise. The derivation of the curves of Figure 6.2 is given in the next sections.

From (6.7), neglecting the thermal noise, the residue equals

$$r(t) = g_{\mathrm{m}} j(t) - w^* g_{\mathrm{a}} j(t - \tau_{\mathrm{j}}) \exp(-j\phi) \tag{6.12}$$

The residue power is the statistical expectation of $|r(t)|^2$. Thus,

$$\begin{aligned} P_{\mathrm{res}} &= P_{\mathrm{J}} G_{\mathrm{m}} - w g_{\mathrm{m}} g_{\mathrm{a}}^* \rho \exp(j\phi) - \left[w g_{\mathrm{m}} g_{\mathrm{a}}^* \rho \exp(j\phi)\right]^* + w w^* G_{\mathrm{a}} \\ P_{\mathrm{J}}\rho &= E[j(t) j^*(t - \tau_{\mathrm{j}})] \end{aligned} \tag{6.13}$$

where $E(\ )$ denotes the statistical expectation operator and ρ denotes the normalized correlation coefficient between the delayed and undelayed waveforms. The value of ρ depends upon the product of the radar bandwidth and the propagation delay. When the bandwidth equals zero, the product equals zero and ρ equals unity. For nonzero products, the magnitude of ρ is less than unity. The optimum weight for minimum residue power is found by evaluating the partial derivative of (6.13) with respect to w^* and setting the derivative equal to zero. Thus,

$$\partial P_{\mathrm{res}} / \partial w^* = P_{\mathrm{J}}\left[-g_{\mathrm{m}}^* g_{\mathrm{a}} \rho \exp(-j\phi) + w G_{\mathrm{a}}\right] \tag{6.14}$$

$$w_{\mathrm{opt}} = (\rho g_{\mathrm{m}} / g_{\mathrm{a}})^* \exp(-j\phi)$$

Note that (6.14) is similar to (6.10) with ρ equal to the ratio of powers. Combining (6.13) and (6.14) gives the minimum residue power, neglecting thermal noise, for the nonzero bandwidth as

$$P_{\min} = P_{\mathrm{J}} G_{\mathrm{a}} (1 - |\rho|^2) \tag{6.15}$$

For zero bandwidth, ρ equals unity and $P_{\min}$ equals zero. For nonzero bandwidth, $P_{\min} > 0$.

6.4.1.1 Bandwidth Effect and Radar System Parameters

For the usual ECM-ECCM condition, the transmitted noise jammer bandwidth is much greater than the radar receiver bandwidth. Thus, the received jammer waveform has a bandwidth equal to the radar bandwidth. A consequence is that radar system design is a constant trade-off between competing desires. As an example, the desire for increased range resolution indicates a need for large radar signal bandwidth. However, the need to maintain good performance in a noise jammer environment indicates a need to obtain good jammer cancellation and, therefore, small radar signal bandwidth. This section modifies the cancellation equation to express the result in terms of radar system parameters in order to highlight the pertinent system trade parameters.

There is a relation between the normalized correlation coefficient and the transmitted radar waveform. For a matched filter processor, the radar receiver transfer function is the complex conjugate of the spectrum of the radar waveform. Thus, for a transmitted waveform spectrum of $S(f)$, the receiver transfer function equals $S^*(f)$. When the noise jammer bandwidth is much greater than the radar bandwidth, the received jammer power spectrum is proportional to the squared magnitude of the receiver transfer function or $|S(f)|^2$. The autocorrelation coefficient is the Fourier transform of the power spectrum. However, the Fourier transform of the squared magnitude of the receiver transfer function also equals the complex envelope of the radar signal waveform at the matched filter output. Thus, as an example, assume the transmit radar waveform is a pulse of duration τ. The matched filter output is a triangle function of total duration 2τ. The value of the normalized correlation coefficient equals the value of the triangle function at the lag time equal to the propagation delay between the two antennas. Thus,

$$\rho = 1 - |\tau_j/\tau| \tag{6.16}$$

Alternately, for an LFM waveform of two-sided bandwidth B, the magnitude of the transmitted spectrum, or matched filter transfer function, is approximately the rectangle function $\mathrm{Rect}(f/B)$. The autocorrelation function is the Fourier transform and evaluates to

$$\rho = \sin(\pi B\tau_j)/\pi B\tau_j \approx 1 - (\pi B\tau_j)^2/6 \tag{6.16a}$$

The approximation is via a power series expansion of the sine function and is valid for values of ρ that are close to unity. Also note that for ρ close to unity

$$1 - \rho^2 \approx (\pi B\tau_j)^2/3 \tag{6.17}$$

Assume a main antenna width equal to D and the auxiliary antenna placed on the edge of the main antenna. From (6.3)

$$\tau_j = D\sin(\theta_J)/(2c) \tag{6.18}$$

Because the azimuth beamwidth, θ_a, is related to the antenna width, this reduces to

$$\tau_j = \sin(\theta_J)/(2f_0\theta_a) \tag{6.19}$$

Combining (6.15), (6.17), and (6.19) gives the cancellation ratio, CR, defined as the ratio of the power received by the main antenna prior to adaptation to the residue power, as

$$CR = P_{main}/P_{min} = \{[(\pi B/f_0)\sin(\theta_J)/(\theta_a)]^2/12\}^{-1} \tag{6.20}$$

where P_{main} is the jammer power received by the main antenna via the sidelobe gain of G_m. It can be seen that the cancellation ratio decreases as the radar bandwidth increases and decreases as the antenna beamwidth decreases. Thus, for both range and angle resolution, improved resolution degrades the ECCM performance of the sidelobe canceler. As a numeric example of the achievable cancellation ratio for typical system parameters, assume an antenna beamwidth of 1/30 radians (1.91 degrees), a bandwidth of 3 MHz, a carrier frequency of 3 GHz, and a jammer angle of 30 degrees. Evaluating (6.20) shows that the cancellation ratio equals 37.3 dB. Thus, for received jammer powers less than about 37 dB, the residue power is close to the thermal noise level and approximately independent of jammer power. However, for received jammer power greater than about 37 dB, the residue power increases linearly (dB) with jammer power. If the system bandwidth increases to 10 MHz, the transition power level decreases to 26.8 dB.

6.4.2 Amplifier Mismatch Effect upon Jammer Cancellation

The propagation delay limitation on jammer cancellation depends upon the jammer's angle of arrival. When the jammer angle is zero, there is no propagation delay. However, perfect jammer cancellation only occurs if the transfer functions in the main antenna and auxiliary antenna paths are identical. When the transfer functions differ (by other than a scalar multiple), the two waveforms at the subtracted inputs are not identical and perfect cancellation will not occur.

Equation 6.15 is general in the sense that it is valid for any two inputs that are adaptively combined to compute the residue. This can be derived by generalizing (6.7) and (6.8) as

$$r(t) = y_{\mathrm{m}}(t) - w^* y_{\mathrm{a}}(t) \tag{6.21}$$

The residue power equals

$$P_{\mathrm{res}} = P_{\mathrm{main}} + |w|^2 P_{\mathrm{aux}} - [w^*\rho + (w^*\rho)^*](P_{\mathrm{main}} P_{\mathrm{aux}})^{1/2} \tag{6.22}$$

where P_{main} and P_{aux} are the total powers received by the main and auxiliary antennas and ρ is the normalized correlation coefficient defined as

$$\rho(P_{\mathrm{main}} P_{\mathrm{aux}})^{1/2} = E\left[y_{\mathrm{m}}^*(t)\, y_{\mathrm{a}}(t)\right] \tag{6.23}$$

Differentiating (6.22) with respect to w^* shows that the weight value that minimizes the residue power is

$$w_{\mathrm{opt}} = \rho(P_{\mathrm{main}}/P_{\mathrm{aux}})^{1/2} \tag{6.24}$$

and combining this with (6.22) shows that the minimum residue power equals

$$P_{\mathrm{min}} = P_{\mathrm{main}}(1 - |\rho|^2) \tag{6.25}$$

Note that this is similar to (6.15) but that this derivation also includes the effects of thermal noise.

When the weight value differs from the optimum weight, direct substitution into (6.22) shows that

$$w = \delta + w_{\mathrm{opt}} \tag{6.26}$$

$$P_{\mathrm{res}} = P_{\mathrm{min}} + |\delta|^2 P_{\mathrm{aux}} = P_{\mathrm{min}} + P_{\mathrm{add}}$$

and P_{add} is the additional residue power due to the difference between the weight value and the optimum weight value.

An analysis of the cancellation ratio limitation due to the amplifier mismatch will be performed by neglecting thermal noise and propagation delay. The result will be an equation for the cancellation ratio as a function of channel match. This limitation is negligible when the propagation delay limitation is more severe; it is of primary importance when it is a more severe limitation than that due to propagation delay.

Equation 6.25 is used to analyze the effects of mismatched amplifiers and filters in the main and auxiliary channels. Denote the impulse response of the main and auxiliary channels as $h_m(t)$ and $h_a(t)$, respectively. Neglecting thermal noise and propagation delay, the input to each channel is $j(t)$. Thus, the channel outputs are

$$y_m(t) = \int j(t-x)h_m(x)dx \qquad y_a(t) = \int j(t-x)h_a(x)\,dx \tag{6.26}$$

When the bandwidth of the noise jammer is much greater than the channel bandwidth

$$\begin{gathered} E[y_a(t)y_m^*(t)] = P_J \int h_a(x)h_m^*(x)\,dx \\ P_{main} = P_J \int |h_m(x)|^2\,dx \qquad P_{aux} = P_J \int |h_a(x)|^2\,dx \end{gathered} \tag{6.27a}$$

Though valid, these equations are only of marginal utility because it is easier to visualize (and specify) channel mismatches in the frequency domain rather than the impulse function domain.

Applying Parsavell's theorem, these integrals can be transformed into the frequency domain. Thus,

$$\begin{gathered} E\left[y_a(t)y_m^*(t)\right] = P_J \int H_a(f)H_m^*(f)\,df \\ P_{main} = P_J \int |H_m(f)|^2\,df \qquad P_{aux} = P_J \int |H_a(f)|^2\,df \end{gathered} \tag{6.27b}$$

where $H_m(f)$ and $H_a(f)$ are Fourier transforms of the respective impulse functions. As an example, assume that both filters are nominally rectangular bandpass functions of gain unity over the bandwidth B. Due to errors, they have gains that vary linearly with frequency but with opposite slopes. Thus,

$$\begin{aligned} H_m(f) &= 1 - \varepsilon f \\ H_a(f) &= 1 + \varepsilon f \end{aligned} \qquad -B/2 < f < B/2 \tag{6.28}$$

Then

$$\begin{aligned} P_{main} &= P_J B(1 + \varepsilon^2 B^2/12) = P_{aux} \\ E\left[y_a(t)y_m^*(t)\right] &= P_J B(1 - \varepsilon^2 B^2/12) \\ \rho = (1 - \varepsilon^2 B^2/12)&/(1 + \varepsilon^2 B^2/12) \approx 1 - \varepsilon^2 B^2/6 \end{aligned} \tag{6.29}$$

and combining this with (6.25) gives

$$CR = P_{\text{main}}/P_{\text{min}} = \varepsilon^2 B^2/3 \tag{6.30}$$

The peak transfer function error, ε_{P}, equals $\varepsilon B/2$ and in terms of the peak error

$$P_{\text{main}}/P_{\text{min}} = 4\varepsilon_P^2/3 \tag{6.31}$$

Transfer function errors are commonly expressed in decibels. For an error of ε_{dB}

$$P_{\text{min}}/P_{\text{main}} = (4/3)(10^{0.05\varepsilon\,\text{dB}} - 1)^2 \tag{6.32}$$

As an example, if ε_{dB} equals 0.1 dB, $P_{\text{min}}/P_{\text{main}}$ equals -37.5 dB. For a peak error of 0.5 dB, it equals -23.3 dB. Comparing these values with those obtained for the propagation delay analysis, it can be seen that amplifier mismatches of more than a few tenths of a decibel would be the limiting factor in achievable jammer cancellation.

Similar results are obtained for other forms of transfer function. As an example, assume that the transfer function error is a sinusoidal ripple. The equivalent of (6.28) is

$$\begin{aligned} H_{\text{m}}(f) &= 1 + \varepsilon_{\text{P}} \sin(\theta_{\text{m}} + 2\pi f/f_{\text{m}}) \\ H_{\text{a}}(f) &= 1 + \varepsilon_{\text{P}} \sin(\theta_{\text{a}} + 2\pi f/f_{\text{a}}) \end{aligned} \tag{6.33}$$

where θ_{m} and θ_{a} are independent random variables uniformly distributed over 2π. The respective ripple frequencies are f_{m} and f_{a}. Note that

$$E(\cos\theta_{\text{m}}) = E(\sin\theta_{\text{m}}) = E(\cos\theta_{\text{a}}) = E(\sin\theta_{\text{a}}) = 0 \tag{6.34}$$

Using (6.34) in (6.27b) shows that

$$\begin{aligned} P_{\text{main}} = P_{\text{aux}} &= P_{\text{J}} B(1 + \varepsilon_{\text{P}}^2/2) \\ \rho P_{\text{ma}} &= P_{\text{J}} B \end{aligned} \tag{6.35}$$

so

$$\rho = (1 + \varepsilon_{\text{P}}^2/2)^{-1} \tag{6.36}$$

Assuming ε_{P} is small

$$P_{\text{min}}/P_{\text{main}} = \varepsilon_{\text{P}}^2 \tag{6.37}$$

which is very close to the result of (6.31).

6.5 Optimum Weight Estimation

The optimum weight depends upon the value of the main antenna voltage gain, amplitude, and phase at the spatial angle of the jammer. Since the sidelobe levels usually depend upon random errors, the gain value is unknown. Thus, the optimum weight cannot be implemented and methods of estimating the optimum weight are required. Early work, published in the classified literature, resulted in the Howells-Applebaum correlation feedback loop.[1] The analytic properties of the algorithm were analyzed by Applebaum.[2] An essentially identical algorithm, usually denoted as the least mean square (LMS) algorithm was developed by Widrow and applied to communication problems.[3] The technique was initially implemented at IF but can also be implemented using digital techniques.

A conceptual block diagram of the LMS technique configured at IF is shown by Figure 6.3. It is also termed a correlation feedback loop because the implementation is a feedback circuit that drives the weight value until the residue voltage has no component in common with the auxiliary input. If there is a common component at the correlation mixer, the value of the weight changes because of the feedback action in a direction to remove the common component from the residue. Thus, the jammer is canceled.

The main and auxiliary antenna inputs are both at a radian carrier frequency of ω_0. Both are down converted (mixed) with an offset oscillator whose radian frequency equals ω_0 - ω_{IF}. The filters pass the difference frequency of ω_{IF}.

6.5.1 Multiplication by Mixing and Filtering

For each mixer shown in the block diagram, the combination of mixing and filtering is intended to implement the mathematical operation of multiplication of the two inputs. The design requires careful choice of mixers, IF frequencies, and filter characteristics in order to obtain a "multiplier" with small errors.

The nonlinear input-output characteristic of the mixer can be represented by a power series as

$$e_{\mathrm{out}} = \sum_k \alpha_{\mathrm{k}} e_{\mathrm{in}}^k \tag{6.38}$$

1. Howells, P. W., Intermediate Frequency Sidelobe Canceler, U.S. Patent 3,202,990, Aug. 24, 1965 (filed May 4, 1959).
2. Applebaum, S. P., Steady State and Transient Performance of the Sidelobe Canceler, Special Projects Laboratory, Syracuse University Research Corporation, April 1966.
3. Widrow, B., et al., "Adaptive Antenna Systems," *Proc. IEEE,* Vol. 55, Dec. 1967.

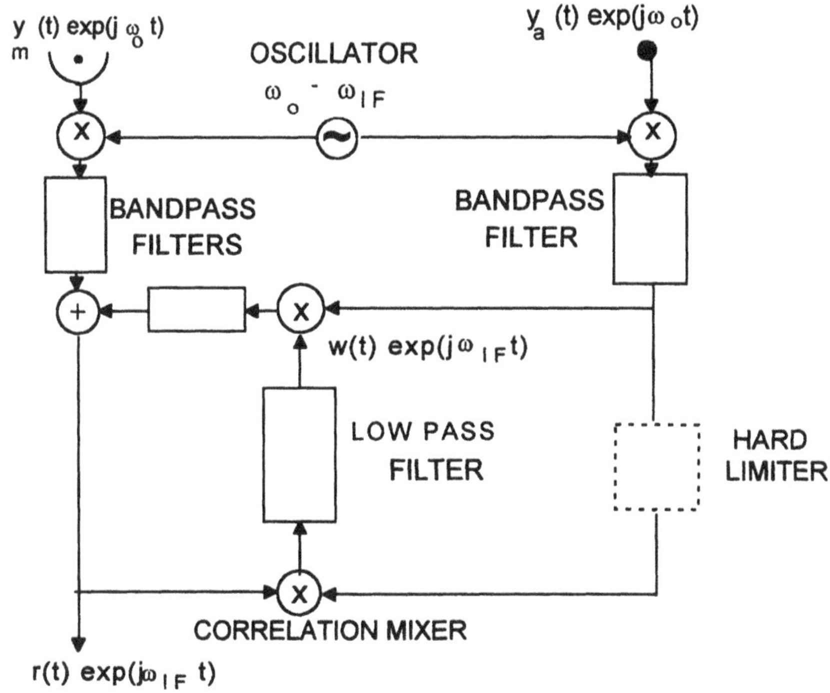

Figure 6.3 Conceptual block diagram of the IF LMS algorithm.

The mixer input, e_{in}, is the sum of the two individual inputs

$$e_{in} = r_1(t)\cos[\omega_1 t + \phi_1(t)] + r_2(t)\cos[\omega_2 t + \phi_2(t)] \tag{6.39}$$

where $r(t)$ denotes envelope modulation and $\phi(t)$ denotes phase modulation. For $k > 1$ in (6.38), the nonlinearity creates output terms whose radian frequencies differ from the input ones of ω_1 and ω_2. For $k = 2$, the cross term of the squaring creates sum and difference frequency terms. The difference term is the desired output and is

$$e_2 = r_1(t)r_2(t)\cos[(\omega_2 - \omega_1)t + \phi_2(t) - \phi_1(t)] \tag{6.40}$$

and $\omega_2 - \omega_1$ equals the IF frequency. Define the complex envelopes

$$\begin{aligned} a_1(t) &= r_1(t)\exp[j\phi_1(t)] \\ a_1(t) &= r_2(t)\exp[j\phi_2(t)] \end{aligned} \tag{6.41}$$

The equivalent of (6.40) is that the complex envelope of the output equals

$a_2(t)$ times the complex conjugate of $a_1(t)$. Thus, the action of each mixer-filter is that if the complex representations of the two inputs are denoted as $a_1(t)\exp(j\omega_1 t)$ and $a_2(t)\exp(j\omega_2 t)$, the complex representation of the output is $a_1^*(t)a_2(t)\exp[j(\omega_2 - \omega_1)t]$. Note that the desired output of (6.40) cannot be attained exactly because the squared terms create frequencies at $2\omega_2$ and $2\omega_1$. The filter, after the mixer, attenuates these terms but cannot totally eliminate them. Similarly, frequency terms due to the other exponents of (6.38) create error terms and it is mandatory that the design chooses mixer and filter characteristics so that the multiplication approximation is done with small error.

6.5.2 SLC Weight Estimation Using LMS Algorithm

Continuing the discussion of the block diagram, the residue is multiplied with the down-converted auxiliary voltage by the correlation mixer. Both of these voltages are at IF. The filter following the correlation mixer is narrow band and centered at zero frequency. The output of the narrow band filter is the weight value for the auxiliary antenna. The weight is mixed with the auxiliary voltage and bandpass filtered. Note that the weight is time dependent and, thus, not equal to the optimum weight. In addition, it is a function of random processes and therefore the weight is a random process. However, it will be shown that the average value of the weight converges to a close approximation to the optimum weights. (The dashed box denotes a hard limiter. This modification is discussed in a later section.)

In terms of the complex envelopes, it can be seen from Figure 6.3 that

$$\begin{aligned} r(t) &= y_{\mathrm{m}}(t) - w^*(t)y_{\mathrm{a}}(t) \\ w(t) &= \int h(t-x)r^*(x)y_{\mathrm{a}}(x)\,dx \end{aligned} \tag{6.42}$$

where $h(\ )$ denotes the impulse response of the narrow band filter after the correlation mixer. Often the filter is a single-pole filter whose operation is described by the differential equation

$$T[dw(t)/dt] + w(t) = gr^*(t)y_{\mathrm{a}}(t) \tag{6.43}$$

where T is the filter time constant and g is the filter's DC gain. Combining (6.43) with the first part of (6.42) gives

$$T[dw(t)/dt] + w(t) = g\left[y_{\mathrm{m}}^*(t)y_{\mathrm{a}}(t) - w(t)y_{\mathrm{m}}^*(t)y_{\mathrm{a}}(t)\right] \tag{6.44}$$

Since $w(t)$ is a random process, (6.44) is a stochastic differential equation. The solution is the joint probability density function for the set of $w(t)$. However, the equation cannot be solved by any known techniques. Instead, the moments of $w(t)$ can be obtained. This section will derive the first moment, or mean, of $w(t)$. It will be shown that the mean is time dependent. That is, if the interference environment is initially solely thermal noise, the weight is a value appropriate for a thermal noise environment. If the environment suddenly changes to thermal noise plus a jammer, the feedback action causes the weight to change to a value appropriate for jammer cancellation. The change of weight value is not instantaneous. The weight value is time dependent, and the time period required to change from the value appropriate to a thermal noise environment to the jammer cancellation value is termed the convergence time.

6.5.3 LMS Convergence Time Analysis

Rearranging (6.44) gives

$$T[dw(t)/dt] + w(t)[1 + g|y_a(t)|^2] = gy_m^*(t)y_a(t) \qquad (6.44a)$$

Defining $w_a(t)$ as the statistical expectation (mean or average value) of $w(t)$ and taking the mean value of both sides of (6.44a) give

$$T[dw_a(t)/dt] + w_a(t) + gE[w(t)|y_a(t)^2] = g\rho(P_{main}P_{aux})^{1/2} \qquad (6.45)$$

where P_{main} and P_{aux} are defined in (6.23) as the main and auxiliary antennas powers, respectively, and ρ denotes the correlation coefficient of the main and auxiliary antennas. It will be assumed that the low-pass filter's bandwidth is very narrow compared to the radar bandwidth. Then, the instantaneous value of the weight is approximately uncorrelated with the instantaneous value of the auxiliary voltage. Thus,

$$E[w(t)|y_a(t)^2] = w_a(t)P_{aux}$$

Substituting this into (6.45) produces the first-order constant coefficient differential equation

$$\begin{gathered} T[dw_a(t)/dt] + w_a(t)[1 + gP_{aux}] = g\rho P_{ma} \\ (P_{main}P_{aux})^{1/2} = \rho P_{ma} \end{gathered} \qquad (6.46)$$

Assuming the average weight value equals zero at $t = 0$, the solution is

$$w_a(t) = (g\rho P_{ma})/[1 + gP_{aux}]\{1 - \exp[-(1 + gP_{aux})t/T]\} \qquad (6.47)$$

The steady-state solution is

$$w_s = (g\rho P_{ma})/[1 + g P_{aux}] \tag{6.48}$$

Comparing this to (6.24), it can be seen that when the filter's DC gain, g, is large, the steady-state average weight value approximately equals the optimum weight.

The transient time constant equals $T/(1 + g P_{aux})$. Thus, the time constant decreases as either the filter gain is increased or the jammer power increases. During the time of the transient decay, the residue power is greater than that achieved at steady state. Neglecting the difference between the steady-state average weight and the optimum weight and assuming the normalized correlation coefficient equals one, the transient time weight error is

$$\delta(t) = (P_{main}/P_{aux})^{1/2} \exp[-t/T(1 + g P_{aux})] \tag{6.49}$$

From (6.26), this weight error causes the residue power to equal the minimum residue power plus the additional residue power

$$P_{add} = P_{main} \exp[-2t/T(1 + g P_{aux})] \tag{6.50}$$

The sketch of Figure 6.4 depicts the convergence as a parametric function of

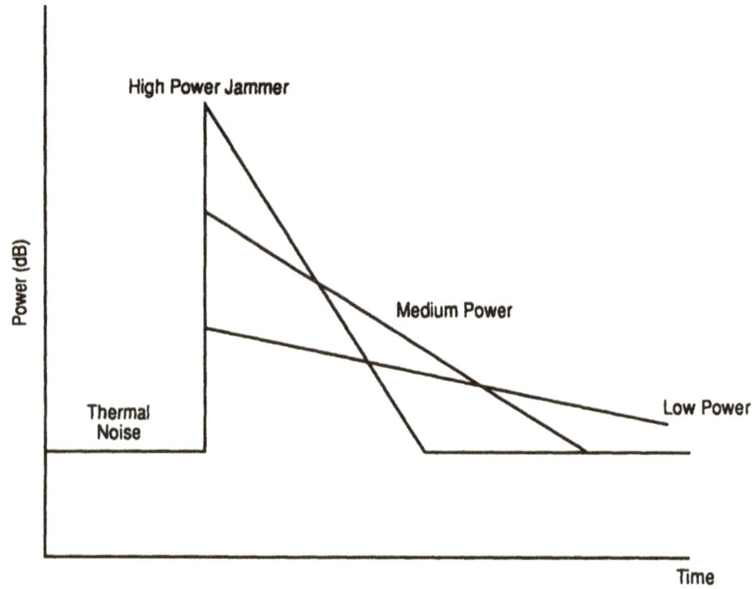

Figure 6.4 LMS single jammer convergence characteristics.

jammer power. As indicated by the figure, initially the residue power is approximately equal to the thermal noise power. Then the environment changes and from (6.50) at time equal to zero the additional residue power equals the main antenna power. As the weight approaches its steady-state value, the residue power (in decibels) decreases linearly with a slope that depends on the jammer power. As can be seen, the convergence is fast for large jammer power and slow for small jammer power. There are two operational constraints on the acceptable values of the time constant. First, the time constant should be longer than the time duration comparable to a range cell so that there is little tendency for self-cancellation of a target return. Second, the time constant should be much shorter than the radar beam's dwell period. For many radar, this implies that the convergence time should be greater than a few microseconds but much less than a few milliseconds. However, variations in the jamming threat can easily cause the received jammer power and, therefore, the time constant to vary by four, or more, orders of magnitude. Thus, long convergence time might limit the applicability of this algorithm.

Note that the solution of (6.47) through (6.50) indicates that the feedback circuit is unconditionally stable. This implies that the low-pass filter gain value can be chosen solely on the basis of achieving the desired convergence time. However, the unconditional stability conclusion depends upon the form of the low-pass filter. Implementations using a single-pole filter are unconditionally stable. In actual implementations, stray capacitance causes the low-pass filter to differ from a single-pole filter and the circuit is not unconditionally stable. It is then necessary to ensure stability by careful design and use of a decreased value for the low-pass filter gain. As will be seen, the equivalent digital single-pole filter implementation is not unconditionally stable and it is necessary to always use a restricted low-pass filter gain value.

Several modifications to the LMS algorithm have been proposed to decrease the variability of convergence time. The next sections analyze two modified LMS algorithms whose convergence time is less dependent upon the power level of the auxiliary antenna. Both modify the auxiliary antenna input to the correlation mixer.

6.5.4 LMS Algorithm with Hard Limiter

This technique modifies the auxiliary antenna input to the correlation mixer using an IF hard limiter. An IF hard limiter is a device whose input-output characteristic is such that when the input is $r(t)\cos[\omega_1 t + \phi(t)]$, the output is $L\cos[\omega_1 t + \phi(t)]$, where the amplitude, L, is a constant. Thus, its action is that all amplitude modulation is removed by the limiting action but the phase modulation is retained.

Equation (6.44a) is modified to

$$T[dw(t)/dt] + w(t)[1 + g_L y_a^*(t) y_L(t)] = g y_m^*(t) y_L(t) \tag{6.51}$$

$$y_L(t) = L y_a(t)/|y_a(t)|$$

where $y_L(t)$ denotes the hard-limited version of $y_a(t)$. As was discussed previously for the LMS algorithm, the resulting equation is a stochastic difference equation and the only available analysis is via moments. Taking the average value of both sides of (6.51) will again produce a first-order constant coefficient differential equation. The time constant, T_L, and steady-state weight value, w_L, are

$$\begin{aligned} T_L &= T/\{1 + g_L E[y_a^*(t)\, y_L(t)]\} \\ w_L &= g_L E[y_m^*(t)\, y_L(t)]/\{1 + g_L E[y_a^*(t)\, y_L(t)]\} \end{aligned} \tag{6.52}$$

To evaluate the indicated expectations, it is necessary to specify the statistics of $y_m(t)$ and $y_a(t)$. For jamming cancellation, a reasonable assumption is that all voltages have a zero-mean, joint Gaussian distribution. Then, the required expectations can be evaluated by transformation to a pair of independent zero-mean, Gaussian-distributed, unit-variance, complex, random variables as was done to develop the required expectations for (4.36). In addition it is necessary to evaluate several integrals.[4] It is shown that when the correlation filter gain is large, the steady-state mean value of the weight is still approximately equal to the optimum weight. However, as will be discussed later, a large gain causes an increased residue power at steady state due to the noisy nature of the steady-state weight. It is also shown that the convergence time varies inversely with the square root of the auxiliary antenna power rather than inversely with the auxiliary antenna power as for the nonlimiter LMS algorithm. An indication of the relative convergence time properties is given by Table 6.1 as a function of jammer power. The tabulated results assume that both the linear and limiter-configured algorithms are designed to have the same convergence time when the jammer power equals 40 dB and the normalized response time is one for both algorithms. For the linear algorithm, the convergence time increases by a factor of 10 for a 10-dB decrease in jammer power. For the limiter-configured algorithm, the convergence time increases by a factor of the square root of 10 for a 10-dB decrease in jammer power.

4. Full details are available in Nitzberg, R., *Adaptive Signal Processing for Radar*, 1992, Norwood, MA: Artech House, p. 95.

Table 6.1
Response Time

Jammer power	Normalized response time: Linear	Normalized response time: Limiter
40 dB	1	1
30 dB	10	$10^{1/2}$
20 dB	100	10

6.5.5 Normalized LMS Algorithm (NLMS)

The use of the hard limiter changes the auxiliary input to the correlation mixer from $y_a(t)$ to $y_a(t)/|y_a(t)|$. The NLMS algorithm changes the correlation mixer auxiliary input to $y_a(t)/|y_a(t)|^2$. It is shown in *Adaptive Signal Processing for Radar*[5] that this causes the convergence time for the single jammer–single auxiliary antenna implementation to be totally independent of the auxiliary power. The circuitry required to implement the algorithm is instantaneous automatic gain control (IAGC), where the IAGC action controls the feedback loop gain such that it is inversely proportional to $|y_a(t)|^2$. The details of the implementation depend upon the IAGC circuitry. The desired IAGC action can be obtained by gain control only of the auxiliary antenna input to the correlation mixer. Alternately, the IAGC action could be split so that the residue and auxiliary antenna voltages were each divided by $|y_a(t)|$.

6.5.6 Digital LMS Algorithm

To obtain improved performance for pulse compression and Doppler processing, for example, some radar system designs use an A/D converter for the main and auxiliary antennas. Digital systems implement the correlation feedback loop via digital filtering and multiplying. The actual implementation can be via a microprocessor or a dedicated special-purpose computer.

Though A/D converters were discussed in Chapter 4, a few additional facts are necessary to discuss and analyze the digital LMS algorithm. There are several implementation techniques for the A/D converter. All of them can be viewed as converting a bandpass waveform into a pair of real-valued discrete sequences, and the pair of sequences can be viewed as a single complex-valued sequence. Figure 1.20 depicts a synchronous demodulation technique. Thus, consider a

5. Nitzberg, R. *Adaptive Signal Processing for Radar*, Norwood, MA: Artech House, 1992, p. 96.

bandpass waveform

$$e(t) = x_r(t)\cos\omega_0 t - x_i(t)\sin\omega_0 t$$

The low-pass voltages $x_r(t)$ and $x_i(t)$ are referred to as the I and Q components, respectively, and I and Q stand for in-phase and quadrature phase. After demodulation, the voltages $x_r(t)$ and $x_i(t)$ are inputs to the pair of A/D converters. The ideal A/D converter performs two cascaded actions. The first is discrete time sampling of the input waveform; the second is conversion of analog samples to digital samples. The first action samples the input at equispaced time instants that are multiples of the sampling period T_s. At time kT_s, where k is an integer, the pair of samplers output the voltages $x_r(kT_s)$ and $x_i(kT_s)$. These real voltages can be regarded as the complex envelope

$$y(kT_s) = x_r(kT_s) + jx_i(kT_s)$$

and the dependence on T_s is usually understood, so the complex output is denoted simply as $y(k)$. For either thermal noise or noise jamming, the statistics of $x_r(t)$ and $x_i(t)$ are that they are zero-mean Gaussian processes with common variance. Then $y(k)$ is a zero-mean complex Gaussian sequence with variance equal to twice the power of $e(t)$. The factor of two in power occurs because of a mathematical peculiarity in the definition of complex random processes and causes no analytic difficulty. The sampling period is usually approximately equal to the reciprocal of the radar bandwidth. Then, samples of $e(t)$ at different time instants are approximately statistically independent. They are exactly independent when $e(t)$ is strictly bandlimited with bandwidth less than or equal to $1/T_s$. The following analyses assume independence at different sampling time instants.

A conceptual block diagram of the digital LMS algorithm is shown in Figure 6.5. The implementation shows a single-pole digital filter used after the correlation mixer. In general, other filters can be used. The analysis of this section will only consider the single-pole filter. As can be seen directly from the figure, the generated adaptive weight and residue satisfy the difference equations

$$\begin{aligned} r(k) &= y_m(k) - w^*(k)y_a(k) \\ w(k+1) &= w(k) + g_d r^*(k) y_a(k) \end{aligned} \tag{6.53}$$

The weight update equation shows that the next weight value equals the current weight value plus a correction term. The correction term equals the product of the conjugate of the current residue and current auxiliary antenna voltage

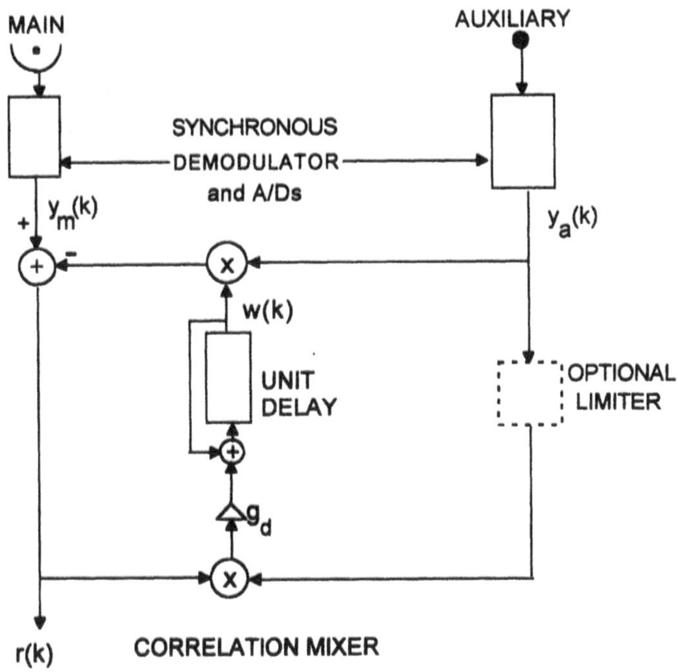

Figure 6.5 Digitally implemented LMS algorithm.

multiplied by the scaling gain factor g_d. Combining the two equations of (6.53) gives

$$w(k+1) = [1 - g_d|y_a(k)|^2]w(k) + g_d y_m^*(k) y_a(k) \tag{6.54}$$

As for the analog LMS algorithm analysis, the time-varying dependence of the average value of the adaptive weight will be derived. Note that the current value of $w(k)$ does not depend upon the current values of $y_m(k)$ and $y_a(k)$ but only upon previous values. Since the sampling rate is such that the current values of $y_m(k)$ and $y_a(k)$ are statistically independent of the previous values, $w(k)$ is statistically independent of $y_a(k)$. Thus, taking the average value of both sides of (6.53), assuming that the initial value of the average weight equals zero, the solution is

$$w_a(k) = [1 - (1 - g_d P_{aux})^k] g_d \rho (P_{main} P_{aux})^{1/2} / (g_d P_{aux}) \tag{6.55}$$

This equation indicates that there are two significant differences in the convergence properties of the digital LMS algorithm as compared to the analog LMS algorithm. One difference is that, at convergence, the average steady-state weight is exactly equal to the optimum weight. This contrasts with the analog implementation where the steady-state weight is only approximately equal to the optimum weight. This difference in steady-state behavior occurs because the DC gain of the narrow band filter after the correlation multiplier is always finite for the analog implementation but is infinite in the digital implementation. A second difference is that for the analog algorithm, the feedback circuit is stable for all positive values of the filter gain g. For the digital algorithm, stability occurs only if

$$|1 - g_d P_{aux}| < 1 \tag{6.56}$$

or equivalently

$$0 < g_d < 2/P_{aux} \tag{6.57}$$

Thus, g_d must be positive and less than a value that is a function of the auxiliary power. A conservative design is to set g_d less than $2/P_{max}$, where P_{max} is the power level that saturates the auxiliary antenna A/D converter. Then defining a normalized gain g_{LMS} as

$$g_d = g_{LMS}/P_{max} \qquad 0 < g_{LMS} < 2 \tag{6.57a}$$

For a specified value of g_d, the functional form of (6.55) is very much like exponential decay. Specifically, for $g_d P_{aux}$ less than unity, (6.55) may be rewritten as

$$w_a(k) = \{1 - \exp[k \ln(1 - g_d P_{aux})]\} g_d \rho (P_{main}/P_{aux})^{1/2} \tag{6.58}$$

and the effective time constant, T_e, is

$$T_e = -1/\ln(1 - g_d P_{aux}) \tag{6.59}$$

For $g_d P_{aux}$ close to unity, the effective time constant is short. As P_{aux} decreases, the time constant increases and tends toward infinity as P_{aux} tends to zero. When $g_d P_{aux}$ is small compared to unity, T_e is approximately

$$T_e = -1/(g_d P_{aux}) \tag{6.59a}$$

and combining with (6.57a) gives

$$T_e = -1/[g_{LMS}(P_{aux}/P_{max})] \tag{6.59b}$$

The weight value for the digital implementation, like that of the analog implementation, converges exponentially with a time constant that varies approximately inversely with the auxiliary power level and inversely with the feedback gain. Thus, faster convergence is obtained by increasing the feedback gain. However, as will be shown later, increased gain causes increased residue power after convergence due to weight noise.

6.5.7 Digital LMS Algorithm with Limiter

When the auxiliary input to the correlation multiplier is modified by limiting, the modification of (6.53) is very much like that caused by limiting in the analog implementation. Specifically, the modified stochastic difference equation is

$$w(k+1) = \left[1 - g_{\mathrm{L}} y_{\mathrm{a}}^*(k) y_{\mathrm{L}}(k)\right] w(k) + g_{\mathrm{l}} y_{\mathrm{m}}^*(k) y_{\mathrm{L}}(k) \tag{6.60}$$

where g_{L} denotes the gain after the correlation mixer and $y_{\mathrm{L}}(k)$ denotes the limiter output. There are two possible methods of implementing a limiting action. One is like the analog limiter implementation in that

$$y_{\mathrm{L}}(k) = y(k)/|y(k)| \tag{6.61a}$$

The other is a digital limiter and

$$y_{\mathrm{L}}(k) = \mathrm{Real}[y(k)]/\mathrm{Real}|y(k)| + j\,\mathrm{Imag}[y(k)]/\mathrm{Imag}|y(k)| \tag{6.61b}$$

The first type of limiting is mathematically identical to the limit action analyzed for the analog implementation of the LMS algorithm. It can be shown that

$$\begin{aligned} E\left[y_{\mathrm{m}}^*(k) y_{\mathrm{L}}(k)\right] &= (\gamma/P_{\mathrm{aux}}) E\left[y_{\mathrm{m}}^*(k) y_{\mathrm{a}}(k)\right] \\ E\left[y_{\mathrm{a}}^*(k) y_{\mathrm{L}}(k)\right] &= (\gamma/P_{\mathrm{aux}}) E\left[y_{\mathrm{a}}^*(k) y_{\mathrm{a}}(k)\right] \end{aligned}$$

with γ equal to $\pi^{1/2}/2$ when the limiting is done using (6.61a) and a slightly different value when implemented using (6.61b). Thus, the action of the limiter is essentially identical to that obtained for the analog implementation. That is, the digital limiter causes the convergence time constant to vary inversely with the square root of the auxiliary antenna power just as was obtained for the analog limiter implementation of the LMS algorithm.

6.5.8 Digital NLMS Algorithm Analysis

The digital version of the NLMS algorithm for a single auxiliary antenna uses the weight update equation

$$w(k+1) = w(k) + [g_N/|y_a(k)|^2]r^*(k)y_a(k) \qquad (6.62)$$

Just as for the analog implementation, the modification as compared to the LMS weight update equation results in a convergence time that is independent of the auxiliary power. Thus, combining the first equation of (6.52) and (6.62) and solving for the average weight value results in

$$w_a(k+1) = (1-g_N)w_a(k) + g_N E\left[y_m^*(k)y_a(k)/|y_a(k)|^2\right] \qquad (6.63)$$

It can be shown that for complex-valued, zero-mean, Gaussian-distributed variates, the expectation of (6.63) evaluates to[6]

$$E\left[y_m^*(k)y_a(k)/|y_a(k)|^2\right] = \rho(P_{main}/P_{aux})^{1/2}$$

The solution of the constant coefficient difference equation is

$$w_a(k) = [1-(1-g_N)^k]\rho(P_{main}/P_{aux})^{1/2} \qquad 0 < g_N < 2 \qquad (6.64)$$

and the inequality constraint on the gain constant is required to ensure stability. It can be seen that the convergence time constant is independent of the auxiliary power level and that the steady-state average weight value equals the optimum weight.

6.6 Weight Noise

For either the analog or digital implementations, the adaptive weight computed by the various algorithms is a function of random processes and therefore is itself a random process. The analyses demonstrated that the average value of the adaptive weight converged to the optimum weight. However, since the weight is a random process, after convergence the weight value deviates randomly from the optimum weight and causes the residue power to be larger than that achieved by the optimum weight.

From (6.26) the total residue power at convergence is

$$P_{res} = P_{min} + P_{aux}E|\delta|^2 \qquad (6.65)$$

6. Nitzberg, R., *Adaptive Signal Processing for Radar*, Norwood, MA: Artech House, 1992, p. 96.

when the deviation is random. The residue power due to weight noise for the LMS and NLMS algorithms is computed in this section. It is shown that the faster convergence characteristic of the NLMS algorithm is coupled with substantially greater weight noise residue power.

6.6.1 LMS Algorithm Weight Noise

In order to evaluate (6.60), (6.53) is rewritten as

$$\begin{aligned} r(k) &= y_{\mathrm{m}}(k) - w^*(k) y_{\mathrm{a}}(k) \\ w(k) &= w_{\mathrm{opt}} + \delta(k) \\ r(k) &= r_{\mathrm{opt}}(k) - \delta^*(k) y_{\mathrm{a}}(k) \\ r_{\mathrm{opt}}(k) &= y_{\mathrm{m}}(k) - w^*_{\mathrm{opt}} y_{\mathrm{a}}(k) \end{aligned} \tag{6.66}$$

where $r_{\mathrm{opt}}(k)$ is defined as the residue voltage obtained when using the optimum weight and $\delta(k)$ is the deviation of the weight from the optimum value. The weight update equation can then be rewritten as

$$w(k+1) = w(k) + g_{\mathrm{d}}\left[r^*_{\mathrm{opt}}(k) y_{\mathrm{a}}(k) - \delta^*(k) |y_{\mathrm{a}}(k)|^2\right] \tag{6.67}$$

and subtracting the optimum weight from both sides results in

$$\delta(k+1) = \delta(k) + g_{\mathrm{d}}\left[r^*_{\mathrm{opt}}(k) y_{\mathrm{a}}(k) - \delta^*(k) |y_{\mathrm{a}}(k)|^2\right] \tag{6.68}$$

Define the variance of the random weight error, $V_\delta(k)$, as $E[\delta(k)\delta^*(k)]$. From (6.68) a difference equation for the weight error variance is

$$\begin{aligned} V_\delta(k+1) = E\{&\left[\delta(k) + g_{\mathrm{d}}\left(r^*_{\mathrm{opt}}(k) y_{\mathrm{a}}(k) - \delta^*(k) |y_{\mathrm{a}}(k)|^2\right)\right] \\ &\cdot \left[\delta(k) + g_{\mathrm{d}}\left(r^*_{\mathrm{opt}}(k) y_{\mathrm{a}}(k) - \delta^*(k) |y_{\mathrm{a}}(k)|^2\right)\right]^*\} \end{aligned} \tag{6.69}$$

Note that $w(k)$ depends only on previous data and the A/D sampling frequency is chosen such that current data are statistically independent of previous data. Thus, $\delta(k)$, $y_{\mathrm{m}}(k)$, and $y_{\mathrm{a}}(k)$ are statistically independent, so

$$\begin{aligned} E[|\delta(k) y_{\mathrm{a}}(k)|^2] &= V_\delta P_{\mathrm{aux}} \\ E\left[\delta(k) r_{\mathrm{opt}}(k) y^*_{\mathrm{a}}(k)\right] &= E\{\left[\delta(k) y^*_{\mathrm{a}}(k)\right]\left[y_{\mathrm{m}}(k) - w^*_{\mathrm{opt}} y_{\mathrm{a}}(k)\right]\} \\ &= E[\delta(k)]\, E\left[y^*_{\mathrm{a}}(k) y_{\mathrm{m}}(k)\right] - w^*_{\mathrm{opt}} |y_{\mathrm{a}}(k)|^2] = 0 \end{aligned}$$

Therefore, (6.69) reduces to

$$V_\delta(k+1) = (1 - 2g_d P_{aux} + g_d^2 E|y_a(k)|^4) V_\delta(k) + g_d^2 P_{min} P_{aux} \tag{6.70}$$

For complex-valued, zero-mean Gaussian variates, a useful result is

$$E|z|^{2n} = n! E(|z|^2)^n$$

so

$$E|y_a(k)|^4 = 2P_{aux}^2$$

Thus, (6.70) is a constant coefficient difference equation, which converges if

$$0 < g_d P_{aux}(1 - g_d P_{aux}) < 1 \tag{6.71}$$

At steady state, the variance, $V_{\delta s}$, is

$$V_{\delta s} = P_{min}(g_d/2)/(1 - g_d P_{aux}) \tag{6.72}$$

so the additional steady-state residue power due to weight noise is

$$p_{add} = P_{min}(g_d P_{aux}/2)/(1 - g_d P_{aux}) \tag{6.73}$$

and the normalized total residue power at convergence equals

$$p_{res}/P_{min} = \{[1 - (g_{LMS}/2)(P_{aux}/P_{max})]/[1 - g_{LMS}(P_{aux}/P_{max})]\} \tag{6.74}$$

Graphs of the normalized residue power are shown in Figure 6.6 versus the normalized gain, g_{LMS}, for several values of the ratio P_{aux}/P_{max}. Note that for small values of g_{LMS} the additional steady-state residue power due to weight noise is small but increases as the received jammer power increases. However, as g_{LMS} increases toward unity, the additional power, and therefore the normalized residue power, tend toward infinity. Thus, even though the first moment analysis indicates that g_{LMS} equal to unity is satisfactory for stability, the second moment analysis indicates it should be much less than unity.

6.6.2 LMS with Limiter Weight Noise

The weight update equation of (6.60) is combined with (6.66), and the resulting difference equation for the variance of the weight error is

$$V_\delta(k+1) = [1 - g_L(\pi P_{aux})^{1/2} + g_L^2 P_{aux}] V_\delta(k) + g_d^2 P_{min} \tag{6.75}$$

At steady state, the solution of the constant coefficient difference equation is

$$V_{\delta s} = (P_{min} g_L)/[(\pi P_{aux})^{1/2} - g_L P_{aux}] \tag{6.76}$$

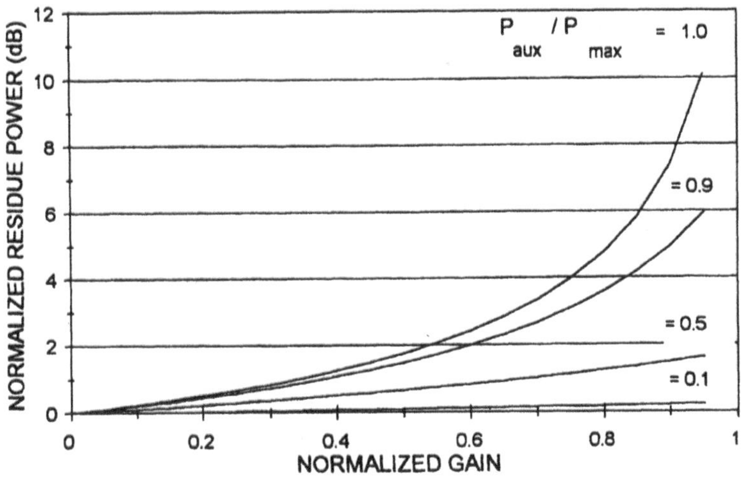

Figure 6.6 Weight noise effect on LMS algorithm.

so the additional steady-state residue power due to weight noise is

$$p_{\text{add}} = (P_{\min} g_{\text{L}} P_{\text{aux}})/[(\pi P_{\text{aux}})^{1/2} - g_{\text{L}} P_{\text{aux}}] \tag{6.77}$$

and the normalized total residue power at convergence equals

$$p_{\text{res}}/P_{\min} = (\pi P_{\text{aux}})^{1/2}/[(\pi P_{\text{aux}})^{1/2} - g_{\text{L}} P_{\text{aux}}] \tag{6.78}$$

The interpretation of (6.78) is identical to the interpretation of (6.74). That is, for small values of the weight update gain, the additional steady-state residue power is small. As the effective update gain increases, the additional residue power tends toward infinity. This is shown by the plots of Figure 6.7. The figure also includes some plots for the linear LMS algorithm. It can be seen that the superior convergence properties of the limiter-configured algorithm are coupled with increased weight noise-induced residue power at convergence.

6.6.3 NLMS Algorithm Weight Noise

Using the procedures of the preceding sections, it can be shown that the difference equation for the weight error variance is

$$V_\delta(k+1) = (1 - g_{\text{N}})^2 V_\delta(k) + g_{\text{N}}^2 P_{\min} E(1/|y_{\text{a}}(k)|^2) \tag{6.79}$$

It can be shown that the expectation in (6.79) is infinite. This implies that the weight variance at steady state is also infinite so that, for this algorithm, the variance is not a good indicator of performance.

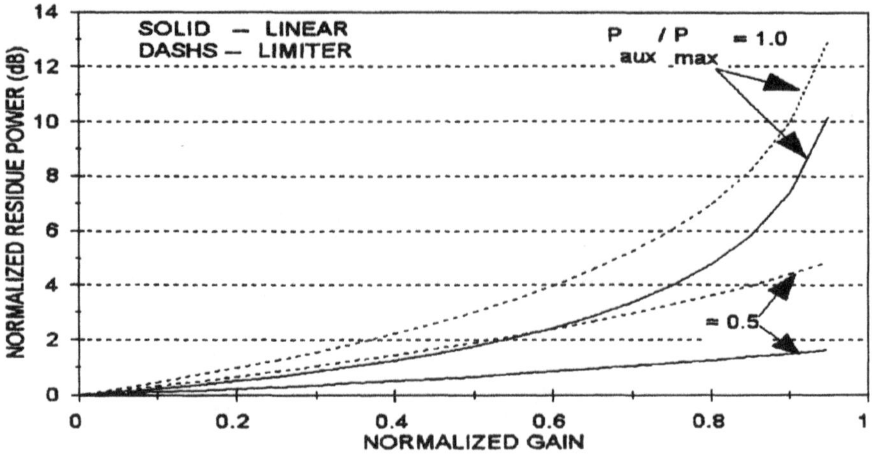

Figure 6.7 Weight noise effect on limiter LMS algorithm.

6.7 Multiple Sidelobe Canceler

As discussed in detail in a later chapter, the SLC can only cancel a single jammer. A multiple-jammer environment requires the multiple sidelobe cancellation (MSLC). A conceptual block diagram of the optimum technique, neglecting algorithm configuration, is shown in Figure 6.8.

As shown in Chapter 8, the number of auxiliary antennas should be at least equal to the number of noise jammers. Also, as for the SLC, the residue exceeds the thermal noise power in a nonpredictable manner, so a CFAR circuit is necessary. The algorithms used to estimate the optimum weights are extensions of the previously discussed algorithms. The relative convergence and weight noise-

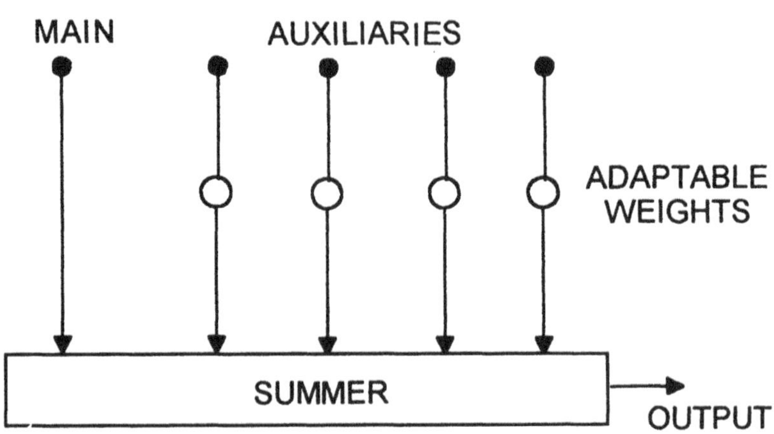

Figure 6.8 Conceptual block diagram of optimum MSLC.

induced residue power is discussed in detail. Additional quantitative examples of the effects of weight noise as a function of feedback gains and algorithms are given. It is shown that the infinite variance condition of the NLMS only occurs for the SLC and not for the MSLC.

6.8 Sidelobe Blanking

The sidelobe canceler works well for rejecting a noise jammer received via the main antenna sidelobes but does not reject a false target-repeater jammer received via the main antenna sidelobes. The difference in performance for the two jammer classes is because the algorithms used to estimate the optimum SLC weights work well with high-duty factor waveforms but poorly for low-duty factor waveforms. As an example, the weights developed by the LMS or other algorithms relax toward the quiescent value of zero during the period between jammer pulses. For multiple false target jammer environments, not only do the weights relax toward zero between received pulses, but the weights can tend toward different values during each received pulse since successive received pulses can be from different jammers. Thus, an alternate ECCM technique is needed to counter false target-repeater jammers. A conceptual block diagram of the fundamental sidelobe blanker is shown in Figure 6.9. The depiction of the two antenna patterns indicates that the gain of the auxiliary antenna exceeds the sidelobe gain of the main antenna. The sidelobe blanker action decides whether a pulse is received via the mainlobe of the main antenna (a true target return) or via the sidelobes of the main antenna (a false target generated by a jammer) by

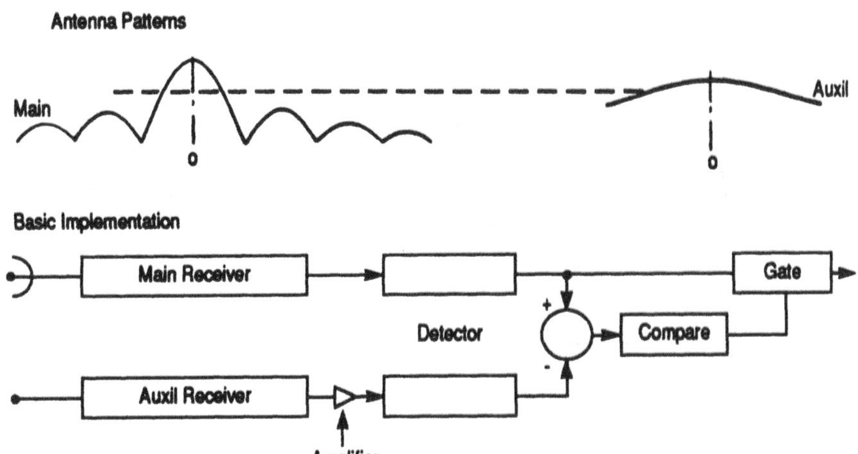

Figure 6.9 Conceptual block diagram of sidelobe blanker.

comparing the relative magnitudes of the voltages received by the two antennas. In the absence of thermal noise, the magnitude of the auxiliary antenna is always greater than that of the main antenna for a false target and the reverse for a true target. As shown in the figure, the comparator circuit gates off when a false target is detected. Otherwise the gate is in the on mode. Two major problems of the circuit of Figure 6.9 are that it performs poorly in either a noise-jamming environment or a clutter environment. The noise-jamming environment problem is that a sidelobe noise jammer causes the blanker to gate off continuously. To prevent this, it is necessary to incorporate the SLC technique in both the main and blanker auxiliary channels prior to the blanking comparator. The clutter environment problem is that the clutter can mask the presence of a sophisticated false target jammer until after Doppler processing and pulse compression circuitry. To prevent this, it is necessary to include Doppler processing and pulse compression circuitry in both the main and auxiliary channels.

A combined sidelobe blanker and canceler is depicted in Figure 6.10. Two auxiliary antennas are shown. One is used for a blanker channel; the other is

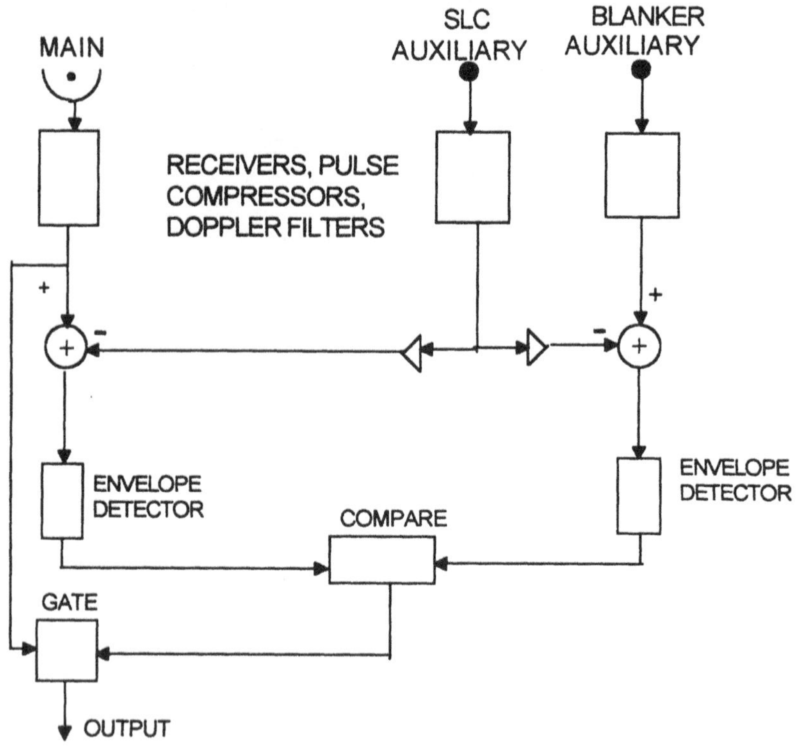

Figure 6.10 Combined noise jammer/repeater jammer ECCM.

used as a noise jammer cancellation channel for both the main antenna and the blanker channel. A pulse compression network and Doppler processor (delay line canceler and/or Doppler filter bank) is shown for the main and blanker channels prior to the blanking comparison point. The noise canceler subtraction point is also shown prior to the blanking comparison point. For best performance, the adaptive weight computer of the noise canceler should obtain data that are not corrupted by clutter or false targets. One way of obtaining these data is to use a time period prior to pulse transmission to sample the noise-jamming environment and compute the adaptive weights. These weights are then held for the duration of the Doppler processing period. Thus, the assumption is that the noise-jamming environment does not change after the adaptive weights are computed. This assumption is usually correct. It is grossly incorrect if the jammers modulate their characteristics. The assumption is slightly incorrect if the radar beam is continuously scanned during the Doppler processing period because the sidelobe antenna gains in the jammers' directions will change.

Appendix 6A: Differentiation Theorem

A mathematically correct method of solving for the complex-valued weight that minimizes (6.9) is to obtain the partial derivatives with respect to the real and imaginary components of the weight and set the partials equal to zero. The theorem actually used to derive the weight that minimizes (6.9) is that differentiating a real-valued quadratic with respect to w^* while assuming it is independent of w gives the same result as differentiating with respect to the real and imaginary components of the weight. The theorem is useful because it is simpler to derive the solution by this procedure.

To prove the theorem, define the general real-valued quadratic

$$q = aww^* + bw^* + b^*w + c \tag{6A.1}$$

Substituting w and b in terms of their real and imaginary components,

$$w = w_r + jw_i \qquad b = b_r + jb_i \tag{6A.2}$$

into (6A.1) gives

$$q = a(w_r^2 + w_i^2) + 2(b_r w_r + b_i w_i) + c \tag{6A.3}$$

Then

$$\partial q/\partial w_r = 2aw_r + 2b_r \qquad \partial q/\partial w_i = 2aw_i + 2b_i \tag{6A.4}$$

Setting these two partial derivatives equal to zero gives

$$\begin{gathered} w_{\mathrm{r}} = -b_{\mathrm{r}}/a \qquad w_{\mathrm{i}} = -b_{\mathrm{i}}/a \\ w = w_{\mathrm{r}} + jw_{\mathrm{i}} = -(b_{\mathrm{r}}/a + jb_{\mathrm{i}}/a) = -b/a \end{gathered} \tag{6A.5}$$

Differentiating (6A.1) with respect to w^* while assuming it is independent of w gives the same result as (6A.5), so the theorem is proved. Also, note that differentiating (6A.1) with respect to w while assuming it is independent of w^* gives the same result as (6A.5).

7

Constant False Alarm Rate Processors

7.1 Introduction

The conceptual basic radar system shown in Figure 1.2 makes a target present or absent decision in each of the range-angle-Doppler frequency resolution cells. In a thermal noise-only environment, either an operator or an automatic decision device makes a target present decision when the input waveform envelope exceeds a prespecified threshold. The threshold is computed from (1.31a) and depends upon the thermal noise power and the desired false alarm probability. However, when the interference environment consists of the combination of thermal noise and noise jamming and clutter, the output interference power is greater than that due to thermal noise alone. This is true even if adaptive processing is used to "cancel" the jamming and/or clutter. As an example, the use of adaptive antennas can achieve significant jammer power cancellation, but even for the narrow band optimum weight implementation, the output residue power is greater than that which occurs for the noise-alone environment. In addition, the residue power is usually even larger due to the effects of nonzero bandwidth and/or channel-to-channel mismatch. Thus, if a constant prespecified threshold is used in a thermal noise-plus-jammer environment, the false alarm rate increases due to the residue power increase and data processor saturation will result. To prevent the saturation, a method of preventing the false alarm increase is mandatory. Similar observations hold for the clutter-plus-thermal noise environment. Again, the output residue power is greater than that due solely to noise, and a method of preventing the false alarm increase is mandatory.

The false alarm rate increase is often unacceptable even when the residue power is only slightly different than that due to thermal noise. A quantitative indication of the severity of the problem, for a prespecified threshold T, is given by Figure 1.9. It shows the attained false alarm probability for typical design false alarm probability values of 1×10^{-6} and $1. \times 10^{-8}$ as functions of the ratio of the design interference power (usually thermal noise power) to the actual residue power. It can be seen from the figure or (1.34) that when the actual residue power exceeds the design interference power by 3-dB power, a positive decibel power ratio, the attained false alarm probability increases to the square root of the desired false alarm probability whereas a negative 3-dB power ratio causes the attained false alarm probability to decrease to the square of that desired. Specifically, when p_{FD} equals $1. \times 10^{-8}$, for the +3- or −3-dB power ratio, the values of p_{FA} are $1. \times 10^{-4}$ and $1. \times 10^{-16}$, respectively. Note that although decreases of the false alarm rate do not cause data processor saturation, it does indicate that the threshold is too high and so target detectability is significantly reduced; this is also unsatisfactory. Clearly, relatively small residue power changes can cause intolerably large performance changes. A form of adjustable or adaptive threshold is required to implement a CFAR processor action.[1]

7.2 CFAR Concept

The basic concept of a CFAR technique is that the voltage of a test cell is compared to the voltages of a set of reference cells. If the test cell voltage is "similar" to those of the reference cells, a target absent decision is made. If the test cell voltage is not similar, a target present decision is made. Given this strategy, several questions must be answered in order to obtain a satisfactory implementation. One is that there is a tacit assumption that it is possible to obtain a set of reference cells whose statistical descriptors are identical to that of the test cell for the target absent condition. How is this done? A second question is how the dichotomy of "similar" or "not similar" is made.

There are several methods of obtaining a set of reference cells. The proper method depends upon the type of interference environment. One common technique, shown in Figure 7.1, uses a range interval bracketing the test cell as the reference cells. Use of this set implies that the characteristics of the interference does not change over the range interval of the reference and test cells. This is true when the interference environment consists of the sum of thermal and constant power jamming noise. It is not true for environments that include ground clutter because the clutter cross section tends to change significantly (and randomly) in relatively small-range increments.

1. This chapter is a condensation of Nitzberg, R. *Adaptive Signal Processing for Radar*, Norwood, MA: Artech House, 1992, pp. 203–252.

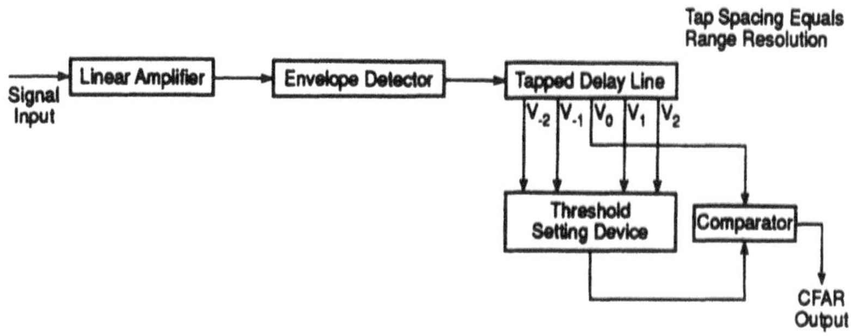

Figure 7.1 CFAR processor using range window for reference cells.

A second technique uses an azimuth interval bracketing the test cell azimuth with all cells at the same range. This technique works well only when the interference characteristics are unchanging over the azimuth interval. It would not work well for a sidelobe noise-jamming environment because the received jamming power changes with azimuth due to the changing sidelobe gain of the antenna pattern. It could work well when the interference is the sum of thermal noise plus mainlobe weather clutter because the weather clutter characteristics are often uniform over azimuth.

A third technique combines the two domains to implement a two-dimensional reference cell configuration. Some radar systems could implement a multidimensional reference set by including Doppler frequency, elevation angle, and polarization, for example.

A technique that is appropriate for ground clutter obtains the reference cells from previous scans of the resolution cell. This technique is called clutter map. In theory, the data from prior scans is used to estimate the clutter-plus-thermal noise power of every resolution cell. When a target, such as an aircraft, enters the cell, the increase in reflected power allows target detection in the presence of the clutter. Often the clutter map is used with a Doppler filter bank and a different map is determined for each filter of the bank.

The following sections discuss several different CFAR processors. The fact that there are so many is indicative of the substantial difficulty in obtaining a single technique that is satisfactory for all interference environments. In general, there is not a totally satisfactory solution.

7.2.1 Cell-Averaging CFAR

Equation (1.32a) relates the threshold value T, the thermal noise power P_{n}, and the design false alarm probability p_{F} when using a square law envelope detector.

The equation is also correct when the noise power is the sum of thermal noise and jamming noise or the residue power after cancellation. Denote the power as P_r for either case. Taking the natural logarithm of both sides of this equation derives that the threshold equals the total interference noise power multiplied by a constant, k_1, which depends only upon the design false alarm probability. Denote the square law envelope detector output for the test cell as q_0. Then a target present decision is made when

$$q_0 > T = k_1 P_r \tag{7.1}$$

and the probability of detection for a specified false alarm probability is given by

$$p_D = p_F^{[1/(1+\gamma_a)]} \tag{7.1a}$$

where γ_a is the average SNR for the Swerling fluctuating target. Equation 7.1a can be solved for γ_a as

$$\gamma_a = \ln(p_f/p_D)/\ln(p_D) \tag{7.1b}$$

When the total interference power is unknown, it seems reasonable to estimate the interference power as $\bar{P}_r$ and make a target present decision when

$$q_0 > T_a = k_2 \bar{P}_r \tag{7.2}$$

where T_a is an adaptive threshold and k_2 is a new multiplier constant that is usually not equal to k_1.

A common method to estimate the interference power is by averaging the outputs of the reference cells. For M reference cells, the estimate is

$$\bar{P}_r = \left(\sum_{m=1}^{M} q_m\right) \Big/ M \tag{7.3}$$

where q_m denotes the square law detector output of the mth reference cell.

An alternate, but similar, procedure shown in Figure 7.1 is to use an envelope detector that outputs the magnitude of the envelope rather than a square-law envelope detector that outputs the magnitude squared of the envelope. The former is often substantially simpler to implement. Then, denoting the envelope of the test and reference cells as e_0 through e_M, the equivalent of (7.2) and (7.3) is to make a target present decision when

$$e_0 > T_a' = k_2'\left(\sum_{m=1}^{M} e_m\right) \Big/ M \tag{7.4}$$

Also, target present or absent decisions could be based on any monotonic function of the magnitude of the envelope. That is, denote the cell voltages for an arbitrary monotonic function as f_0 through f_M. The decision is then based on

$$f_0 > k_2''\left(\sum_{m=1}^{M} f_m\right) \Big/ M \tag{7.5}$$

The question of which is the statistically best procedure requires detailed consideration of the statistics of the cell voltages. It is shown in the reference that for the usual case of complex Gaussian statistics, implementations using square law detectors maximize the probability of detection for a specified false alarm probability. Thus, although easier to implement, a processor based upon (7.2) and (7.3) has a larger loss. It is also shown that for a Swerling 1 target, the probability of detection for the statistically optimum CFAR processor is given by

$$p_D = \left\{(1 + \gamma_a)/\left[1 + \gamma_a + \left(p_f^{-1/M} - 1\right)\right]\right\}^M \tag{7.6}$$

Figure 7.2 shows the probability of detection for a false alarm probability of $1. \times 10^{-6}$ for the Swerling 1 target parametric with the number of reference cell (normalizing) samples. The curve labeled infinite is computed based upon (7.1a) for the known power level case. It can be seen that as the number of reference cells increases, the curves approach the known power level curve. Thus the CFAR loss decreases as the number of reference cells increases. The next section develops equations for the CFAR loss. It is shown that the CFAR loss does approach 0 dB for a homogeneous environment as the number of reference cells approaches infinity. Later sections consider the performance in nonhomogeneous environments.

The reference cells are sometimes denoted as normalizing cells because of a variation of implementation for the CACFAR processor. Thus, (7.2) and (7.3) can be combined and rearranged to the equivalent target present/absent decision of

$$q_0 \Big/ \left(\sum_{m=1}^{M} q_m\right) \Big/ M > k_2 \tag{7.7}$$

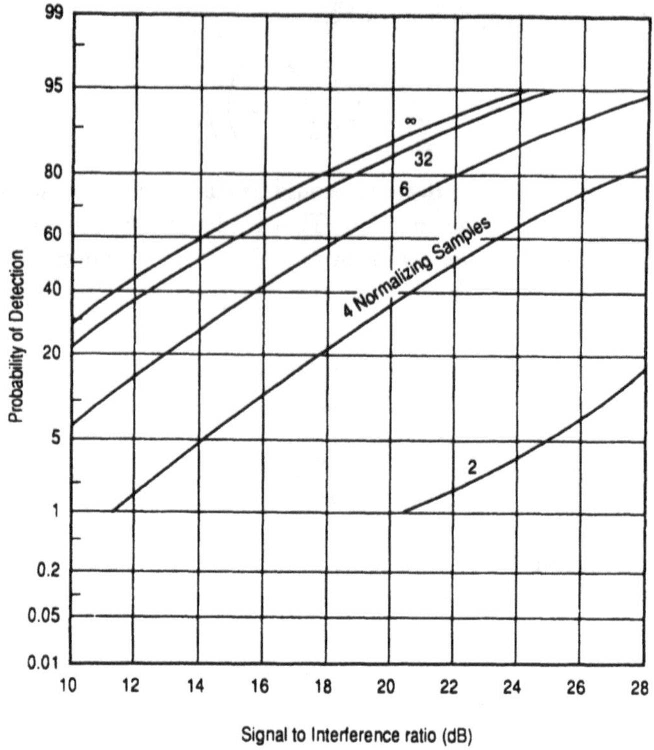

Figure 7.2 Performance of optimum CACFAR in homogeneous environments.

and in this form it can be seen that the test cell voltage is normalized by the average of the reference cell voltages and compared to the threshold multiplier constant. This form is also useful in developing a concept of why the technique produces a CFAR action for a homogeneous environment and why it fails to produce a CFAR action for a nonhomogeneous environment.

The set of detector output voltages, $\{q_0, q_1, q_2, \ldots, q_M\}$, have the common power level P_0 and the power level is unknown. This set can be thought of as formed by "nature" generating a set of unit power random variates $\{\mu_0, \mu_1, \mu_2, \ldots, \mu_M\}$ and multiplying each component of the set by the common factor P_0. Thus, each q_m equals $P_0\mu_m$. Substituting this into (7.7) gives

$$P_0\mu_0 \Big/ \left(\sum P_0\mu_m\right) \Big/ M > k_2 \tag{7.7a}$$

It can be seen that the left-hand side of the inequality, the decision random variable, is independent of P_0. Thus, the probability that the inequality is satisfied is independent of the unknown power level, and the false alarm probability in a homogeneous environment processor is constant.

7.2.2 Homogeneous Environment CACFAR CFAR Loss

The value of SINR required for a specified p_D and p_F can be obtained from (7.6). Thus,

$$\gamma_a(M) = [(p_D/p_f)^{1/M} - 1]/(1 - p_D^{1/M})] \tag{7.8}$$

where the notation $\gamma_a(M)$ is introduced to emphasize the dependence upon M. The limiting value of $\gamma_a(M)$ as M approaches infinity can be determined via L'Hospital's rule. It can be shown to be

$$\gamma_a(\infty) = \ln(p_f/p_D)/\ln(p_D) \tag{7.9}$$

A comparison of (7.9) with (7.1b) shows that the limiting value is identical to that for the known level environment. Thus, the CFAR loss tends to zero as the number of reference cells tends to infinity.

The CFAR loss equals the ratio of (7.9) and (7.1b). This loss is graphed in Figure 7.3 and is approximately independent of the probability of detection. Note that both scales are logarithmic and that the CFAR loss, in decibels, plots as approximately straight lines. Thus, as an example, the decibel loss is reduced from 1 dB to 1/2 dB by doubling the number of reference cells. An approximate analysis that develops the straight line relationship is developed next.

First note the identity

$$(p_D/p_f)^{1/M} - 1 = \exp[(1/M)\ln(p_D/p_f)] - 1$$

For large M, the argument of the exponential is small and well approximated by

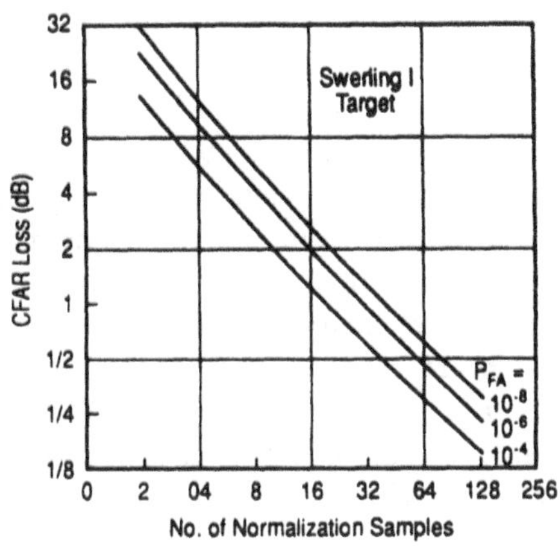

Figure 7.3 Decibel CFAR loss for CACFAR.

the first few terms of a power series. Thus,

$$(p_D/p_f)^{1/M} - 1 \approx (1/M)\ln(p_D/p_f)[1 + \ln(p_D/p_f)/(2M)]$$

A similar approximation for $(1 - p_D^{1/M})$ can be developed. Then combining the approximations with (7.9) gives

$$\gamma_a(M)/\gamma_a(\infty) \approx [1 + \ln(p_D/p_f)/(2M)]/[1 + \ln(p_D/p_f)/(2M)]$$

Since for small x, $1/(1+x) \approx 1 - x$, this can be further approximated as

$$\gamma_a(M)/\gamma_a(\infty) \approx 1 - \ln(p_f)/(2M)$$

where only terms involving first powers of $1/M$ have been retained. The decibel loss is

$$L_{dB} = 10\log_{10}[\gamma_a(M)/\gamma_a(\infty)]$$

and from the preceding approximations, this may be evaluated as

$$L_{dB} \approx -5\ln(p_f)/[M\ln(10)]$$

Thus, the decibel loss, to a first-order approximation, does not depend upon the probability of detection. Also, note that doubling the number of reference cells causes the decibel loss to be halved.

7.3 Nonhomogeneous Environments

The cell-averaging CFAR computes an adaptive threshold to maintain a constant false alarm rate. The adaptive threshold equals a gain constant (determined by the design false alarm probability) multiplied by the sum of the instantaneous voltages of the cells in the range window bracketing the test cell. However, the attained false alarm probability is only equal to the design value when the interference power (thermal noise plus clutter plus jamming plus additional targets) is constant (homogeneous) over the total range window including the test cell. For clutter, the required homogeneity often does not occur. Then the attained false alarm probability differs considerably from the design value. This section considers several nonhomogeneous environments and alternate "CFAR" techniques that are better able to cope with these environments. The word CFAR is in quotes because usually the false alarm rate is not exactly constant in a clutter environment.

7.3.1 CACFAR Performance in Mismatched Environments

The synthesis and prior analysis of the CACFAR circuit are done under the assumption that, for the target absent condition, the test cell and reference cell powers (interference-residue power) are all equal. This assumption constitutes the design environment. The actual environment is usually the same as the design environment when the environment is the sum of thermal and jamming noises. However, in the presence of clutter, the actual environment differs considerably from the design environment. One consequence of this mismatch is that the actual false alarm rate is not equal to the design false alarm rate. A second consequence is that the probability of target detection does not necessarily increase monotonically as the number of reference cells is increased.

The first portion of the following sections is a qualitative discussion of the false alarm probability in a mismatched environment. Later sections discuss equations to quantify the effect of the environment mismatch upon the false alarm probability and detection probability. Still later sections discuss and analyze alternative CFAR techniques that can attain a constant false alarm rate even though all cells do not have the same power. It is shown that attaining a constant false alarm rate for these more severe environments can be done, but the CFAR loss increases. It should be noted that the "CFAR loss" is a loss only with respect to performance in a known power level environment. It is not an avoidable loss but instead is a cost of operating in a nonthermal noise environment.

7.3.2 Mismatched Environment Example

One example of a nonhomogeneous environment is a step function environment. A step function environment is one where the clutter cross section (reflection power) varies as a step function versus range. This is an approximate model of reality when there is a sharp boundary between two different types of clutter (such as at a coastline) or a sharp boundary between thermal noise only and the sum of thermal noise plus clutter. For the CACFAR, the false alarm probability is not the design value if the range window encompasses the step transition. When the test cell is in the high-power region, the adaptive threshold is too low since it is partially formed from the low-power interference region and the false alarm probability is high. When the test cell is in the low-power region, the false alarm probability is too low and as a consequence the probability of target detection is low.

As a qualitative indication of the false alarm probability attained in a step-like nonhomogeneous environment, consider the environment sketched in Figure 7.4(a). This depicts an interference environment whose power is initially at some constant power level (possibly the thermal noise condition) and then increases linearly, when measured in decibels, between R_1 and R_2. This

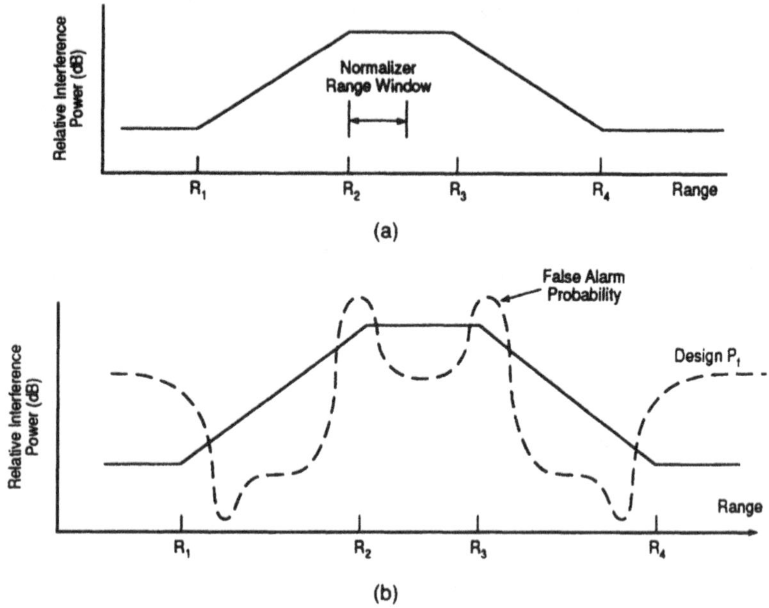

Figure 7.4 Mismatched environment effect.

transition is possibly a more realistic indication of a change from one class of environment to another than a sharp transition step. The power level is then constant until R_3, when it starts decreasing. The decrease terminates at range R_4, when the power level attains its initial value. Note that the total range over which the normalizer obtains reference cell samples (indicated as the range window) is less than either the plateau extent or the ramp extent. Consider the false alarm probability that will be obtained for the CACFAR when operating in this environment. First note that the attained false alarm probability of the circuit equals the design value when the range window is contained completely within a region where the interference power is constant. Thus, at the extreme left of the sketch, the attained false alarm probability equals the design false alarm probability. As the range window shifts to the right, a portion of the window overlaps the ramp. Note that the average adaptive threshold voltage increases as the region of the ramp is entered. Since the adaptive threshold voltage is increasing, the false alarm probability decreases from its design value. Next, consider the case where the range window has moved completely off the ramp and is on the plateau. Again, the attained false alarm probability equals the design value. Conceptually shifting the range window to the left until the left-hand portion of the window enters the ramp region, it can be seen that the average adaptive threshold voltage is decreased as compared to the value attained when completely in the plateau region. This lower value of average adaptive threshold voltage causes the attained false alarm probability to be higher than the design value.

Another region of interest is when the normalizer (reference cells) window is completely on the ramp. In this region, the average adaptive threshold voltage will tend to be high, so the false alarm probability will be low. To prove that the average adaptive threshold voltage is high, note that any CFAR technique is essentially estimating the value of the interference power of the test cell. Consider a cell pair symmetrically located about the test cell. That is, if the test cell is identified as the jth cell, a symmetrically located cell pair of interest is $j+i$ and $j-i$ for any value of i. Then, in the ramp region, denoting the residue power at the center tap as P_r, the interference powers of the other taps would be P_r/b^i and $P_r b^i$. The constant b is related to the slope (in decibels) of the interference power versus range. The CACFAR averages these two power values in order to estimate the value of the test cell's interference power. The difference between the average and the actual value is

$$\delta = P_r[(b^i + 1/b^i)/2 - 1] = P_r(b^{i/2} - 1/b^{-i/2})^2/2 \qquad (7.10)$$

It can be seen that δ is always positive. Thus, the adaptive threshold voltage is always high and the attained false alarm probability is always low. Since (7.10) has the same value whether b is positive or negative, the low false alarm probability occurs whether the interference power variation ramp has positive or negative slope. Thus, the false alarm probability has the same value on the up ramp as on the down ramp. In addition, the values are the same at the corners of the plateau. A sketch of the expected form of variation of the false alarm probability as a function of range for the CACFAR normalizer is shown in Figure 7.4(b). It can be seen that the environment symmetry causes similar symmetry in the range variation of the attained false alarm probability.

A more quantitative indication of the attained false alarm probability for a ramp environment when using the CACFAR technique requires a detailed analysis. An equation for the probability of detection for a generalized environment is derived in the reference. It is shown that a generalization of (7.6) is

$$p_D = \prod_{m=1}^{M} \{[1 + P_m(p_f^{-1/M} - 1)]/P_0(1+\gamma_a)\}^{-1} \qquad (7.11)$$

where P_0 is the total interference power of the test cell and P_m is the total interference power of the mth reference cell. Note that (7.11) reduces to (7.6) when the test cell and reference cells all have the same power. The equation can be used to evaluate the false alarm probability by setting γ_a equal to zero. Thus, Figure 7.5 shows that the attained false alarm probability, computed using (7.11), for normalizing with 4, 8, or 16 cells for a design false alarm probability, is

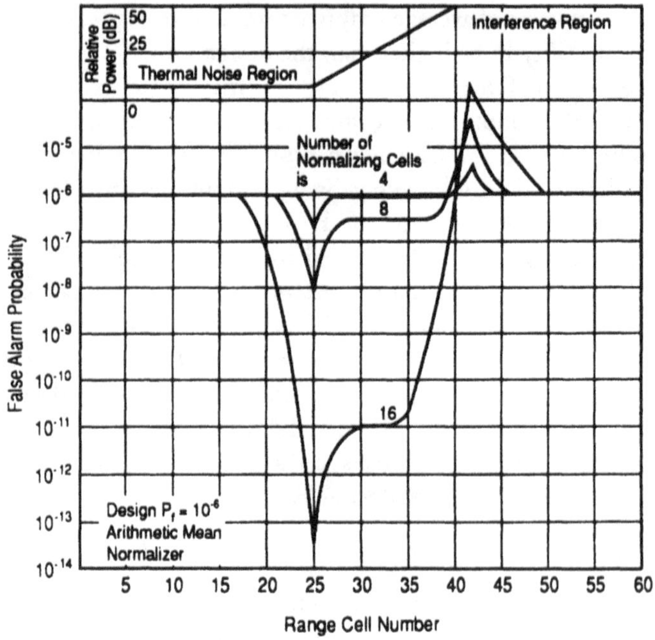

Figure 7.5 False alarm probability attained for a ramp environment.

1×10^{-6}. As previously indicated, the attained false alarm probability equals the design value only when the normalization window is completely within a region of constant power level. This occurs both at low-range cell numbers and at high-range cell numbers. As the test cell increases from low-range cell numbers toward high-range cell numbers, the normalization (reference cells) window encounters a region where the interference power is higher than that of the test cell. This tends to increase the adaptive threshold voltage and the attained false alarm probability decreases from the design value. Note that the departure from the design value occurs earlier as the number of normalization cells increases. This occurs because the increased width of the normalization window encounters the discontinuity earlier. It can be seen that the departure from the design false alarm probability increases as the number of normalization cells increases. Since the adaptive threshold voltage will be high in the ramp region, the probability of target detection is low. Thus, although performance increases monotonically with the number of normalization samples in a homogeneous environment, this is not true for all environments.

The difficulty of detecting targets in a ramp environment when using the CACFAR circuit is shown by the curves of Figure 7.6. The SINR required for target detection is shown for four different slopes of a ramp environment. The

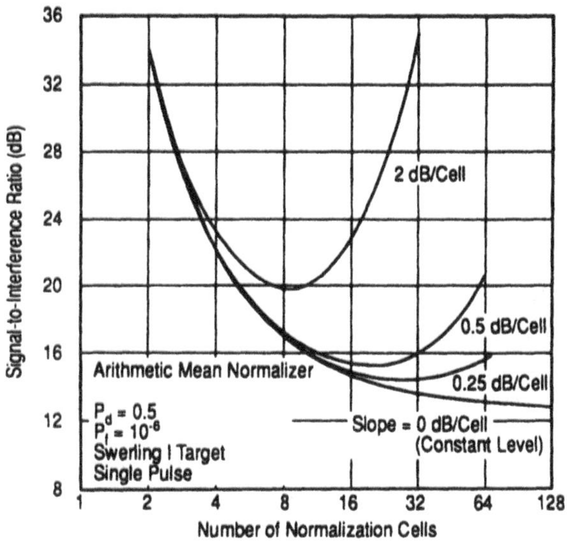

Figure 7.6 SINR required for target detection for ramp environments.

0-dB slope is a constant power (homogeneous) environment. As previously discussed, and as shown by the curve, the required SINR decreases monotonically with the number of normalization cells. For nonzero slopes, the required SINR does not decrease monotonically with the number of normalization cells. Instead there is an optimum number of cells that minimizes the required SINR. The optimum number decreases as the slope of the ramp increases. As an example, for a slope of 0.5 dB per cell, the optimum number of normalization cells for this environment when using the CACFAR technique is approximately 20. The minimum CFAR loss for this processor is then 2.4 dB. As another example, for a 2.0-dB slope, the minimum CFAR loss for the CACFAR technique is 7.2 dB.

In both Figures 7.5 and 7.6, the CACFAR is referred to as *an arithmetic mean normalizer.* This terminology is to emphasize that there are other ways to combine the reference cell voltages to generate an adaptive threshold. One example is to base the adaptive threshold on the geometric mean of the reference cell voltages rather than the arithmetic mean. The implementation of the geometric mean is done using logarithmic amplifiers and is therefore relatively simple. Specifically, extrapolating from (7.7), the target present/absent decisions using the geometric mean normalization are implemented by the inequality

$$q_0 \Big/ \left(\prod_{m=1}^{M} q_m \right)^{1/M} > k_g \tag{7.12}$$

and it is equivalent to that of

$$\ln(q_0) - \left[\sum_{m=1}^{M} \ln(q_m)\right] \Big/ M > \ln(k_g) \qquad (7.13)$$

7.4 Geometric Mean CFAR

The CACFAR forms the adaptive threshold from the arithmetic mean of the reference cells. This technique works well for a homogeneous environment. That is, when the residue power of the test cell and the reference cells all are the same value of P_0, the arithmetic mean of the reference cells is a good estimator for the power P_0. An alternate explanation of its merits is given by (7.7b) where it is shown that the probability of target declaration is independent of the power level of the interference. However, for a nonhomogeneous environment, the arithmetic mean is a poor estimate of P_0. Specifically, as discussed in the previous section, the arithmetic mean is a poor estimate for a ramp environment. It is now demonstrated that a geometric mean is a satisfactory estimator for a ramp environment.

Assume the M reference cells are bracketing the test cell. Denote the test cell indices as zero and the reference cell indices as ± 1 through $M/2$ and (7.12) is modified to

$$q_0 \Big/ \left(\prod_{m=1}^{M/2} q_m q_{-m}\right)^{1/M} > k_g \qquad (7.12a)$$

The residue power of the cells is given by

$$P_{\text{residue}}(m) = P_0(P_s)^m \qquad (7.14)$$

The interfence power expressed in decibels is obtained by taking the logarithm of (7.14), and it can be seen that this expression defines a ramp environment. Thus, P_0 denotes the test cell residue power and that the logarithm of P_s is the slope of the ramp.

As was previously described, the observed set of detector voltages, $\{q_{-M/2}, \ldots, q_0, q_1, q_2, \ldots, q_{M/2}\}$ can be thought of as formed by "Nature" generating a set of unit power random variates $\{\mu_{-M/2}, \ldots, \mu_0, \mu_1, \mu_2, \ldots, \mu_{M/2}\}$. For a ramp environment the mth component of the set is multiplied by the factor $P_0(P_s)^m$. Thus, each q_m equals $[P_0(P_s)^m]\mu_m$. Substituting this into (7.12a) demonstrates that the value of the decision random variable is independent of the values of P_0 and P_s and depends only upon the values of the set $\{\mu_{-M/2}, \ldots, \mu_0, \mu_1, \mu_2, \ldots, \mu_{M/2}\}$. Thus, when the normalization voltage equals the geometric

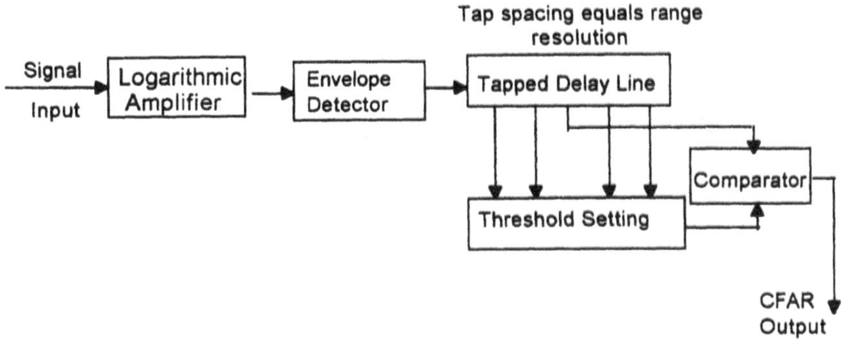

Figure 7.7 Geometric mean CFAR.

mean, the false alarm probability is independent of the value of test cell interference power and of the slope of the ramp.

A conceptual block diagram of a geometric mean CFAR is shown in Figure 7.7. Comparing it and Figure 7.1, it can be seen that the only difference is that the linear amplifier has been changed to a logarithmic amplifier. Curves of the SINR required for target detection are shown in Figure 7.8 for a ramp environment when the CFAR is implemented using an arithmetic mean normalizer or a geometric mean normalizer. As expected, the required SINR for the geometric mean normalizer does not depend on the environment's slope whereas that for the arithmetic mean normalizer does. Note that the required

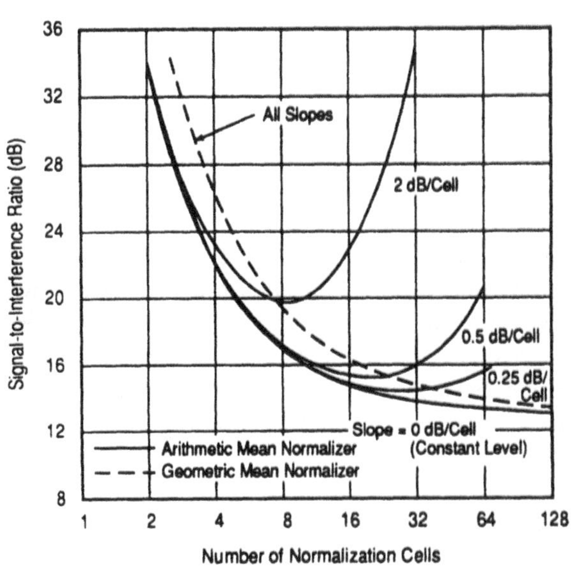

Figure 7.8 Required SINR for ramp environments.

SINR for the geometric mean normalizer exceeds that of the arithmetic normalizer when the environment is homogeneous. This is expected since the CACFAR is optimum for a homogeneous environment. Also, the increased SINR is indicative of a general CFAR circuitry property. That is, when a circuit is designed such that its false alarm probability tends to remain constant over more general classes of environment, it also becomes more difficult for the circuit to detect targets and the CFAR loss increases.

7.5 CFAR Problems and "Solutions"

The type of environment determines the preferred form of CFAR circuit. For a noise-jamming environment, a homogeneous environment, the CACFAR is the statistically optimum technique. There are several candidate techniques for a clutter environment (nonhomogeneous environments), and the preferred technique depends upon the specific type of clutter environment. One example is the previously discussed ramp environment. Another is a step transition. This latter clutter environment occurs at a sharp transition from one type of clutter to another such as at a coastline where there is land on one side of the transition and ground on the other side. Some clutter environments are neither of these and are more realistically described as a random cross section varying as a function of range and angle. A different type of nonhomogeneity occurs when there are multiple, closely spaced targets. Then, even for a noise-jamming environment, the interference power of the reference window cells is not equal to that of the test cell because some of the reference cells also include the power due to a target. One example of multiple, closely spaced targets is when an aircraft launches a missile. When the large cross-section aircraft is in the reference window, its presence will substantially increase the adaptive threshold so that detecting the small cross-section missile in the test cell becomes almost impossible. Another example is when closely spaced aircraft are in an approach pattern for landing on an airport runway.

Though it is known that the CACFAR is the optimum technique for the noise-jamming environment, no optimality properties are known for any other environment. As a specific example, it is not known whether the geometric mean CFAR is statistically optimum for a ramp environment. Therefore, many different techniques have been suggested and evaluated. Some examples, in addition to the arithmetic and geometric mean implementations previously discussed, include

1. Greatest-of;
2. Order statistics;

3. Censored;
4. Composite.

Note that many alternate techniques can be rapidly synthesized.

7.5.1 Greatest-of CFAR

The step transition environment is a generalization of the transitions depicted in Figure 7.4. As shown in Figure 7.4, a major problem with a step transition is that the false alarm probability becomes very high and tends to overload the data processor. The split window greatest-of CFAR (GOCFAR) is a circuit modification that is often used to avoid the high values of false alarm probability. The circuit modification is to split the range window into the half leading the test cell and the half lagging the test cell. The voltages of the reference cells of each portion are summed, and the larger sum is multiplied by a gain constant to form the adaptive threshold. This circuit will decrease the high false alarm rate that occurs for the step environment. However, when the test cell is in the low clutter cross-section portion of the step transition, the adaptive threshold is even larger than when a CACFAR is used and thus is much too high. Therefore, this modification does not alleviate the low false alarm probability/low detection probability condition but tends to worsen it. This CFAR is used when the data processor overload problem is deemed much more serious than the low target detection problem.

The modification slightly increases the CFAR loss in a homogeneous environment relative to that incurred by the CACFAR. The additional loss is a function of the false alarm probability and the total number of reference cells. The additional loss values, additional relative to the CACFAR technique, are shown in Figure 7.9 for a detection probability of 0.9. It can be seen that the additional loss is of the order of a few tenths of a decibel and decreases as the number of reference cells increases.

A modification of the GOCFAR is an example of a composite technique. Instead of always basing the adaptive threshold on the greater of the leading and lagging sums, the two sums are compared. If they are very different, implying that the environment does contain a step, the adaptive threshold is based upon the greater of the two. If, however, the two sums are close in value, implying that the environment is homogeneous, the two sums are added and the adaptive threshold is based on the total sum. This is the CACFAR! Thus, a composite technique is configured to first classify the environment based upon a function of the reference cell voltages and then configure a CFAR technique based upon the environment classification.

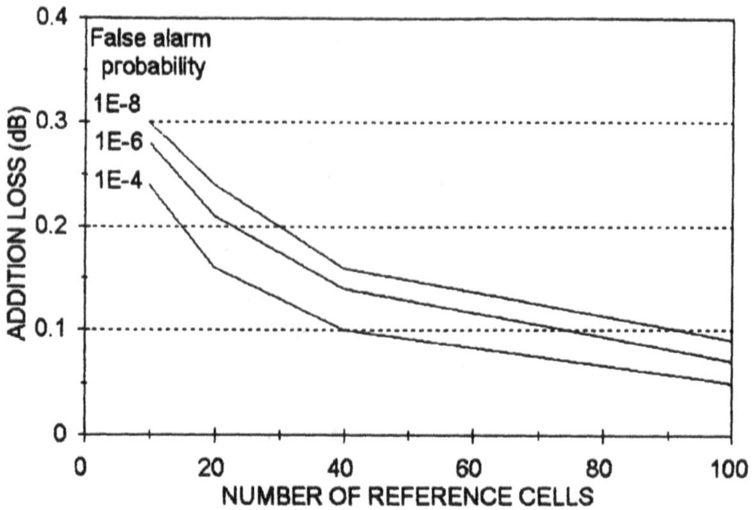

Figure 7.9 Additional CFAR loss for GOCFAR.

7.5.2 Multiple Target Effect upon CACFAR

As previously stated, if an aircraft is in the reference window, its presence can substantially increase the adaptive threshold so that detecting the presence of another aircraft in the test cell is much more difficult. An example is when closely spaced aircraft are in an approach pattern for landing on an airport runway. As a quantitative indication of the frequency of the problem, assume multiple aircraft are in a landing approach pattern at a speed of 250 knots (128.7 m/s) and the radar's LFM bandwidth is 1 MHz. This bandwidth implies a range resolution of 150m but using Taylor weighting for sidelobe control stretches the range resolution cell size to 211.5m. A 64-cell reference window therefore has a total range extent of 13.54 km. The required aircraft time separation to prevent multiple-target CFAR problems equals 105.2 sec. Since many large airports land aircraft with as little as 30-sec time separation, the multiple-aircraft CFAR problem is sometimes a significant problem.

It is shown in the reference that when all targets are statistically independent with Swerling type fluctuation, the probability of detection for an M-reference cell CACFAR, and K additional targets in the reference window, is given by

$$p_D = [1 + k_a/(1 + \gamma_A)]^{-(M-K)}[1 + k_a(1 + \gamma)/(1 + \gamma_A)]^{-K} \quad (7.15)$$

where k_a is the adaptive threshold multiplier, γ is the SINR of the test cell target, and γ_A is the SINR of each additional targets.

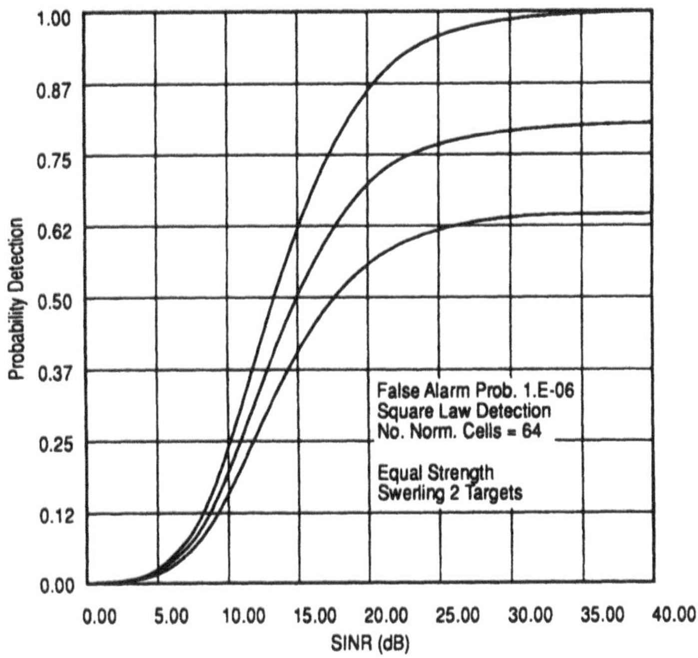

Figure 7.10 CACFAR target detection suppression by additional targets.

The probability of detection is shown in Figure 7.10 for a 64-reference cell CACFAR. The threshold multiplier is chosen for a false alarm probability of $1.\times10^{-6}$ in a homogeneous environment. The upper curve is the probability of detecting a target in the test cell when the normalization cells contain only homogeneous interference. The next lower curve is for the case when one of the reference cells also contains a target of average cross section equal to the test cell's target cross section. The lowest curve is when two of the reference cells contain targets of average cross section equal to the test cell's target cross section. The curves show that when there are additional targets in the reference cells, the probability of target detection asymptotically approaches a value of less than unity as the SINR increases. Thus, the additional CFAR loss due to additional targets depends upon the specified detection probability. For a detection probability close to unity, the loss is essentially infinite.

There are several alternate CFAR techniques that substantially reduce the multiple-target additional CFAR loss. One technique is to use the geometric mean of the reference cells rather than the arithmetic mean used in the CACFAR technique. Another is the censoring CACFAR with or without the split window greatest-of modification. The censoring technique deletes the largest cell voltage

Table 7.1
CFAR Loss(dB) in a Homogeneous Environment

CACFAR	0.97
GOCFAR	1.13
OSCFAR (75% rank)	1.45
GO-OSCFAR (75% rank)	1.66
CACFAR; one-cell censoring	1.01
CACFAR; two-cell censoring	1.06

(or two largest, for example) from the window and sums the remaining cells to form the adaptive threshold. For the split window version, the technique deletes the largest (or two largest, for example) from each half window, sums the remaining cells of each half window, and uses the larger sum for the adaptive threshold voltage. One penalty of using the censoring technique is that it slightly increases the CFAR loss in a homogeneous environment. Also, the technique will fail if the reference window contains more additional targets than the design assumes.

A third technique that has relatively small CFAR loss in a homogeneous environment but can cope with many additional targets in the reference window is based upon using the median of the reference cells to form the adaptive threshold. This works well because though the mean is significantly increased when a few of the reference cells contain additional targets, the median is essentially unaffected. Note that this technique can also be combined with the split window technique in order to cope with step function environments. More generally, instead of using the median, the adapted threshold can be chosen as a multiple of an arbitrary order statistic of the reference cells (OSCFAR). As an example, the order statistic can be chosen as the 75% rank or the 50% rank (the median). The CFAR loss in a homogeneous environment depends upon the order statistic used to form the adaptive threshold. It can be shown that the loss is minimized when the chosen order statistic is in the range of the 75 to 80 percentile.

To aid in comparing the various techniques, the loss for several CFAR processors for 32 reference cells in a homogeneous environment and a false alarm probability of 1×10^{-6} is indicated in Table 7.1. The losses are approximately independent of the detection probability.

7.6 Clutter Map CFAR

Ground clutter is extremely nonhomogeneous and is often modeled by a random cross section versus range. The prior CFAR techniques are not applicable because they explicitly assume a functional variation over range. However, the radar cross section of each cell of the ground clutter tends to change very slowly with time. Thus, the estimate of the cell's interference power can be done by scan-to-scan averaging. Note that though this technique is quite applicable to ground clutter, it is less applicable to weather or sea or chaff clutter because for these latter clutter types the radar cross section of each resolution cell changes relatively rapidly with time due to the wind-induced motion of the basic scatterers.

The prior CFAR techniques are implemented using a moving window integrator. Specifically, extrapolating this technique, the power estimate for the kth resolution cell using the last M scans of data $\varrho_{\mathrm{I}k}$ is

$$\varrho_{\mathrm{I}k} = \left(\sum_{m=1}^{M} q_m(k) \right) \Big/ M \qquad (7.16)$$

where $q_m(k)$ denotes the detector output. Note that this technique requires long-term storage of the voltages of the M cells used to compute the estimate. Since there can be many resolution cells, the total memory required for storing the data from M scans would be excessive. To reduce this storage requirement, the CFAR map technique often uses a background estimate derived by exponential smoothing of each resolution cell. Thus, for the mth scan and the kth resolution cell, the background estimate is formed from previous scans and the current measurement, $q_m(k)$, as

$$\varrho_{\mathrm{I}k}(m) = (1 - w)\varrho_{\mathrm{I}k}(m-1) + w q_m(k) \qquad (7.17)$$

It is shown in the reference that the probability of detection is given by

$$p_{\mathrm{D}} = \prod_{r=0}^{\infty} [1 + k_{\mathrm{C}} w (1 - w)^r / (1 + \gamma_{\mathrm{a}})]^{-1} \qquad (7.18)$$

where k_{C} is the threshold multiplier for a specified false alarm probability. Its value may be determined by iteratively evaluating (7.18) with $\gamma_{\mathrm{a}} = 0$. The probability of detection is then determined by evaluating (7.18) for each specific value of γ_{a} with the now known value of k_{C}.

Curves of the probability of detection versus SINR for several values of the weight, w, are shown in Figure 7.11 for a false alarm probability of $1. \times 10^{-6}$. The weight value equal to zero is equivalent to the case where the total interference power is known exactly a priori so that CFAR is not required. The difference between the zero weight curve and the other curves is the CFAR loss. It can be seen that the loss decreases monotonically as w decreases. This occurs

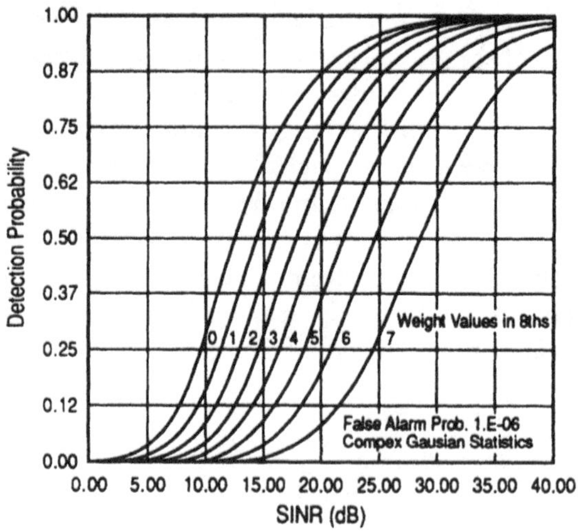

Figure 7.11 Clutter map probability of detection.

because decreasing w corresponds to using longer data windows to estimate the resolution cell power. The CFAR loss is graphed in Figure 7.12 for a detection probability equal to 0.9. Figure 7.12(a) shows the CFAR loss as the weight value ranges over zero to unity. Figure 7.12(b) expands the scale with the range of weight value of zero to 1/8. For the latter figure, the loss is plotted for false alarm probabilities of $1. \times 10^{-6}$ and $1. \times 10^{-8}$. It can be seen that the loss increases as the false alarm probability decreases. However, the curves are computed for the tacit assumption that the interference power is constant over all time. In a real environment, this tacit assumption is false and the value of w cannot be too small because it must be chosen commensurate with the rate of change of the environmental power.

7.6.1 Clutter Map Time Constant

As indicated by the previous figures, small values of the update weight are desirable in order to decrease the clutter map CFAR loss. However, note that (7.17) is a difference equation and, like a digital filter, has a transient response time. The response time is important because, though the clutter cross section of the ground changes slowly, it does change. As examples, the cross section changes substantially from dry to wet conditions caused by rain or snow. On a slower time scale, the clutter cross section usually changes substantially from summer to winter.

The solution to the homogeneous equation form of (7.17) is

$$\varrho_{Ik}(m) = C(1 - w)^k = C \exp[k \ln(1 - w)]$$

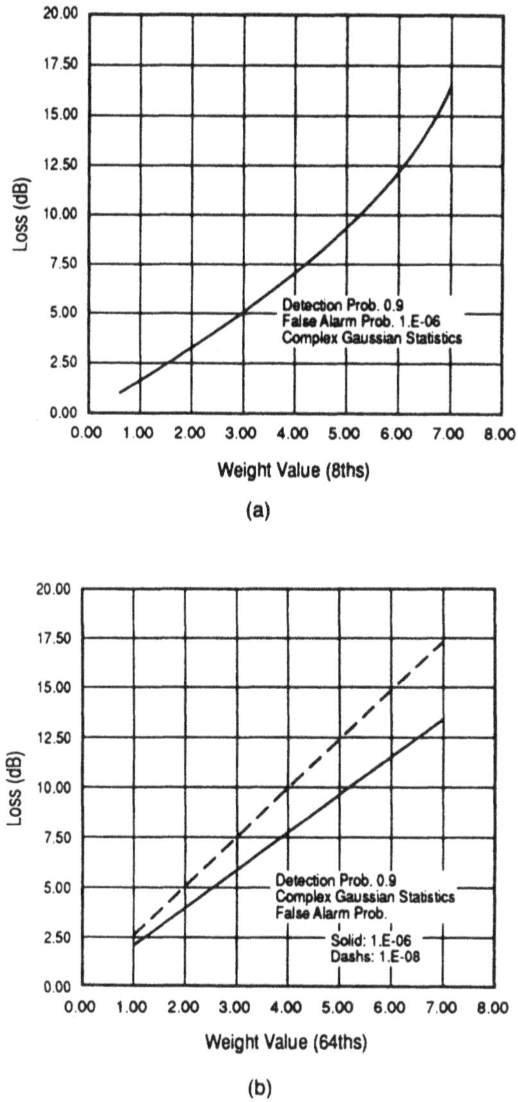

Figure 7.12 Clutter map CFAR loss.

where C is an arbitrary constant. This is like an exponential decay with a time constant equal to $-1/\ln(1-w)$. Assuming the response time to a step change is approximately three time constants, the number of scans required to adapt to an environmental change is approximately $-3/\ln(1-w)$. Table 7.2 indicates the time period to respond to a step clutter cross section change when the antenna rotates at 6 rpm.

Table 7.2
Clutter Map Response Time

w	Number of scans	Time @ 6 rpm rotation
4/8	4.33	43.3 sec
1/8	22.47	224.7 sec (3.74 min)
1/64	190.50	1905 sec (31.75 min)

It can be seen that there is a significant trade-off between low CFAR loss associated with low weight values and fast response time associated with large weight values.

7.7 Composite CFAR Techniques

The goal of composite techniques is to reduce the CFAR loss. One example is to recognize that often the environment is only thermal noise. When this is true, the optimum procedure is to not use any CFAR but to use a precomputed fixed threshold. A composite technique designed to exploit this condition uses both a fixed threshold value, T, and an adaptive threshold, T_a, computed using a CACFAR configuration or other variant. The fixed threshold is used unless the adaptive threshold is much larger. The concept of the technique is illustrated by Figure 7.13.

The concept of using an adaptive threshold only when it is much larger than the fixed threshold is equivalent to using a threshold T', defined as

$$T' = \begin{cases} T & \alpha T_a < T \\ T_a & \alpha T_a > T \end{cases}$$

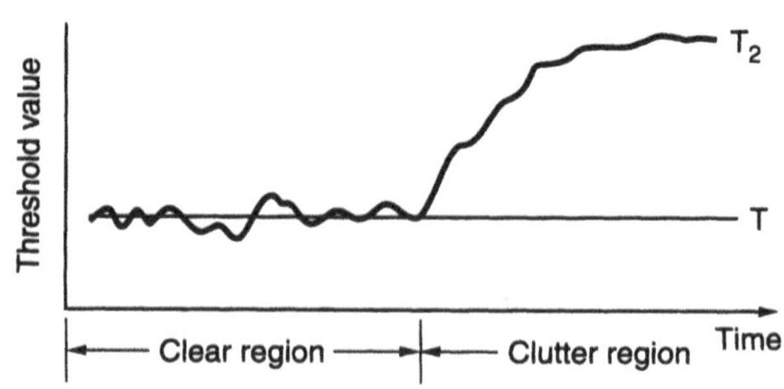

Figure 7.13 A composite CFAR technique.

where α is a parameter that controls the dichtomy of large and small. Note that when $\alpha = 0$ the fixed threshold is always used and when α tends to infinity the adaptive threshold is always used. The intermediate values of α control the percentage of time that either threshold is used. In a thermal noise environment, the CFAR loss can be made as small as desired using a value of α close to zero. However, the cost of decreased loss is the increased false alarm rate in a nonthermal noise environment.

As previously mentioned, another composite technique is based on the GOCFAR. The modification is that the individual split window sums are combined to one total sum, resulting in a CACFAR circuit, when it is decided that the two sums are very similar so that the environment is homogeneous rather than a step. There are many other possible composite techniques.

8

Multiple Sidelobe Cancelers

The combination of a radar antenna, usually a parabolic reflector and a single auxiliary antenna configured as a sidelobe canceler, creates a system antenna pattern that is approximately null at the spatial angle of a single jammer. When the environment contains more than a single jammer, it is necessary to use additional auxiliary antennas. Figure 8.1 shows a conceptual block diagram of a MSLC.

As shown, the output of each antenna passes through mixers, amplifiers, and filters before the adaptive weights are applied. The adaptive weights are chosen to satisfy the optimality criterion of minimizing the output residue power. This is approximately the condition of creating a system antenna pattern that has nulls at the spatial angles of the jammers. A defect of this optimality criterion is that it ignores the system antenna gain at the target's angle. Specifically, the weighted auxiliary antennas form a perturbation beam. The subtraction of the perturbation beam from the main beam creates a system beam whose gain in the directions of the jammers is approximately zero. However, the subtraction of the perturbation beam also affects the gain in the target direction. When the gain of the main antenna is much greater than the product of the number of auxiliary antennas and the gain of an auxiliary antenna, the perturbation of the main beam gain in the target direction is usually sufficiently small so that the residue power minimization criterion is reasonable. However, for array antennas where all elements have approximately the same gain, this is an inappropriate criterion. An alternate optimality criterion is to maximize the output SINR, and this is more appropriate for arrays. Also, as will be discussed in detail later, arrays and Doppler filters are mathematically equivalent and the maximum SINR criterion is appropriate for both.

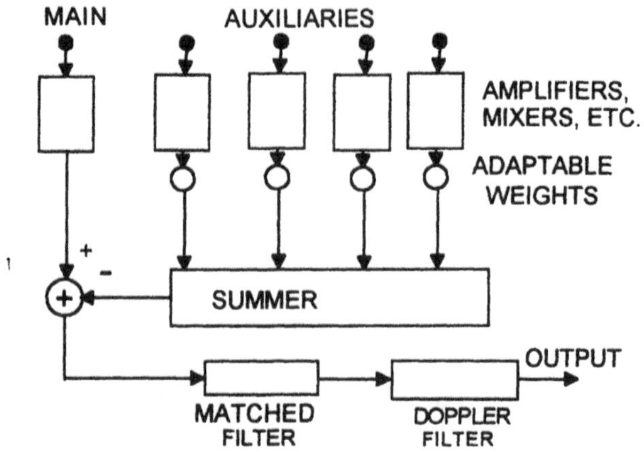

Figure 8.1 Conceptual block diagram of MSLC.

8.1 Optimum MSLC Weight Derivation

Denote the voltage on the main antenna as $y_0(t)$ and that on the nth auxiliary antenna as $y_n(t)$. For N auxiliary antennas the residue voltage equals

$$r(t) = y_0(t) - \sum_{n=1}^{N} w_n^* y_n(t) \tag{8.1}$$

where w_n denotes the adaptable weight for the nth auxiliary antenna and the conjugation is used for convenience in later representations. Denote the N-dimensional vectors W and Y with nth components w_n and $y_n(t)$, respectively. Then the vector notation equivalent of (8.1) is

$$r(t) = y_0(t) - W^{\mathrm{H}} Y \tag{8.1a}$$

The thermal noise and jammer noises are all zero-mean, Gaussian-distributed random processes. Thus, the residue voltage is also a zero-mean Gaussian process. Its power is defined as

$$P_{\text{residue}} = E\left[r(t)r^*(t)\right]$$

and from (8.1) this is[1]

$$\begin{aligned} P_{\text{residue}} &= E\left\{\left[y_0(t) - \sum w_n^* y_n(t)\right]\left[y_0(t) - \sum w_n^* y_n(t)\right]^*\right\} \\ &= E\Big\{y_0(t)y_0^*(t) - \sum w_n y_0(t) y_n^*(t) - \sum w_n^* y_n(t) y_0^*(t) \\ &\quad + \sum\sum w_n^* y_n(t) w_m y_m^*(t)\Big\} \end{aligned} \tag{8.2}$$

1. All summations without indicated ranges are over the range of 1 to N.

The power received by the main antenna, P_{main}, equals $E[y_0(t)y_0^*(t)]$. The correlations of the antenna-to-antenna voltages are

$$E\left[\, y_n(t)y_m^*(t)\right] = r_{nm} \tag{8.3}$$

Substituting (8.3) into (8.2) gives

$$P_{\text{residue}} = P_{\text{main}} - \sum w_n r_{n0}^* - \sum w_n^* r_{n0} + \sum\sum w_m w_n^* r_{nm} \tag{8.4}$$

There is a set of weights, the optimum MSLC weights, that minimizes the residue power. These weights can be computed only if all the correlation values are known. This knowledge is equivalent to knowing the spatial angles and powers of all the jammers and the voltage gains of all antenna at the jammer angles. This knowledge is never available. Thus, the optimum weights and the performance that is obtained using the optimum weights is not achievable. The optimum weight theory is primarily useful in determining a performance upper bound so as to have a reference for comparing the performance of candidate estimation algorithms.

The optimum weight solution is found from (8.4) by evaluating the $2N$ partial derivatives of P_{residue} with respect to the real and imaginary parts of the weights and setting the derivatives equal to zero. As shown in the appendix of Chapter 5, an equivalent procedure is forming the N partial derivatives with respect to each w_n^* and equating the N partials to zero. This results in N linear equations for the N unknown weights given by

$$\begin{aligned}
r_{11}w_1 + r_{12}w_2 + r_{13}w_3 + \cdots + r_{1N}w_N &= r_{10} \\
r_{21}w_1 + r_{22}w_2 + r_{23}w_3 + \cdots + r_{2N}w_N &= r_{20} \\
r_{31}w_1 + r_{32}w_2 + r_{33}w_3 + \cdots + r_{3N}w_N &= r_{30} \\
&\vdots \\
r_{N1}w_1 + r_{N2}w_2 + r_{N3}w_3 + \cdots + r_{NN}w_N &= r_{N0}
\end{aligned} \tag{8.5}$$

The matrix equivalent of (8.4) and (8.5) is obtained by defining the N by N covariance matrix R whose nth row, mth column entry is the correlation between the nth and mth auxiliary antennas, r_{nm}, and the correlation vector R_0 whose nth column entry is the correlation between the nth auxiliary and the main antennas r_{n0}. Thus,

$$P_{\text{RESIDUE}} = P_{\text{MAIN}} - R_0^{\text{H}} W - W^{\text{H}} R_0 + W^{\text{H}} R W \tag{8.4a}$$

In matrix notation, the optimum weight vector is determined by computing the partial derivative of (8.4a) with respect to W^{H} and setting it equal to zero.

This results in the equation for the optimum weight vector W_{OPT}. Substituting W_{OPT} into (8.4a) gives the associated minimum residue power

$$\begin{aligned} R\,W_{\mathrm{OPT}} &= R_0 \\ P_{\mathrm{MIN}} = P_{\mathrm{MAIN}} - R_0^{\mathrm{H}} R^{-1} R_0 &= P_{\mathrm{MAIN}} - R_0^{\mathrm{H}} W_{\mathrm{OPT}} \end{aligned} \tag{8.5a}$$

8.2 Covariance Matrix and Correlation Vector Evaluation

The values of the antenna-to-antenna correlations depend on the jammers' angle and powers and the gains of the antennas to the jammers. Thus, the voltage of the main antenna at the subtraction point due to the thermal noise and J jammers is

$$y_0(t) = n_0(t) + \sum_{k=1}^{J} g(\psi_k)\, j_k(t) \tag{8.6}$$

where $n_0(t)$ denotes the thermal noise of the main antenna and $g(\psi_k)$ denotes the antenna voltage gain for the kth jammer at the jammer angle ψ_k and so $g(\psi_k)\, j_k(t)$ denotes the voltage received from the kth jammer. Similarly, the voltage of the nth auxiliary antenna at the subtraction point due to thermal noise and the J jammers is

$$\begin{aligned} y_n(t) &= n_n(t) + \sum_{k=1}^{J} j_k(t - \tau_{nk}) \exp[-j\phi_{nk}] \\ \tau_{nk} &= d_n \sin(\psi_k)/c \\ \phi_{nk} &= 2\pi f_0 \tau_{nk} \end{aligned} \tag{8.7}$$

where d_n is the spacing between the main and the nth auxiliary antennas. The spacing induces the time delay τ_{nk}, which induces the phase shift ϕ_{nk}. The auxiliary antennas are assumed to be unity gain omnidirectional. The jammers are assumed to be statistically independent. Then the received power for the main antenna is

$$P_{\mathrm{MAIN}} = E|y_0(t)|^2 = P_n + \sum_{k=1}^{J} |g(\psi_k)|^2 P_{Jk} \tag{8.8}$$

where P_n is the thermal noise power and P_{Jk} is the kth jammer power received by a unit gain antenna. From (8.6) and (8.7) the nth component of the correlation vector can be evaluated as

$$r_{n0} = E\left[\, y_n(t) y_0^*(t)\right] = \sum_{k=1}^{J} g^*(\psi_k) \exp[-j\phi_{nk}]\, P_{Jk}\, \rho(\tau_{nk}) \tag{8.9}$$

where $\rho(\tau_{nk})$ denotes the correlation coefficient corresponding to the time delay τ_{nk} and it is assumed that the thermal noises of different antenna are statistically independent. The nth row mth column entry of the covariance matrix can be evaluated from (8.7) as

$$r_{nm} = E\left[y_n(t)y_m^*(t)\right] = P_n\,\delta(n-m) + \sum_{k=1} \exp[-j2\pi(d_n - d_m)\,f_0 \sin(\psi_k)/c]\,P_{Jk}\,\rho(\tau_{nk} - \tau_{mk}) \qquad (8.10)$$

and $\delta(\,)$ is Dirac's delta function equal to unity when the argument is zero and equal to zero for all other arguments.

8.3 Optimum Weight Performance Evaluation

The minimum residue power depends upon several factors in a chaotic manner. As examples, it depends upon the number of auxiliary antennas; the number, powers, and spatial angles of all the jammers; the voltage gains of all antennas in the directions of the jammers; the radar system bandwidth; and equipment implementation errors such as the channel-to-channel filters passband matches. The reference to chaotic dependence is because small changes in any one of the previous variables can cause enormous, and unexplainable, changes in the residue power. Thus, it has been necessary to evaluate the performance using computer evaluation of many special cases. The evaluation methodology was to

1. Select the number of auxiliary antennas (N);
2. Select the number and powers of the jammers;
3. Select the number of random experiments for the evaluation;
4. For each experiment
   a. Randomly and independently choose the jammer angles, which were restricted to the angular region of 8 sidelobes bracketing the mainlobe but excluding the mainlobe region;
   b. Compute all antenna voltage gains for each jammer;
   c. Compute the covariance matrix and the correlation vector;
   d. Solve for optimum weights;
   e. Compute residue power for optimum weights;
5. Repeat the experiment using a new set of random jammer angles;
6. Derive performance statistics.

The performance statistics are presented as histograms. Initially the evaluation assumes that the transfer function of all the channels due to the mixers and filters, for example, are perfectly matched. After, the effect of errors is evaluated.

8.3.1 MSLC Evaluation Examples

The first example is for a 64-equispaced element array with element-to-element spacing equal to 0.5 wavelength. Each element is omnidirectional and has unity gain. The array's elements are combined to form a main antenna, and the sidelobes are weighted with small errors so that the average sidelobe gain is also equal to unity. Some of the array elements are also used as auxiliary elements. For this example the elements of the array that are used for auxiliary elements are approximately uniformly spaced. They are the end elements, 1 and 64, and elements 21 and 42. Three jammer cases are considered. One is for a single jammer whose power is such that the power level received at each array element is 40 dB greater than the thermal noise. The second case is for two equal power jammers with total power equal to 40 dB above the thermal noise power. The third case is for four equal power jammers with a total power equal to 40 dB above the thermal noise power. The evaluation is done for three different radar bandwidths.

The bandwidth affects the normalized correlation coefficient, ρ, in (8.9) and (8.10). The normalized correlation coefficient is evaluated via the Fourier transform of the radar's receiver's power spectrum. Assuming an LFM type waveform, the spectrum is approximately rectangular, so the correlation coefficient is given by

$$\rho(\tau_{nk}) = \sin(\pi B\tau_{nk})/(\pi B\tau_{nk}) \tag{8.11}$$

The antenna's phase center is at the center of the equispaced element array. Thus, the distance from the main antenna phase center to the nth array element equals $|31.5 - n|\lambda/2$ and noting that $f_0\lambda = c$ gives the distance as $|31.5 - n|c/(2f_0)$. Substituting this into (8.7) and then into (8.11), gives

$$\begin{aligned}\rho(\tau_{nk}) = {} & \sin[\pi B|31.5 - n|\sin(\psi_k)c/(2f_0)]/ \\ & [\pi B|31.5 - n|\sin(\psi_k)c/(2f_0)]\end{aligned} \tag{8.11a}$$

Thus, the normalized correlation coefficient depends upon the ratio of the radar bandwidth to the carrier frequency. The dependence on the ratio is true for any type of radar spectrum. A similar equation is obtained to evaluate $\rho(\tau_{nk} - \tau_{mk})$.

Each of the four-jammer cases has been evaluated for three radar normalized bandwidths for a total of six cases and the six histograms are plotted in Figure 8.2. The vertical axis is the histogram probability. The horizontal axis is the normalized residue power, the ratio of the residue power to the thermal

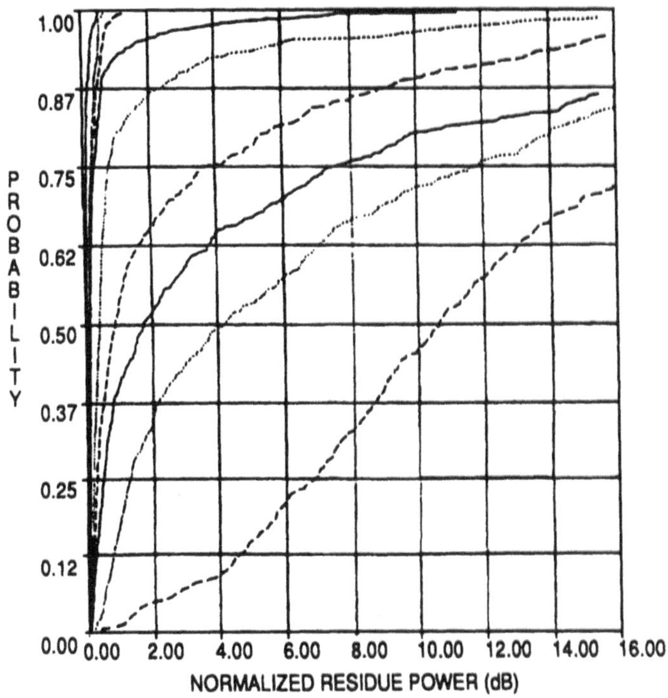

Figure 8.2 40 dB jammers; 4 uniformly spaced auxiliaries; no errors.

noise power, expressed in decibels. The solid curve graphs are for the very narrow normalized bandwidth case of B/f_0 approaching zero; the short-dashed cases are for a normalized bandwidth equal to 0.001 (say a 1-MHz bandwidth and a carrier frequency of 1 GHz); the long-dashed curves are for a normalized bandwidth equal to 0.004. The three curves grouped at the upper left of the figure are the single-jammer cases. The three curves grouped at the lower left are the four-jammer cases, and the intermediate curves are for the two-jammer cases. It can be seen that the residue power for the one-jammer case is much less than that of the four-jammer cases. For all number of jammers the residue power increases as the radar bandwidth increases. There is approximately a probability of 0.87 that the normalized residue power is less than 1 dB for all three bandwidths of the single-jammer cases. However, the normalized residue power is much greater for the four-jammer cases. For the narrow bandwidth case the probability that the normalized residue power exceeds 12 dB equals 0.18. For the normalized bandwidth of 0.004 the probability that the normalized residue power exceeds 12 dB equals 0.4. The poor performance exhibited by the previous example is primarily caused by a poor choice of which elements are used as auxiliary elements. Specifically, uniformly spaced auxiliaries create grating lobes and the grating lobes limit

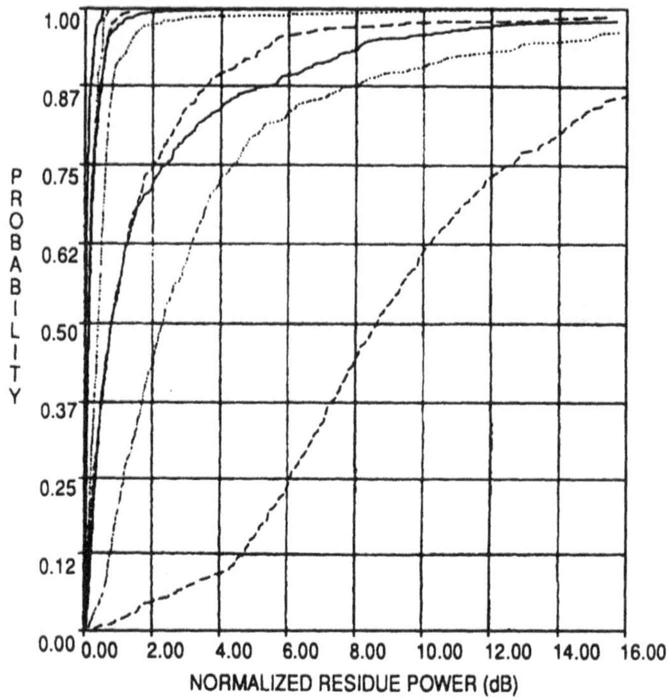

Figure 8.3 40 dB jammers; 4 non-uniformly spaced auxiliaries; no errors.

performance. This is demonstrated by the curves of Figure 8.3. This case differs from the prior case only by the selection of auxiliary elements such that the element-to-element spacing is nonuniform. Comparing this figure with the previous figure it can be seen that the normalized residue power is decreased for every case by using nonuniformly spaced auxiliary antennas. In particular, the probability of very high residue power for the two-jammer cases has been substantially reduced as compared to the uniformly spaced auxiliary antenna configuration. Note that the probability that the normalized residue power exceeds 12 dB for all bandwidths of the two-jammer case is approximately zero. Though the normalized residue power for the four-jammer cases is also reduced by nonuniformly spacing the auxiliary antennas, the residue power is still large. This demonstrates that the four-auxiliary MSLC does not adequately cancel four jammers.

8.3.2 Channel-to-Channel Mismatch Effect

If there is no thermal noise and the transfer functions of the channels of each antenna to the summing point are perfectly matched, there are values for the

N MSLC adaptive weights such that the residue voltage equals exactly zero when the number of jammers is not greater than N. However, in a real world implementation, the filters and mixers, for example, of the main and auxiliary channels of the MSLC implementation are not exactly identical and exact cancellation cannot be achieved even in the absence of thermal noise. Thus, mismatch errors limit cancellation. The analysis of this section quantifies the performance degradation caused by channel-to-channel mismatches.

The analysis of Section 8.2 assumed the channels were perfectly matched, and in particular it was assumed that the center frequency channels gains are identical. It is first shown that the residue power is independent of the center frequency gains. It is then shown that mismatched frequency-dependent gain errors do affect the residue power, and an equation relating the upper bound of cancellation to the frequency-dependent mismatch is derived.

The set of voltages $\{y_n\}$ were defined as both the input and output of the transfer function boxes of Figure 8.1. Denote the center frequency voltage gain for the nth channel as g_n. The previous intrinsic assumption is that $g_n = 1$ for all n. When they differ from unity the effect is that the previous values of $y_n(t)$ are modified to $g_n y_n(t)$. The power received by the main antenna is increased by the factor $|g_0|^2$. The previous values of r_{n0} and r_{nm} are modified to $g_n g_0^* r_{n0}$ and $g_n g_m^* r_{nm}$, respectively. Thus, the correlation vector and covariance matrix are modified to $g_0^* D_g R_0$ and $D_g R D_g^H$, where D_g denotes a diagonal matrix with main diagonal entries of g_n. Substituting these results into (8.5b) gives the minimum residue power as

$$\begin{aligned} P_{\text{MIN}} &= |g_0|^2 P_{\text{MAIN}} - |g_0|^2 (D_g R_0)^H (D_g R D_g^H)^{-1} (D_g R_0) \\ &= |g_0|^2 (P_{\text{MAIN}} - R_0^H D_g^H (D_g^H)^{-1} R^{-1} D_g^{-1} (D_g R_0) \\ &= |g_0|^2 (P_{\text{MAIN}} - R_0^H R^{-1} R_0) \end{aligned}$$

Thus, the residue power is independent of the auxiliary antenna center frequency gains.

The effect of mismatched transfer functions is analyzed by denoting the transfer functions box output as z_n for a filter input y_n. Thus,

$$z_n(t) = \int_{-\infty}^{\infty} h_n(x) y_n(t-x)\,dx$$

where $h_n()$ denotes the filter impulse response of the nth channel. From the analyses of Sections 8.1 and 8.2, the optimum weights and minimum residue power depend on the set of correlations $E[z_n(t)z_m^*(t)]$. Then[2]

$$\begin{aligned} r_{nm}^{(z)} &= E\left[z_n(t)z_m^*(t)\right] = r_{nm}^{(z)}(0) \\ &= E\left[\iint h_n(x_1)y_n(t-x_1)h_m^*(x_2)y_m^*(t-x_2)dx_1\,dx_2\right] \end{aligned} \tag{8.12}$$

The correlations of the inputs are denoted as $r_{nm}^{(y)}$, and the previous section denoted them as r_{nm} without any superscript. Since

$$\begin{aligned} E\left[y_n(t-x_1)y_m^*(t-x_2)\right] &= r_{nm}^{(y)}(x_2-x_1) \\ r_{nm}^{(z)} &= \iint E\left[h_n(x_1)h_m^*(x_2)\right]r_{nm}^{(y)}(x_2-x_1)dx_1\,dx_2 \end{aligned} \tag{8.13}$$

The filter errors are more easily analyzed in the frequency domain. Substituting

$$h_n(x_1) = \int H_n(f_1)\exp(j2\pi f_1 x_1)df_1$$

and a similar expression for $h_m(x_2)$ into (8.13) gives

$$\begin{aligned} r_{nm}^{(z)} = \iiiint & E[H_n(f_1)H_m^*(f_2)]\exp[j\,2\pi(f_1x_1-f_2x_2)] \\ & \times r_{nm}^{(y)}(x_2-x_1)dx_1\,dx_2\,df_1\,df_2 \end{aligned} \tag{8.14}$$

For a magnitude error of $\varepsilon_n(f)$ and a phase error of $\phi_n(f)$, the errored transfer function equals

$$H_n(f) = \gamma_n\{1+\varepsilon_n(f)\exp[j\,2\pi\phi_n(f)]\}$$

where γ_n is a normalization factor such that each filter has unity center frequency gain. The normalization is done for later mathematical convenience and has no effect on the residue power since it has been shown that residue power is independent of center frequency gain factors. Both $\varepsilon_n(f)$ and $\phi_n(f)$ are assumed to be zero-mean, independent random processes. For small errors, the transfer functions are further approximated as

$$H_n(f) = \gamma_n[1+\varepsilon_n(f)+j\phi_n(f)]$$

The filter gain to thermal noise is proportional to $|H_n(f)|^2$ and denoting the variances of $\varepsilon_n(f)$ and $\phi_n(f)$ as V_ε and V_ϕ, respectively, the transfer function

2. Unless otherwise indicated, the limits of integrals are $-\infty$ to $+\infty$.

approximation reduces to

$$H_n(f) = [1 + \varepsilon_n(f) + j\phi_n(f)]/(1 + V_\varepsilon + V_\phi)^{1/2}$$

For small errors this is further approximated as

$$H_n(f) = (1 - V_\varepsilon - V_\phi)^{1/2}[1 + \varepsilon_n(f) + j\phi_n(f)] \tag{8.15}$$

Thus, for $n = m$, $E|H_n(f)|^2$ equals unity and (8.14) reduces to

$$r_{nn}^{(z)} = \iiiint \exp[j\,2\pi(f_1 x_1 - f_2 x_2)]\, r_{nn}^{(y)}(x_2 - x_1)\, dx_1\, dx_2\, df_1\, df_2 \tag{8.16}$$

Recognizing that the Fourier transform of unity is a delta function, integrating on f_1 and f_2 gives

$$r_{nn}^{(z)} = \iint \delta(x_1)\delta(x_2)\, r_{nn}^{(y)}(x_2 - x_1)\, dx_1\, dx_2 \tag{8.17}$$

and integrating on $dx_1\, dx_2$ gives

$$r_{nn}^{(z)} = r_{nn}^{(y)}(0) = r_{nn}^{(y)}$$

Thus, the diagonal terms of the covariance matrix are unchanged.

The off-diagonal terms of the covariance matrix and the terms of the correlation vector are obtained via

$$\begin{aligned} E[H_n(f_1)H_m^*(f_2)] = (1 - V_\varepsilon - V_\phi)E\{&[1 + \varepsilon_n(f) + j\phi_n(f)] \\ &\times [1 + \varepsilon_m(f) - j\phi_m(f)]\} \end{aligned}$$

Assuming the errors are independent from channel to channel results in the same integrals as for (8.16) and (8.17), so

$$r_{nm}^{(z)} = (1 - V_\varepsilon - V_\phi)\, r_{nm}^{(y)}$$

Defining ρ_{H} as $(1 - V_\varepsilon - V_\phi)$, the matrix formulation for the correlation vector is

$$R_0^{(z)} = \rho_{\mathrm{H}} R_0^{(y)} \tag{8.18}$$

For the covariance matrix, the off-diagonal terms are multiplied by ρ_{H}, so the square matrix $R_z - \rho_{\mathrm{H}} R_y$ is diagonal with diagonal elements equal to $(1 - \rho_{\mathrm{H}}) P_{\mathrm{AUX}}$, where P_{AUX} is the total power received by each auxiliary element. Thus,

$$R_z = \rho_{\mathrm{H}} R_y + (1 - \rho_{\mathrm{H}}) P_{\mathrm{AUX}} I \tag{8.19}$$

where I denotes the identity matrix. From (8.5a)

$$P_{\text{MIN}} = P_{\text{MAIN}} - R_0^{\text{H}} R^{-1} R_0 \tag{8.20}$$

Combining the last three equations gives

$$P_{\text{MIN}}^{(z)} = P_{\text{MAIN}} - \rho_{\text{H}}^2 R_0^{(y)\text{H}} [\rho_{\text{H}} R_y + (1 - \rho_{\text{H}}) P_{\text{AUX}} I\,]^{-1} R_0^{(y)}$$

Expanding $P_{\text{MIN}}^{(z)}$ in a Taylor series about $\rho_{\text{H}} = 1$ gives

$$\begin{aligned} P_{\text{MIN}}^{(z)} &= P_{\text{MIN}}^{(y)} + (1 - \rho_{\text{H}})(d/d\rho_{\text{H}})\{P_{\text{MIN}}^{(z)}[1]\} \\ &= \rho_{\text{H}} P_{\text{MIN}}^{(y)} + (1 - \rho_{\text{H}})[P_{\text{MAIN}} + P_{\text{AUX}}[W_{\text{OPT}}^{(y)}]^H W_{\text{OPT}}^{(y)} \end{aligned} \tag{8.21}$$

Note that

$$P_{\text{MIN}}^{(z)} > (1 - \rho_{\text{H}}) P_{\text{MAIN}} \tag{8.22}$$

Thus, (8.22) gives a lower bound for the residue power. The residue power is greater than $(1 - \rho_{\text{H}}) P_{\text{MAIN}}$ for any geometry and any number of auxiliary elements. An upper bound for the cancellation ratio is

$$CR = (P_{\text{MAIN}}/P_{\text{MIN}}^{(z)}) < 1/(1 - \rho_{\text{H}}) \tag{8.22a}$$

As a numeric example, (8.22a) indicates that the cancellation ratio is less than 40 dB for $\rho_{\text{H}} = 0.9999$.

8.3.3 Channel-to-Channel Match Specification

The gain and phase errors will be considered separately. Assume that the probability density function of the zero-mean error $\varepsilon(f)$ is uniform over a range of δ. Then the error variance equals

$$V_\varepsilon = \delta^2/12 \tag{8.23}$$

The maximum peak-to-peak of the transfer function's magnitude equals $(1 + \delta/2)/(1 - \delta/2)$. In decibels this is

$$\text{dB(error)} = 20 \log[(1 + \delta/2)/(1 - \delta/2)]$$

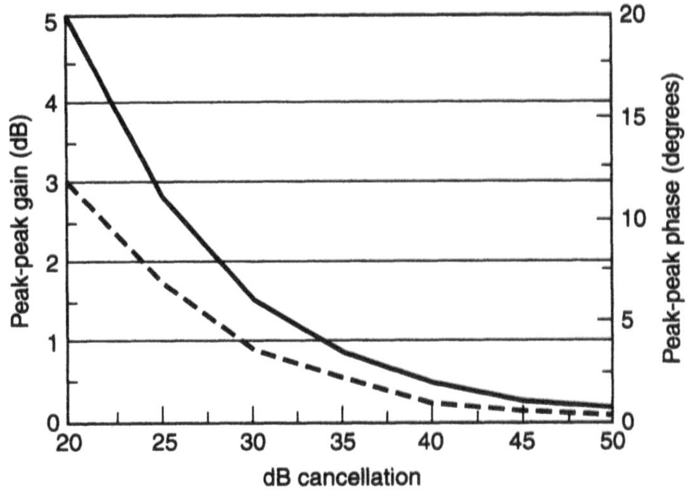

Figure 8.4 40 dB jammers; 4 non-uniformly spaced auxiliaries; no errors.

which can be manipulated to

$$\mathrm{dB(error)} = [20/\ln(10)]\ln[(1+\delta/2)/(1-\delta/2)] \tag{8.24}$$

For small errors (8.23) and (8.24) combine to give

$$V_\varepsilon = [\mathrm{dB(error)}/30]^2 \tag{8.25}$$

As an example, from (8.22), for a cancellation of 40 dB, V_ε equals 0.0001; so from (8.25) the peak-to-peak gain variation across the filter band should be less than 0.3 dB.

Similarly, for the phase error uniformly distributed over a range Δ, $V_\phi = \Delta^2/12$. Since the peak-to-peak phase equals Δ, this indicates that the peak-to-peak phase error should be less than approximately 2 degrees for a cancellation of 40 dB. Figure 8.4 graphs the relation between the upper bound on the achievable cancellation and the filter mismatch errors. The dotted curve is for the amplitude error and the solid curve is for the phase error. Note that the decibel cancellation assumes only a phase or magnitude error. If both exist, then the decibel cancellation values decrease by 3 dB.

8.3.4 Channel-to-Channel Match Evaluation Examples

Figures 8.5 and 8.6 are histograms of the evaluations of the effects of channel-to-channel mismatch on the residue power. The figures are for the 64-equispaced

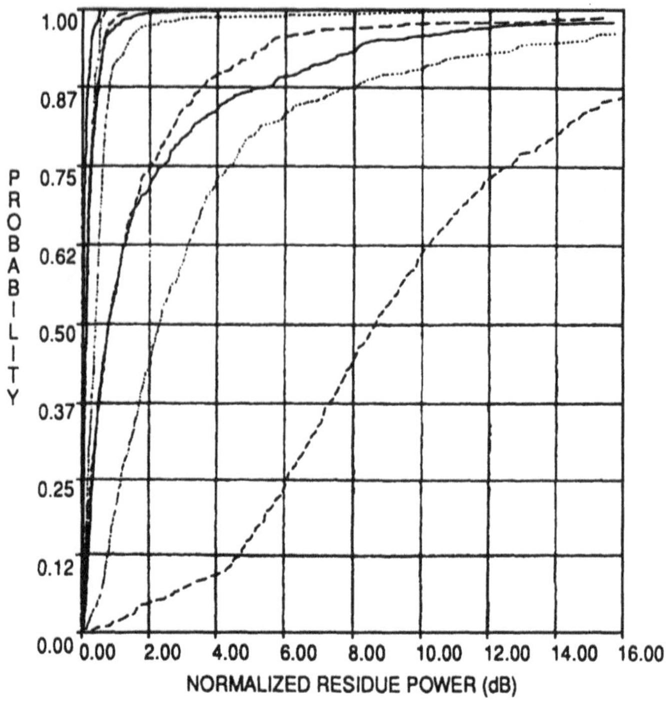

Figure 8.5 40 dB jammers; 4 non-uniformly spaced auxiliaries; 0.1 dB errors.

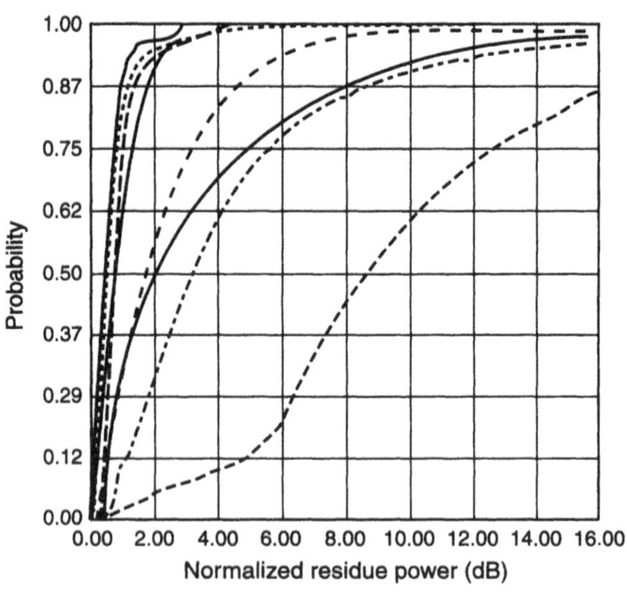

Figure 8.6 40 dB jammers; 4 non-uniformly spaced auxiliaries; 0.1 dB errors.

element array considered in the prior examples, and the four auxiliary elements are the nonuniformly spaced elements previously considered. The three jammer cases are again for one, two, and four jammers. The total jammer power at each array element equals 40 dB for each of the three jammer cases. The normalized bandwidth values are different than the prior examples and equal 0, 0.001, and 0.004. The solid curves are for the zero bandwidth, the dotted curves are for the 0.001 normalized bandwidth, and the dashed curves are for the 0.004 bandwidth. The figures are evaluated using (8.18) and (8.19) for the correlation vector and covariance matrix. Figure 8.5 is for no channel-to-channel mismatch. Figure 8.6 is for a channel-to-channel magnitude mismatch error equal to 0.1 dB. The data for both are evaluated with the same random angles for the jammer locations, so the differences are only due to the mismatch error.

The three leftmost curves of Figure 8.6 are for the single jammer and the three bandwidths. It can be seen that for all three bandwidths the normalized residue power is approximately equal to the thermal noise power. The three rightmost curves are for the four-jammer cases and the three bandwidths. Comparing the curves of Figure 8.7, it can be seen that all of the curves are displaced to the right, indicating a higher probability of large normalized residue power. Note that the least affected curve is for that of four jammers and the 0.004 normalized bandwidth. In this case the performance is almost totally limited by the number of jammers and the large bandwidth, so the mismatch errors have negligible effect. However, for the other cases, the mismatch error has an effect. The normalized residue power for the single-jammer cases are significantly affected. These results are consistent with the previous analysis and the curves of Figure 8.4. The curve of Figure 8.4 indicates that the cancellation upper bound is 40 dB for a 0.3-dB peak-to-peak magnitude error. This implies that errors of this magnitude should affect cancellation for 40-dB jammers but not affect the cancellation for 30-dB jammers.

Figures 8.7 and 8.8 are comparable to Figure 8.5 and 8.6 except that the jammer power is reduced to 30 dB. Comparing the two figures it can be seen that the respective curves are almost identical. This is as expected since 0.1-dB equipment implementation errors should not degrade performance for 30-dB jammer power.

8.3.5 Bandwidth Effect Analysis

This section develops a partial explanation of the dependence of the residue power upon bandwidth and why increasing the number of auxiliary antennas for a fixed number of jammers significantly decreases the residue power.

There is an ambiguity between carrier frequency and spatial angle. Thus, from (8.7), the relative phase shift of a received waveform at the nth auxiliary

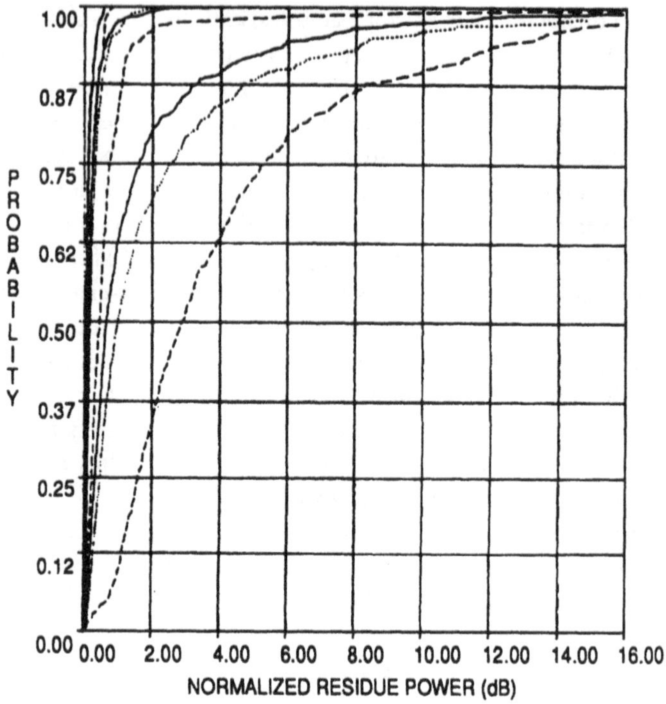

Figure 8.7 30 dB jammers; 4 non-uniformly spaced auxiliaries; no errors.

from the kth jammer is

$$\phi_{nk} = 2\pi f_0 d_n \sin(\psi_k)/c \tag{8.26}$$

Note that the phase shift depends on the product of the carrier frequency and the sine of the arrival angle. Thus, a change of carrier frequency is ambiguous with a change of angle of arrival. The effect of the radar bandwidth, B, upon the phase shift is equivalent to a change of carrier frequency, so a point source of nonzero bandwidth is equivalent to a spatially distributed zero-bandwidth source of spatial width β. The relation between B and β can be obtained from differentiation of (8.26). Thus

$$d\phi_{nk} = [\partial\phi_{nk}/\partial f_0](\Delta f_0) + [\partial\phi_{nk}/\partial\psi_k](\Delta\psi)$$

For the phase shift to be constant, due either to bandwidth or angle change, $d\phi_{nk}$ must equal zero. Define Δf_0 as B and $\Delta\psi$ as β. Thus

$$\beta = -\{[\partial\phi_{nk}/\partial f_0]/[\partial\phi_{nk}/\partial\psi_k]\}B = -(B/f_0)\cos\psi_k \tag{8.27}$$

Note that β depends upon the normalized bandwidth.

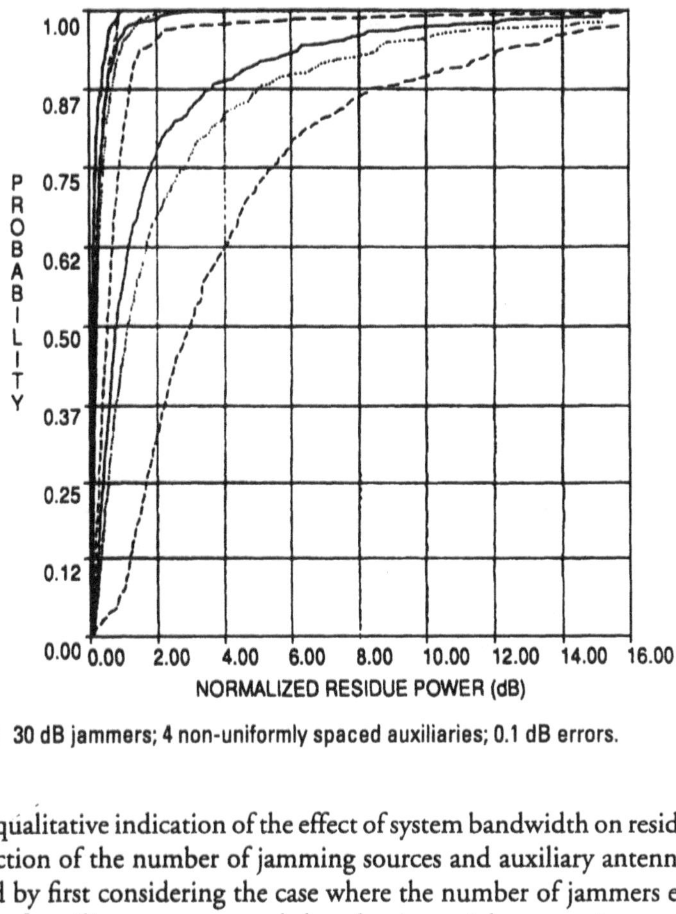

Figure 8.8 30 dB jammers; 4 non-uniformly spaced auxiliaries; 0.1 dB errors.

A qualitative indication of the effect of system bandwidth on residue power as a function of the number of jamming sources and auxiliary antennas can be obtained by first considering the case where the number of jammers equals the number of auxiliary antennas and the adaptive weight vector is set so that the response of the adapted antenna pattern is exactly equal to zero at the jammer angles. This is approximately the case for the residue power minimization. Since there are N auxiliaries and N jammers, the N zero gain setting specifies a set of N equations similar to (8.5). The adapted voltage gain at the kth jammer spatial angle neighborhood can be approximated by the first term of a series as

$$g(\psi_k) = \alpha_k(\psi - \psi_k) \tag{8.28}$$

where α_k is the first derivative of the gain at the jammer angle. Denote the power of the jammer as P_{Jk}; this power is effectively distributed over the angle β. As the spatial extent encompasses a region of nonzero gain, the residue power does not equal zero.

The residue power caused by the voltage gain pattern of (8.28) is given by the integral

$$P_{\text{res}}(k) = |\alpha_k|^2 (P_{Jk}/\beta) \int (\psi - \psi_k)^2 d\psi$$

and the integration limits are from $-\beta/2 + \psi_k$ to $\beta/2 + \psi_k$. Combining with (8.27) this evaluates to

$$P_{\text{res}}(k) = (|\alpha_k|^2/12) P_{Jk}[(B/f_0) \cos \psi_k]^2 \quad (8.29)$$

Thus, the residue increases as the square of the relative bandwidth.

Now, consider the case of N auxiliary antennas and $N/2$ jammers. Again the adapted antenna pattern is specified as being exactly equal to zero at the jammer angles. Since there are only half as many jammers as the previous example, additional conditions can be imposed. Assume that the optimum adaptive weight is approximately that which also causes the derivative of the gain pattern to equal zero at the jammers' angles. Note that in the previous case the gain pattern is a single zero at the jammers' angles. This latter specification creates a double zero at each jammer angle. For this case, the equivalent of (8.28) is

$$g(\psi_k) = \varepsilon_k (\psi - \psi_k)^2 \quad (8.30)$$

and the equivalent of (8.29) is

$$P_{\text{res}}(k) = (|\varepsilon_k|^2/80) P_{Jk}[(B/f_0) \cos \psi_k]^4 \quad (8.31)$$

Thus, the residue increases as the fourth power of the relative bandwidth. Since the relative bandwidth is much less than unity, the residue power is substantially decreased when the number of auxiliary antennas equals twice the number of jammers.

8.4 Algorithms

The optimum MSLC weight minimizes the residue power, but these weights can be computed only if all the correlation values are known. This knowledge is equivalent to knowing the spatial angles and powers of all the jammers and the voltage gains of all antennas at the jammer angles. This knowledge is never available. Thus, the optimum weights and the performance that is obtained using the optimum weights is not achievable and represents a performance upper bound. Algorithms[3] are used to estimate the optimum weights. As was done for CFAR circuits, the estimation data are obtained from the range cells that are adjacent to the test cell. There are three major algorithms that are used as estimators. Two algorithms, LMS and NLMS, were discussed as estimators for the SLC configuration. These two algorithms can be implemented using either analog or digital techniques. In addition, a third algorithm can be implemented for digital systems. This algorithm is denoted as the sample matrix inversion (SMI) algorithm. In concept, it uses the range cell data to form estimates of the covariance matrix and correlation vector and then solves for the weight vector that satisfies the matrix estimates. There is a trade-off between performance and implementation cost for the three algorithms. The SMI has the best performance but the highest cost. The LMS and NLMS algorithms are of lesser cost and lesser performance. The implementation cost of the two algorithms is approximately the same. Just as for the SLC implementation, their performances differ in terms of convergence time and weight noise. That is, the NLMS algorithm converges faster but at the cost of increased steady-state residue power due to weight noise.

8.4.1 LMS Algorithm

A conceptual block diagram of a digital implementation of the LMS algorithm for MSLC is shown in Figure 8.9. The configuration is an extension of the SLC single correlation feedback loop shown in Figure 5.5. Each auxiliary antenna feeds a feedback loop connected to the antenna's adaptive weight. The feedback action causes the weights to change in a direction such that the residue voltage is uncorrelated with each antenna voltage. The uncorrelated condition corresponds to the minimum residue power condition. Because the algorithm uses feedback loops, it does not achieve the uncorrelated condition instantaneously but instead has a convergence period. The convergence analysis is complicated and involves eigenvectors and eigenvalues of the covariance matrix. It can be shown that the average weight vector converges to the optimum vector. The time-dependent residue power is given by

$$P_{\text{RES}}(t) = P_{\text{MIN}} + \sum_{n=1}^{N} P_n \exp(-t/\tau_n) \tag{8.32}$$

where P_n denotes the power of the nth convergence mode and τ_n is the mode's time constant. Each of the N values of P_n and τ_n depend upon the powers and

3. A more detailed discussion is contained in Nitzberg, *Adaptive Signal Processing for Radar*, Norwood, MA: Artech House, 1992, Chapter 7.

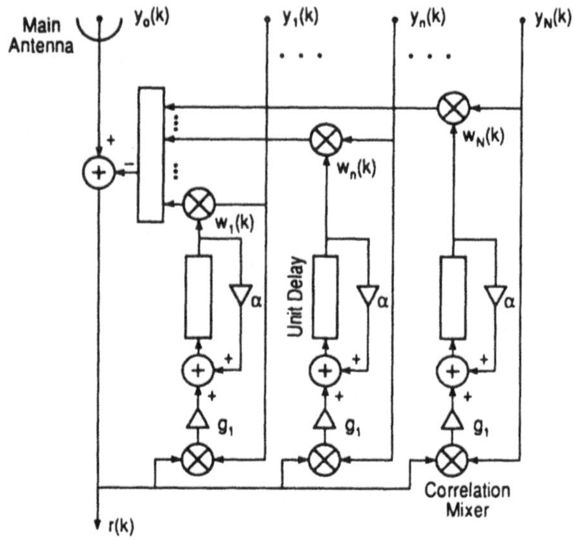

Figure 8.9 Conceptual block diagram of MSLC LMS algorithm.

spatial angles of all of the jammers in a complicated and chaotic manner. That is, small changes in a single jammer power or angle often cause large changes of all parameters of (8.32). The analysis of convergence times was done in a manner similar to that used to evaluate the optimum weight performance. Thus, a number of experiments were performed. For each experiment the jammer angles were randomly and independently chosen. The eigenvectors and eigenvalues of the resultant covariance matrix and correlation vector were computed and from these, the convergence time was computed. Histograms of the convergence times were obtained.

The results are that when the number of jammers equals the number of auxiliary antennas, there is a high probability that the transient response time exceeds the radar's dwell time. Fast convergence is obtained when the number of auxiliaries equals twice the number of jammers. Just as for the SLC implementation, the NLMS algorithm tends to have faster convergence time but at the expense of increased residue power after convergence due to weight noise. Thus, there is a significant trade-off between convergence time, residue power after convergence, and implementation cost.

8.4.2 SMI Algorithm

As previously discussed, the optimum weight vector is the solution of the equation

$$RW_{\mathrm{OPT}} = R_0$$

Because the covariance matrix and correlation vector are unknown, maximum likelihood estimators can be used for these. Denote the estimates as R_{EST} and $R_{0,\mathrm{EST}}$. In concept, the estimates are formed using data from the range cells bracketing the test cell. However, as discussed later, this technique is computationally poor and other indirect methods are used. The conceptual discussion is useful primarily as a tutorial device. Thus, in concept, the optimum weight vector is given by

$$R_{\mathrm{EST}} W_{\mathrm{EST}} = R_{0,\mathrm{EST}} \qquad W_{\mathrm{EST}} = R_{\mathrm{EST}}^{-1} R_{0,\mathrm{EST}} \tag{8.33}$$

To form the estimates of the matrices, denote the voltage of the nth auxiliary antenna for the kth range cell as y_{kn}. Define the vector Y_k with nth component y_{kn} as

$$Y_k = (y_{k1}, y_{k2}, y_{k3}, \ldots, y_{kN})^{\mathrm{T}}$$

where N is the number of auxiliary antennas. Similarly, define the voltage of the kth range cell of the main antenna as y_{k0}. Then for K range cells, the estimates are

$$R_{\mathrm{EST}} = \left[\sum_{k=1}^{K} Y_k Y_k^{\mathrm{H}}\right] \Big/ K \qquad R_{0,\mathrm{EST}} = \left[\sum_{k=1}^{K} y_{k0}^{*} Y_k\right] \Big/ K \tag{8.34}$$

It can be shown that if the statistics of the interference, the sum of the thermal noise and all jammers, is Gaussian, these estimators are the maximum likelihood estimators. The estimate of the optimum weight is then obtained from (8.33).

The algorithm name is a misnomer because the computationally efficient procedure neither estimates a matrix nor inverts a matrix. The preferred procedure is to form a data matrix, Y, from the vector measurements Y_k as

$$Y = (Y_1, Y_2, Y_3, \ldots, Y_K)$$

This matrix is rectangular with N rows and K columns. Similarly define the K-dimensional vector, Y_M with kth component, y_{k0}. In general, K is greater than N and typically K is at least equal to $2N$. Then (8.34) can be expressed as

$$R_{\mathrm{EST}} = YY^{\mathrm{H}}/K \qquad R_{0,\mathrm{EST}} = YY_M^{*}/K$$

and (8.33) can be expressed as

$$YY^{\mathrm{H}} W_{\mathrm{EST}} = YY_M^{*}$$

There are transformations—a particularly convenient one is denoted as Householders—that convert Y^{H} into the product of a unitary matrix and K by N trapezoidal matrix. That is,[4]

$$Y^{\mathrm{H}} = QR \qquad QQ^{\mathrm{H}} = I$$

so Q is a square K by K matrix and R is a K by N matrix. The first N rows of R equal a nonsingular upper triangular matrix and the remaining rows are all equal to zero. As an example a 4 by 2 R is

$$R = \begin{bmatrix} x & x \\ 0 & x \\ 0 & 0 \\ 0 & 0 \end{bmatrix}$$

where x denotes a nonzero entry. In partitioned form .

$$R = \begin{bmatrix} R_{\mathrm{U}} \\ 0 \end{bmatrix}$$

where R_{U} is an upper triangular matrix. Substituting this into (8.33a) and noting that R_{U} is nonsingular gives

$$R_{\mathrm{U}}\, W_{\mathrm{EST}} = Q_1^{\mathrm{H}} Y_M^*$$

where Q_1 is the rectangular matrix equal to the first N columns of Q. Since R_{U} is a triangular matrix, it is simple to solve for W_{EST} without inversion. Note again that for this computational procedure it is not necessary to form an estimate of the covariance matrix nor form an inverse.

Note that the SMI implementation requires a digital implementation. Specifically, an A/D converter is required for each auxiliary antenna and for the main antenna. This requirement is relatively expensive as contrasted with the analog implementation of the LMS or NLMS algorithms. Thus, there is a trade between implementation cost and algorithm convergence time.

8.4.3 SMI Performance Analysis

The performance analysis of the LMS and NLMS algorithms evaluated the convergence time of the algorithms and the additional residue power due to

4. Unfortunately the classical notation of R is the same as that used for a covariance matrix. Here R is not a covariance matrix.

weight noise effects after convergence. These performance criteria are pertinent because the estimated weight vector is random and does not converge to the optimum weight vector. Only the mean value of the estimated weight vector converges to the optimum weight vector. The SMI performance analysis differs from that of the iterative algorithms for several reasons. One reason is that the estimated weight vector converges with probability one to the optimum weight vector. That is, given a sufficient number of data vectors [large enough K in (8.34)], the weight noise effect becomes arbitrarily small. The convergence time is defined as the value of K such that the performance loss compared to that for the optimum weights is small, say 3 dB. It is convenient to analyze the convergence time of the SMI for a different, but related, optimality criterion.

The MSLC algorithm minimizes the residue power without regard to the gain of the adapted antenna to the target direction. This can cause poor performance. It is often desirable to maximize the output SINR rather than simply minimizing the output residue power and ignoring the effect upon antenna gain in the target's direction. It can be shown that MSLC power minimization is a special case of SINR maximization, so the performances of the two criteria are related. The optimum weights for SINR maximization is developed and evaluated in the next chapter. The LMS, NLMS, and SMI algorithms are also used to estimate the weights for this criterion. The performance evaluation for SINR maximization was developed by Reed, Mallet, and Brennan[5] and is presented here as representative of the algorithm's performance.

A major difference between SMI and the iterative algorithms is how the performance depends upon environmental conditions. As previously discussed, the convergence time and weight noise effects of the LMS and NLMS algorithms depend strongly upon the number, powers, and spatial locations of the jammers. However, the convergence time of the SMI algorithm is independent of the jammer environment. Specifically, Reed, Mallet, and Brennan showed that the convergence time is the same for any covariance matrix. With the optimum weights, the attained SINR is maximized and denoted as SINR_{OPT}. With the estimated weights, the attained SINR is denoted as SINR_{SMI}. For any particular set of voltage measurements the value of SINR_{SMI} is random. Define η as the ratio of SINR_{SMI} to SINR_{OPT}

$$\eta = \text{SINR}_{\text{SMI}}/\text{SINR}_{\text{OPT}}$$

Thus, η is the signal processing loss when using the algorithm. Note that η is a

5. Reed, I. S., J. D. Mallet, and L. E. Brennan, "Rapid Convergence Rate in Adaptive Arrays," *IEEE Trans on Aerospace and Electronic Systems*, Vol. AES-10, Nov. 1974, pp. 853–863.

random variable. Its probability density function is given by

$$p(\eta) = [K!/(N-2)!(K+1-N)!](1-\eta)^{N-2}\eta^{K+1-N}$$

and note that the density function does not depend upon the covariance matrix. From the density function the average loss can be evaluated as

$$E(\eta) = (K+2-N)/(K+1)$$

When K is much greater than N this can be approximated as

$$E(\eta) = (K-N)/K$$

From the last equation, the average loss is 3 dB when K equals $2N$. The loss is reduced to 1.25 dB when K equals $4N$.

9

Signal-to-Interference Plus Noise Ratio Maximization

As previously mentioned, the MSLC optimality criterion minimizes the residue power without constraining the gain in the target direction. This criterion is usually satisfactory if the main antenna gain is much greater than that of the perturbation beam. However, a theoretically superior criterion is to maximize the output SINR. When the statistics of the interference are Gaussian, it can be shown that maximizing the SINR maximizes the target detectability.

The SINR criterion was used to derive optimal processors for three different processing areas. The first two are adaptive arrays and adaptive Doppler processors. In addition, two forms of combined array and Doppler adaptivity can be configured. One is a cascade of an adaptive array and an adaptive Doppler; another is simultaneous adaptation in both domains. The first three configurations are sketched in Figure 9.1. The top configuration is a conceptual block diagram of the adaptive array, the second is the adaptive Doppler, and the third is the cascade of these two processors. The figure indicates that the array has N array elements and therefore uses N adaptive weights. The Doppler filter processes M pulses and therefore uses M adaptive weights. The cascade of the two techniques uses $N + M$ adaptive weights. Figure 9.2 shows simultaneous adaptation in both domains. It uses NM adaptive weights.

The bottom of Figure 9.2 indicates the adaptive response of the two techniques for a single jammer and narrow bandwidth clutter. The darkened portion of either figure depicts a region of approximately zero gain. The lower left portion of the figure is the response for the cascaded processors. It shows a null at the jammer angle for all Dopplers and a null at the clutter Doppler frequency for all spatial angles. The two circles represent locations of targets. Neither of

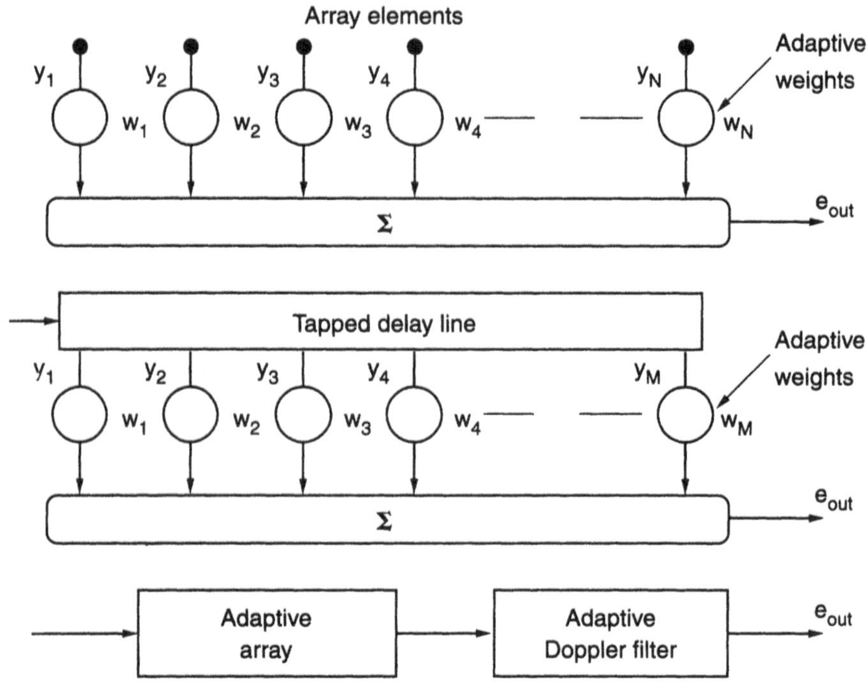

Figure 9.1 Adaptive arrays and Doppler processing.

the two targets is detectable because they lie within a zero response region. The lower right portion of the figure indicates the null response achieved by simultaneous adaptation. The null region is much smaller and it can be seen that neither target is in a null region. Thus, the simultaneous adaptivity technique has potentially superior target detection capability. However, the improved performance is achieved by substantially increasing the number of adaptive weights and therefore the implementation cost.

9.1 SINR Weight Maximization Derivation

The optimum weights for the MSLC residue power minimization criterion were derived in terms of the correlations of the antenna voltages due to the interference.[1] The optimum weights that maximize the output SINR are in terms of the correlations of the antenna voltages due to the interference and also in terms of the voltages induced by the target. Thus, referring to the upper

1. Interference refers to the additive sum of thermal noise and jammers.

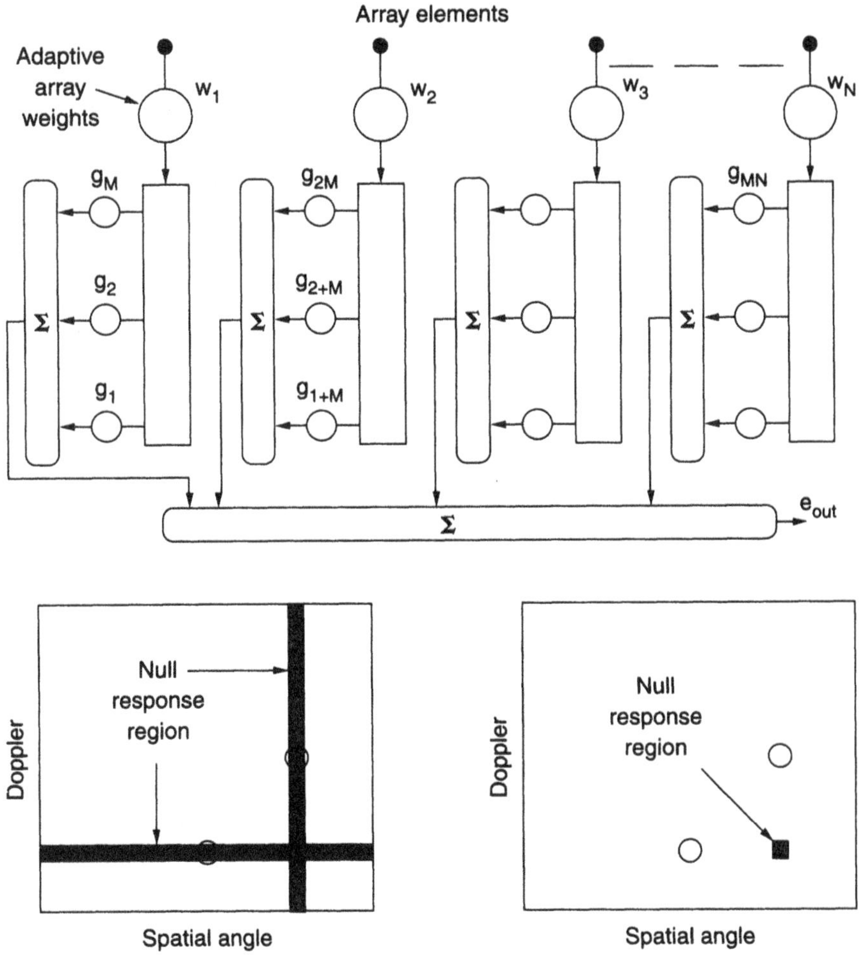

Figure 9.2 Simultaneous two domain adaptivity.

portion of Figure 9.1, the voltage induced on the nth array element is y_n. Since each array element is weighted by the gain w_n, the output voltage is[2]

$$e_{\mathrm{OUT}} = \sum_{n=1}^{N} w_n^* y_n \tag{9.1}$$

2. The standard notation is confusing because it uses the symbol w for both the residue power minimization criterion and the SINR maximization criterion. The values of the weight are different and are derived by different equations.

For the target absent case, the output power due to the interference is

$$P_{\mathrm{OUT}} = E\left|\sum_{n=1}^{N} w_n^* y_n\right|^2 \tag{9.2}$$

Denote the correlation between the nth and mth array element voltages as r_{nm}, that is,

$$r_{nm} = E\left(y_n y_m^*\right)$$

Therefore, (9.2) equals

$$P_{\mathrm{OUT}} = \sum_{n,m=1}^{N}\sum w_n^* r_{nm} w_m \tag{9.3}$$

Denote the portion of y_n due to the target reflection as s_n. The output voltage due to the target equals

$$s_{\mathrm{OUT}} = \sum_{n=1}^{N} w_n^* s_n \tag{9.4}$$

The output SINR equals

$$\mathrm{SINR} = |s_{\mathrm{OUT}}|^2 / P_{\mathrm{OUT}}$$

which may be rewritten in matrix notation as

$$\mathrm{SINR} = |W^{\mathrm{H}} S|^2 / W^{\mathrm{H}} R W = W^{\mathrm{H}} S S^{\mathrm{H}} W / W^{\mathrm{H}} R W \tag{9.5}$$

where the nth entry of S and W are s_n and w_n, respectively. The n, mth entry of the covariance matrix R is r_{nm}. The vector S is denoted as the steering vector since if W equals S, as indicated in (3.26) and reiterated in Section 9.5, this results in a beam that is directed toward the target angle.

One method of finding the W that maximizes the output SINR is to use the Lagrange multiplier technique. Specifically, note that the SINR is independent of the magnitude of W. That is, the SINR value for any particular vector W_1 is the same for any weight vector $\alpha\, W_1$ for any nonzero value of α. Then constrain W to the set such that $W^{\mathrm{H}} R W$ equals one. The Lagrange multiplier technique then finds the W in the set that maximizes $W^{\mathrm{H}} S S^{\mathrm{H}} W$. For the Lagrange multiplier λ, the technique derives W and λ that maximize F, defined as

$$F = W^{\mathrm{H}} S S^{\mathrm{H}} W - \lambda(W^{\mathrm{H}} R W - 1) \tag{9.6}$$

Differentiating with respect to W^{H} and setting the derivative equal to zero yields

$$SS^{H}W - \lambda RW = 0 \tag{9.7}$$

Premultipling by W^{H} and noting that $W^{H}RW = 1$ gives

$$W^{H}SS^{H}W = \lambda$$

and substituting this back into (9.7), recognizing that $S^{H}W$ is a scalar common to both sides, gives the equation for the optimum weight vector that maximizes the output SINR

$$RW_{\text{OPT}} = gS \tag{9.8}$$

for any nonzero value of g.

An alternate derivation is based on matrix properties of the covariance matrix. Note that all covariance matrices are Hermitian. That is, the nth row mth column entry, r_{nm}, is related to the mth row nth column entry, r_{mn}, as

$$r_{nm} = r^{*}_{mn} \tag{9.9}$$

It can be shown that any nonsingular covariance matrix has a Hermitian square root

$$R^{1/2}R^{1/2} = R = (R^{1/2})^{H}R^{1/2} \tag{9.10a}$$

and a Hermitian inverse square root such that

$$R^{1/2}R^{-1/2} = I = (R^{1/2})^{H}R^{-1/2} \tag{9.10b}$$

where I is the identity matrix. Using (9.9) and (9.10), (9.5) can be rewritten as

$$\begin{aligned} \text{SINR} &= |W^{H}S|^{2}/\,W^{H}RW \\ &= |W^{H}(R^{1/2})^{H}R^{-1/2}S|^{2}/\,W^{H}(R^{1/2})^{H}R^{1/2}W \end{aligned} \tag{9.11}$$

Equation (9.11) can be rewritten by defining the vector V as

$$\begin{aligned} V &= R^{1/2}W \\ \text{SINR} &= V^{H}R^{-1/2}S/\,V^{H}V \end{aligned} \tag{9.12}$$

This can be recognized as a form of the Schwarz inequality. The vector V that maximizes the SINR is

$$V_{\text{OPT}} = gR^{-1/2}S$$

for any nonzero g; combining this with (9.12) gives

$$R^{1/2} W_{\mathrm{OPT}} = R^{-1/2} S$$

which is essentially identical to (9.8).

9.2 Array Covariance Matrix Computation

The voltage on the nth array element, $y_n(t)$, for J jammers is given by an equation similar to (7.7) as

$$\begin{gathered} y_n(t) = n_n(t) + \sum_{k=1}^{J} g_n(\psi_k)\, j_k(t - \tau_{nk}) \exp[-j\phi_{nk}] \\ \tau_{nk} = d_n \sin(\psi_k)/c \qquad \phi_{nk} = 2\pi f_0 \tau_{nk} \end{gathered} \tag{9.13}$$

where $n_n(t)$ is the thermal noise on the nth array element, d_n is the spacing between the first and the nth elements, and $g_n(\psi_k)$ is the gain of the nth element in the direction of the kth jammer. The elements of the covariance matrix equal

$$r_{nm} = E[y_n(t) y_m(t)^*]$$

and for statistically independent thermal noise and jammers, this evaluates to

$$\begin{aligned} r_{nm} = {} & P_n \delta(n - m) + \sum_{k=1}^{J} |g_n(\psi_k)|^2 \exp[-j2\pi(d_n - d_m) f_0 \sin(\psi_k)/c] \\ & \times P_{Jk}\, \rho(\tau_{nk} - \tau_{mk}) \end{aligned} \tag{9.14}$$

which is very similar to (7.10).

Similarly, the components of the vector S are

$$s_n(t) = g_n(\psi_{\mathrm{T}}) s(t - \tau_{n\mathrm{T}}) \exp[-j\phi_{n\mathrm{T}}] \tag{9.15}$$

where the subscript T denotes target and the complex envelope of the transmitted waveform is a scalar multiple of $s(t)$. When the product of the radar bandwidth and the maximum value of $\tau_{n\mathrm{T}}$ is much less than one, the target reflection dispersion is negligible and (9.15) is approximated as

$$s_n(t) = g_n(\psi_{\mathrm{T}}) s(t) \exp[-j\phi_{n\mathrm{T}}] \tag{9.15a}$$

In addition, mutual coupling of the array elements is neglected, so the evaluations are restricted to the special case of identical unit gain, omnidirectional, equispaced elements with spacing of d. Then (9.14) and (9.15a) further simplify to

$$r_{nm} = P_n \delta(n-m) + \sum_{k=1}^{J} \exp[-j2\pi(n-m)(d/\lambda)\sin(\psi_k)] \times P_{Jk}\,\rho(\tau_{nk} - \tau_{mk}) \qquad (9.14b)$$

$$s_n(t) = \sigma(t)\exp[-j2\pi(n-1)(d/\lambda)\sin(\psi_T)] \qquad (9.15b)$$

9.3 Optimum Weighting of a Doppler Processor

Note that the preceding optimum weights derivation was in terms of the voltage set $\{y_n\}$ and the steering vector S. From (9.15b) the steering vector components have a constant phase shift differential between array elements. Also note that the Doppler processor shown as the middle portion of Figure 9.1 denotes the tap voltages as y_n. An equivalent to the steering vector phase shift differential is the tap-to-tap phase progression for the target Doppler. Thus, for a target Doppler frequency f_D and an interpulse period of T, the signal at the nth tap is

$$s_n = \exp[-j2\pi(n-1) f_D T] \qquad (9.16)$$

which is completely analogous to (9.15b). Thus, adaptive arrays and adaptive Doppler processors are mathematically equivalent and the optimum weights equation, (9.1), is applicable for either an array or a Doppler processor. The only difference is in determining the form of the covariance matrix and steering vector.

For either type of processor, the nth observed voltage is of the form

$$y_n(t) = n_n(t) + s_T \exp[j(n-1)\phi_T] + c_n(t)\exp[j(n-1)\phi_I] \qquad (9.17)$$

where $n_n(t)$ denotes the thermal noise, s_T denotes the target voltage, and $c_n(t)$ denotes the interference (either jamming or clutter) voltage. The steering vector, S, is

$$S^T = [1, \exp(j\phi_T), \exp(j2\phi_T), \ldots, \exp(j(N-1)\phi_T)] \qquad (9.18a)$$

It is termed the steering vector since if the steering weight vector W is of the form of (9.18a)

$$W^{\mathrm{T}} = [1, \exp(j\phi_{\mathrm{S}}), \exp(j2\phi_{\mathrm{S}}), \ldots, \exp(j(N-1)\phi_{\mathrm{S}})] \tag{9.18b}$$

from (9.1) the target reflection output is

$$e_{\mathrm{OUT}} = \sum_{n=1} w_n^* y_n = \sum \exp[-j(\phi_{\mathrm{T}} - \phi_{\mathrm{S}})(n-1)]$$

When ϕ_{T} equals ϕ_{S}, the beam is steered to the target's angle/Doppler and the output voltage is the maximum possible value of N.

9.3.1 Doppler Covariance Matrix Computation

The tap voltages are the sum of clutter and noise. Since the clutter and noise are statistically independent, the covariance matrix is the sum of a covariance matrix due to noise alone, R_{n}, and one due to clutter, R_{C}. Thus,

$$R = R_{\mathrm{n}} + R_{\mathrm{C}} \tag{9.19}$$

Since the noise is statistically independent from tap to tap,

$$R_{\mathrm{n}} = P_{\mathrm{noise}} I \tag{9.20}$$

where P_{noise} is the noise power and I is the identity matrix.

The entries of the clutter covariance matrix can be developed as follows. Denote the voltage at the delay line input of Figure 9.1 due to clutter as $c(t)$. The voltage at the nth delay line tap equals $c[t-(n-1)\mathrm{T}]$. A particular value of t denotes a particular range cell and the set of voltages at the delay taps corresponds to the set of returns for the particular range cell due to the returns of the N-pulse transmission. The nth row, mth column entry of R_{c} is

$$r_{nm}^{(\mathrm{c})} = E\{c[t-(n-1)T]c^*[t-(m-1)T]\} = r_{nm}^{(\mathrm{c})}[(n-m)T] \tag{9.21}$$

where $r_{nm}^{(\mathrm{c})}$ denotes the clutter's autocovariance function, which equals the Fourier transform of the clutter's power density spectrum. In general, the spectrum varies with range (the time t). As a simple example, the cross section of clutter varies with range, making the magnitude range dependent. In addition, for weather clutter, due to wind speed changing with altitude, the clutter bandwidth varies with range.

In order to unambiguously evaluate (9.21), it is necessary to clarify the role of the variable t since there are two entirely different time variables involved.

One is a "delay time," which really means range. The second is the variation of voltage with time at a specified range. Thus, the mathematical model is that at each range, the return due to clutter is a stationary random process. The power of the random process varies with range because the clutter cross section varies with range. The spectral shape and, in particular, the bandwidth of the random process can also vary with range. For some cases, the power density spectrum of weather clutter is well approximated by a rectangular spectrum of total width, B_c, centered at the Doppler frequency of f_c and of total power, P_c. Thus, the clutter's power density spectrum, $P(f)$, is

$$P(f) = (P_c/B_c)\text{Rect}[(f - f_c)/B_c] \tag{9.22}$$

and the covariance function can be evaluated as

$$r_c(\tau) = P_c \exp(j2\pi f_c \tau) \sin(\pi B_c \tau)/(\pi B_c \tau) \tag{9.23}$$

Combining this with (9.19) gives

$$r_{\text{nm}}(c) = P_c \exp[j2\pi f_c T(m - n)] \sin[\pi B_c T(m - n)]/[\pi B_c T(m - n)] \tag{9.24a}$$

An alternate assumption is that the clutter power density spectrum is Gaussian. For this assumption

$$r_{nm}(c) = P_c \exp[j2\pi f_c T(m - n)] \exp\{-2[\pi B_c T(m - n)]^2\} \tag{9.24b}$$

9.3.2 Optimum Array/Doppler Examples

For a thermal noise environment, P_c equals zero, so the covariance matrix reduces to

$$R = P_{\text{noise}} I \qquad R^{-1} = P_{\text{noise}}^{-1} I$$

and the optimum weight vector equals

$$W_{\text{OPT}} = g R^{-1} S = (g/P_{\text{noise}}) S \tag{9.25}$$

Thus, from (9.18a) and (9.18b), the beam is steered to the target's angle/Doppler. This is the "classical" matched filter discussed previously.

Another example is for a thermal noise plus a single interferer environment of power, P_I. For the adaptive array application it is a jammer at a spatial angle ψ_I; for the adaptive Doppler it is clutter at a Doppler frequency f_I. Both cases induce the differential phase shift of ϕ_I in (9.23). The nth row, mth column entry of R is

$$r_{nm} = P_{\text{noise}}\delta(n-m) + P_I \exp[j(n-m)\phi_I]$$

Then R equals

$$R = P_{\text{noise}} I + P_I S_I S_I^{H} \tag{9.26}$$

where the nth entry of the vector S_I, the interference's angle/Doppler steering direction, equals $\exp[j(n-1)\phi_I]$. The target's steering vector is unchanged and is given by (9.24a). Direct substitution verifies that the inverse of R is given by

$$\begin{aligned} R^{-1} &= P_{\text{noise}}^{-1}\{I - [p/(1+pN)]\, S_I S_I^{H}\} \\ p &= P_I / P_{\text{noise}} \end{aligned} \tag{9.27}$$

so

$$W_{\text{OPT}} = g\, P_{\text{noise}}^{-1}\{S - [(S_I^{H} S)\, p/(1+pN)]\, S_I\} \tag{9.28}$$

Thus, the optimum weight creates a beam pointed at the target minus a beam pointed at the interferer.

To determine the resultant gain to the target direction and to the interferer direction, note that the output voltage equals

$$e_{\text{OUT}} = W_{\text{OPT}}^{H} Y$$

The voltage vector induced by the target is S, so the output voltage for an input from the target direction is

$$\begin{aligned} e_{\text{OUT}} &= W_{\text{OPT}}^{H} S \\ &= g[S^{H}S - [(S_I^{H}S)^{H} p/(1+pN)1]\, S_I^{H} S] \\ &= g[S^{H}S - |S_I^{H}S|^2 p/(1+pN)] \end{aligned} \tag{9.29}$$

Note that for the matched filter of the previous example the sidelobe gain relative to the peak gain, in the direction of the interference is

$$G_I = |S_I^{H}S|^2/(S^{H}S)^2 \tag{9.30}$$

so (9.29) reduces to

$$e_{\text{OUT}} = g(S^{\text{H}}S)[1 + pN(1 - G_{\text{I}})]/[1 + pN] \tag{9.31}$$

For interference absent, $p = 0$, the output voltage equals $S^{\text{H}}S$. Since G_{I} is much less than unity, the output voltage is approximately equal to $S^{\text{H}}S$ even in the presence of the interference.

The gain in the direction of the interference is computed in a similar manner as

$$e_{\text{OUT}} = W_{\text{OPT}}^{\text{H}} S_{\text{I}}$$

and results in

$$e_{\text{OUT}} = g S^{\text{H}} S_{\text{I}}/(1 + pN)$$

Thus, the power gain to the interferer is reduced by the factor $(1 + pN)^2$ as compared to its gain when using the matched filter weights of (9.25).

9.3.3 Multiple-Jammers/Clutter Frequencies

The steering vector S does not depend on the number of interference sources. The only change due to multiple sources is that the covariance matrix changes. For J statistically independent interference sources, the covariance matrix given by (9.26) generalizes to

$$R = I + \sum_{m=1}^{J} P_m S_m S_m^{\text{H}}$$

Though it was shown that the single source narrowband case causes the optimum weight vector to develop a beam directed at the target minus a beam directed at the interferer, it is difficult to develop similar physical insight for the multiple-source case. The optimum weight vector does not develop beams directed at the target and interferers. Also, the nonzero bandwidth case causes additional interpretation difficulties. It is necessary to analyze several special cases in order to develop a physical interpretation of the performance.

An example of the attained performance for multiple interferers is given by Figure 9.3. The antenna is a uniformly spaced 16-element line array, and each element is omnidirectional. The optimum weight set is computed from (9.8), (9.14b), and (9.15b). The array pattern is computed using (9.4) combined with (9.15b). The pattern is periodic and the horizontal axis range equals the

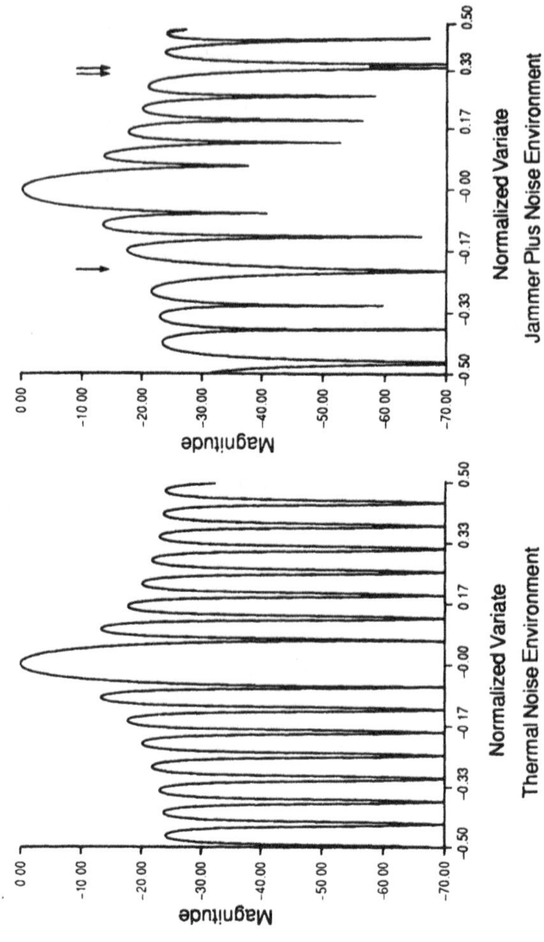

Figure 9.3 Unadapted and adapted responses of array/Doppler processor.

period. When the thermal noise is statistically independent on each element and there is no jamming, as discussed in the first example of Section 9.3.2, the covariance matrix is proportional to the identity matrix. The optimum weight equals the steering vector, and the resulting pattern is as shown on the left side of the figure. The adapted pattern on the right side is for the case of three jammers. Each jammer induces an array element power such that the SNR equals 50 dB. The jammers' angular locations are depicted by the arrows. Note that the pattern has deep nulls at these angles.

The horizontal axis is denoted as a normalized variate to emphasize that the plots are also applicable to Doppler processing. For this interpretation the left-hand plot is the Doppler response of the optimum processor for one filter of the bank when there is no clutter. The right-hand plot is the response of the optimum processor for one filter of the bank when there is clutter at the three Doppler frequencies denoted by the arrows.

9.4 Relative Performance of Several Doppler Filters

The target detection performance of a Doppler filter depends upon the clutter characteristics (total power, center frequency, bandwidth, spectral shape), the number of pulses processed, the target energy, the target fluctuating characteristics, and the effect of the CFAR circuitry. In order to avoid confusion due to the intermingling of effects, the performance results are obtained for the assumption that a perfect lossless CFAR is implemented and the false alarm probability equals $1. \times 10^{-6}$. The results are for eight-pulse processor and a Swerling 1 target with SNR equal to 20 dB. The SNR is the total target energy for the eight pulses. That is, 20 dB corresponds to a numeric value of 100, so the numeric per-pulse SNR equals 12.5. The clutter-plus-noise spectrum is shown in Figure 9.4. The filter is steered to the target's Doppler frequency. The horizontal axis relative to

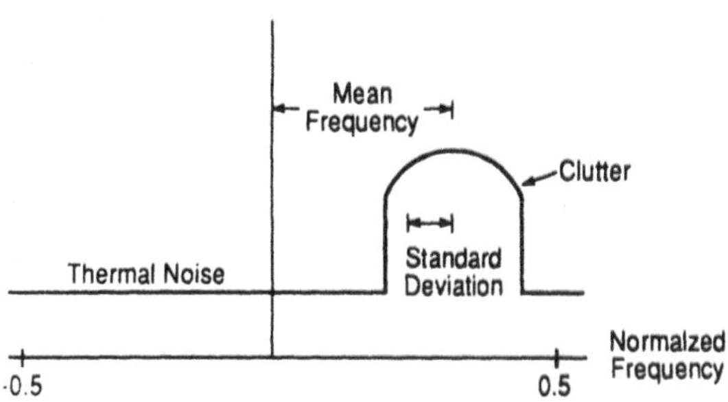

Figure 9.4 Clutter plus noise spectrum.

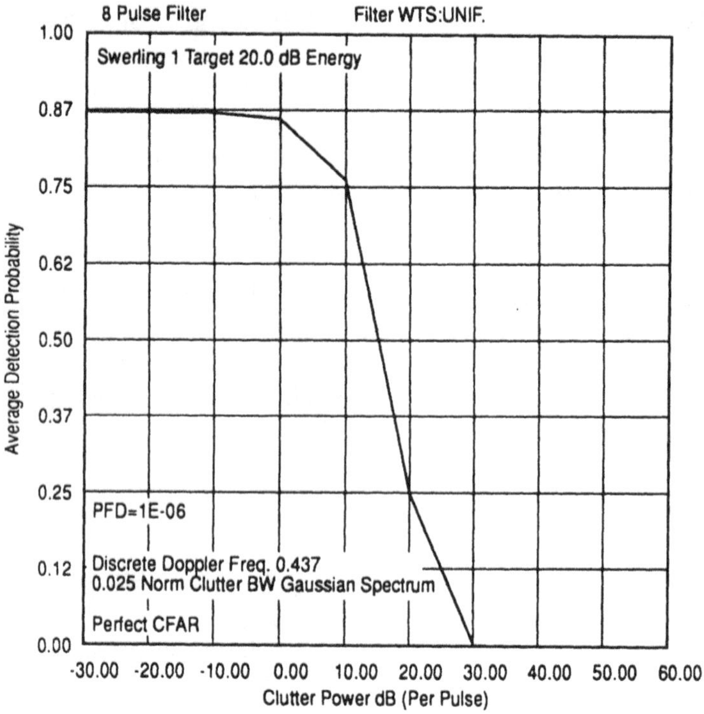

Figure 9.5 Uniform weights.

the steered Doppler is normalized as the ratio of Doppler frequency to the radar PRF. The clutter center frequency is shifted with respect to the target's Doppler. The probability of detection in a noise environment equals 0.872 for a single pulse with 20-dB SNR.

Figure 9.5 shows the probability of target detection in a combined clutter and noise environment when the filter weights are of equal magnitude. Equal-magnitude weights are optimum for a noise-only environment. The horizontal axis is the per-pulse clutter-to-noise ratio (CNR). When the per-pulse CNR equals the extreme left-hand horizontal axis value of −30 dB, this is almost a pure noise environment. It can be seen that for this clutter level, the probability of detection is approximately equal to the maximum possible value of 0.872. It can also be seen that the probability of target detection decreases as the clutter power increases. The probability of detection is computed for a Gaussian-shaped clutter spectrum when the ratio of the clutter bandwidth and the pulse repetition frequency equals 0.025. For the first set of curves, the relative center frequency of the clutter equals 0.4375 times the pulse repetition frequency. Thus, the clutter center frequency is approximately at a sidelobe peak and is far

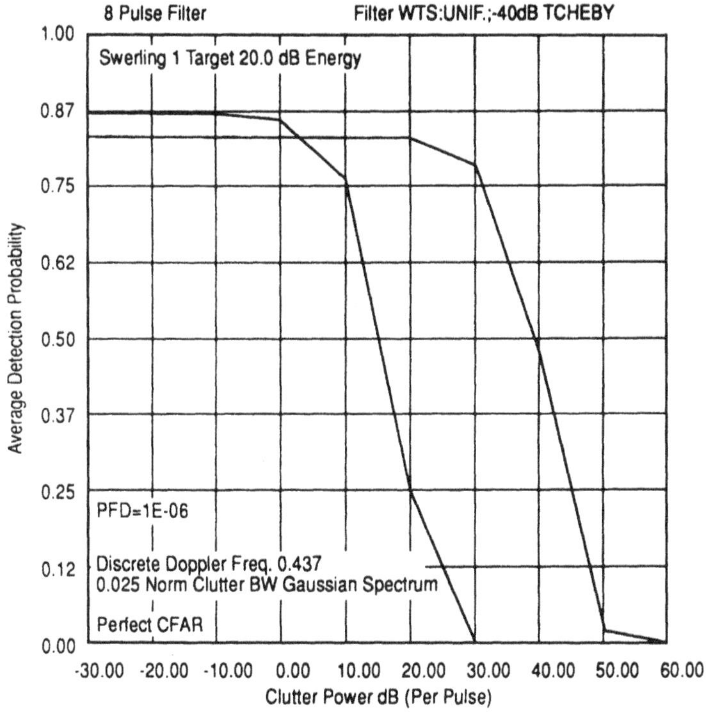

Figure 9.6 Plus Tchebychev weights.

from the mainlobe region. It can be seen that the probability of target detection starts significantly decreasing when the CNR is as small as 0 dB. Figure 9.6 shows the probability of detection for two Doppler filters. One is a repeat of the uniform magnitude weight's filter of the previous figure. The other is for a −40-dB relative sidelobe Tchebychev weighted filter. Note that at the −30-dB CNR, the probability of detection for the Tchebychev filter is less than that of the uniform weight's filter. This demonstrates the well-known signal processing loss associated with a low sidelobe design. The Tchebychev weighted filter has a higher detection probability than the uniform weight's filter when the per-pulse CNR exceeds approximately 0 dB. Figure 9.7 shows the detection probability for the previous two Doppler filters plus the Doppler filter configured using optimum weights. Note that for the low CNR values, the detection probability of the uniform magnitude weight's filter and the optimum filter are identical. This occurs because the optimum weights are the uniform weight set for the low clutter power. Also note that the performance of the optimum weight's filter is much better than that of the other two filters for the high CNR values.

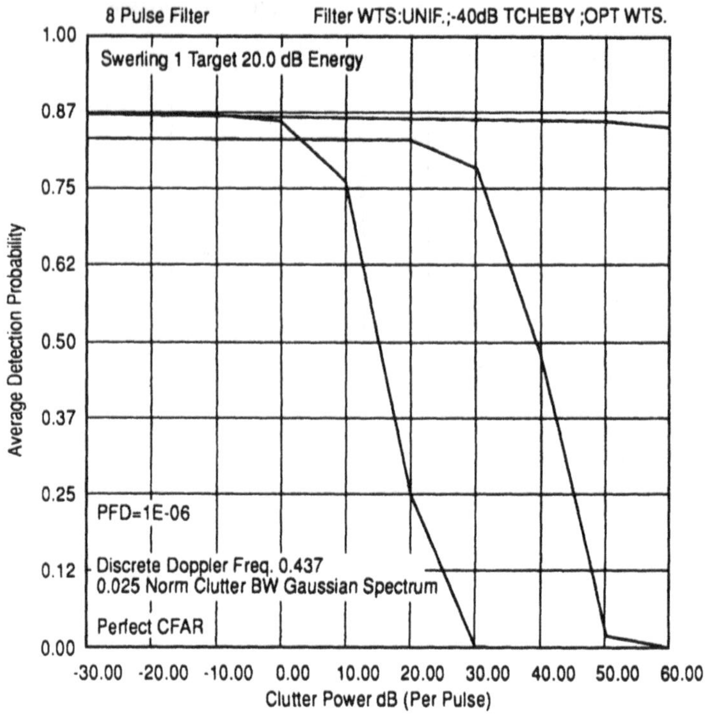

Figure 9.7 Plus optimum weights.

As previously discussed, a defect of Tchebychev weighted filters is that the filter's mainlobe widens as compared to uniform weights, causing difficulty in detecting targets whose Doppler frequency is close to that of the clutter. This defect was not significant for the curves of Figures 9.5 to 9.7 since the clutter center frequency is not close to the mainlobe center frequency. Figure 9.8 shows the probability of detection for the three filters when the clutter's center frequency is close to the mainlobe. For this figure, it equals 0.1875 times the pulse repetition frequency. The curves for the three filters are easily distinguishable. That for Tchebychev weighting has the lowest detection probability at the low CNR value. That for optimum weighting has the highest detection probability at the high CNR value. Comparing Figures 9.7 and 9.8, it can be seen that the detection probability is significantly decreased for all three filters. Also note that there is almost no advantage to using Tchebychev weighting as compared to uniform weighting.

The last case considered is for the condition that the Doppler frequency difference between the target and clutter center frequency is a random variable uniformly distributed over the unambiguous Doppler range. Note that this assumption can cause the clutter spectrum to be within the filter's mainlobe

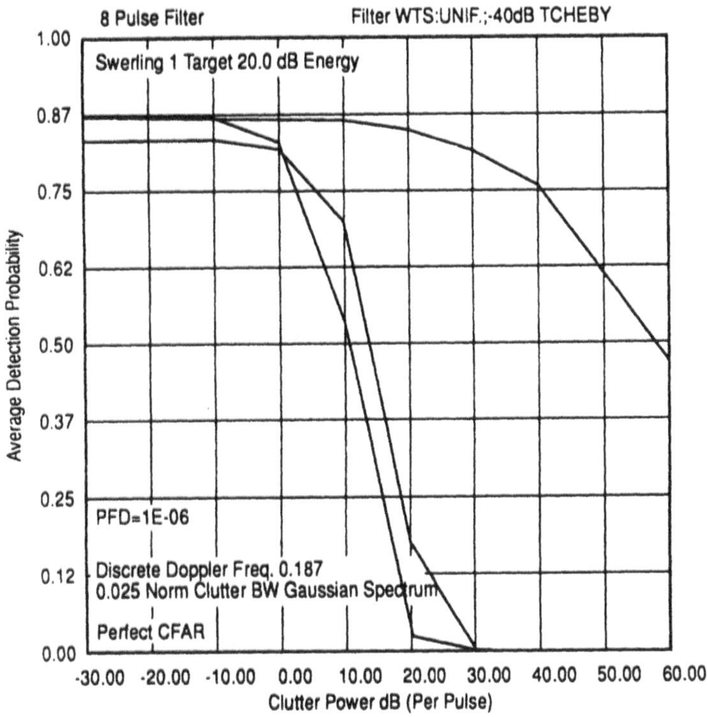

Figure 9.8 Closer clutter spectrum.

region and the increased mainlobe width for Tchebychev weighting to become more significant.

As discussed earlier, high-speed targets and the relatively low PRFs for ground-based search radar cause the target's Doppler frequency to be highly ambiguous, so the uniformly distributed random variable model is appropriate. The method used to evaluate the average detection probability for the random model was to set the clutter center frequency equal to 50 different values uniformly spaced over the PRF for each CNR value. At each of the 50 values, the detection probability was computed for the two deterministic weight sets by computing the residue power and, thus, the filter's output SINR and, thus, the detection probability. The average of the 50 detection probabilities was computed. For the optimum filter, the optimum weight set was computed at each of the 50 frequency values at each CNR value. For each optimum weight set, the resultant output SINR and, thus, the detection probability were computed. Then the 50 detection probability values were averaged. The detection probability for the three filters for this condition is shown by Figure 9.9. Again, optimum weighting is substantially superior to fixed weighting. Figure 9.10 is similar, except that it is for a 16-pulse filter and the SNR value is unchanged

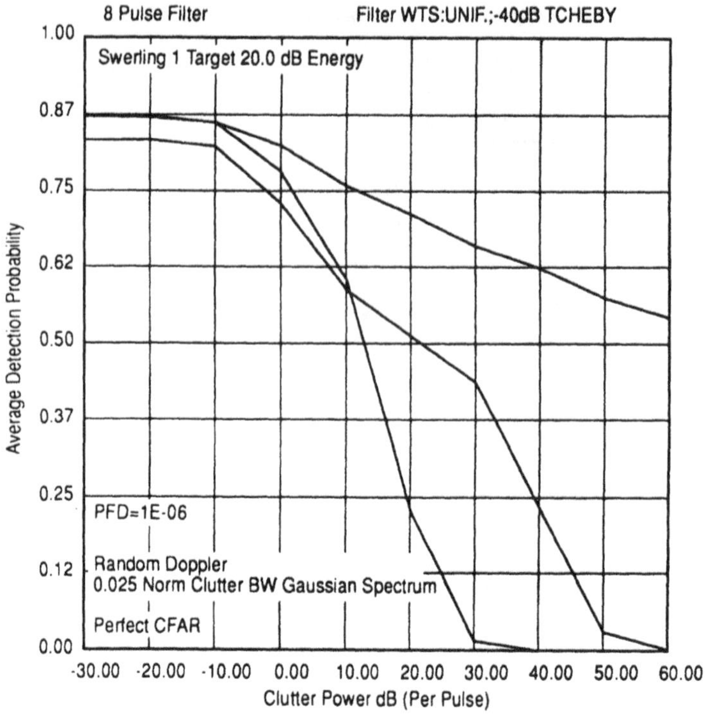

Figure 9.9 Random center frequency clutter spectrum—8 pulses.

at 20 dB. Note that the performance improves significantly at the high CNR values for all three filters as the number of pulses doubles.

9.5 Algorithms

The previously discussed algorithms may be used to estimate the optimum weights. In addition, two special variants of the SMI algorithm can be used. Both involve estimates of the covariance matrix that take advantage of special symmetries in the covariance matrix.

All covariance matrices are Hermitian; that is, they all satisfy the symmetry condition given by (9.9). An example for a 4×4 matrix is

$$R = \begin{pmatrix} a_1 & b^* & c^* & d^* \\ b & a_2 & e^* & f^* \\ c & e & a_3 & g^* \\ d & f & g & a_4 \end{pmatrix} \tag{9.32}$$

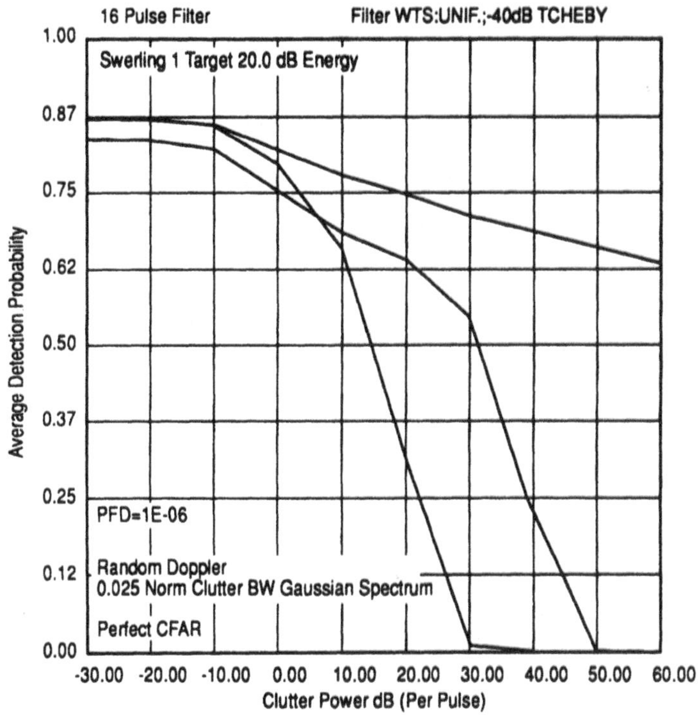

Figure 9.10 Random center frequency clutter spectrum—16 pulses.

The main diagonal entries—a_1, a_2, a_3, and a_4—are real valued. Note that the covariance matrix has conjugate symmetry about the main diagonal as specified by (9.9).

From (9.23) and (9.24), the covariance matrix entries can have a much higher degree of symmetry. The n, mth entries depend only on the difference $n - m$. Thus, the entries of any diagonal are identical and display conjugate symmetry. These matrices are called Hermitian-Toeplitz. For example,

$$R = \begin{pmatrix} r_0 & r_1^* & r_2^* & r_3^* \\ r_1 & r_0 & r_1^* & r_2^* \\ r_2 & r_1 & r_0 & r_1^* \\ r_3 & r_2 & r_2 & r_0 \end{pmatrix} \tag{9.33}$$

Estimators for this class of matrix are associated with power spectrum estimators and are discussed in the next chapter. As is shown, a significant aspect of these estimators is that good estimators can be obtained using as few as a single range cell of data. This contrasts sharply with the estimator of the Hermitian covariance matrix of (8.34), where the number of range cells of data required for good performance is approximately equal to two times the number of array elements (or delay line taps).

The Hermitian-Toeplitz covariance matrix is appropriate for Doppler processors but is not really applicable for arrays because the assumption that all array elements have the same gain is erroneous due to mutual coupling. Thus, (9.17) is modified so that the observed voltage on the nth array element due to interference is

$$y_n(t) = g_n c_n(t) \exp[j(n-1)\phi_I] \tag{9.34}$$

The nth row, mth column entry of R_c is

$$r_{nm}^{(c)} = g_n g_m^* E\{c[t-(n-1)T]c^*[t-(m-1)T]\} \exp[j(n-m)\phi_I] \tag{9.35}$$

and from (9.23) this evaluates to

$$g_n g_m^* r_C[(n-m)T] \exp[j(n-m)\phi_I]$$

As an example, for a four-element array, accounting for element gain modifies (9.33) to

$$R = \begin{pmatrix} g_1 g_1^* r_0 & g_2 g_1^* r_1^* & g_3 g_1^* r_2^* & g_4 g_1^* r_3^* \\ g_1 g_2^* r_1 & g_2 g_2^* r_0 & g_3 g_2^* r_1^* & g_4 g_2^* r_2^* \\ g_1 g_3^* r_2 & g_2 g_3^* r_1 & g_3 g_3^* r_0 & g_4 g_3^* r_1^* \\ g_1 g_4^* r_3 & g_2 g_4^* r_2 & g_3 g_4^* r_2 & g_4 g_4^* r_0 \end{pmatrix} \tag{9.36}$$

However, since the mutual coupling gain modification is caused by the effect of neighboring elements and the array has symmetry when viewed from either end, for this example $g_4 = g_1^*$ and $g_3 = g_2^*$. This symmetry occurs because the linear array or Doppler filter delay line looks the same when viewed from the left to the right or from the right to the left. Note that the conjugate reverse order of voltages is equivalent to defining

$$z_n(t) = y_{N-n+1}^*(t) = g_{N-n+1}^* c^*[t-(N-n+1)T] \exp[-j(N-n+1)\phi_I]$$

and

$$E[z_n(t) z_m^*(t)] = g_n g_m^* r_C[(n-m)T] \exp[j(n-m)\phi_I]$$

Thus, the covariance matrix of the forward data is the same as that of the conjugated backward data. The symmetry may be exploited by modifying the covariance estimate of (8.34) to

$$R_{\mathrm{EST}} = \left[\sum_{k=1}^{K} Y_k Y_k^{\mathrm{H}} + Z_k Z_k^{\mathrm{H}} \right] \Big/ 2K$$

where the nth component of Z is y^*_{N-n+1}. It can be shown that the performance result is approximately identical to that obtained using $2K$ vector samples in the conventional estimator of (8.34).

10

Super-Resolution

Objects are conventionally defined as resolved when the processor output displays two distinct peaks. This definition of resolution ignores the effects of noise on the processed waveform. Always, in the absence of noise and processing errors, proper processing allows objects that are much more closely spaced to be resolved. Thus, in the presence of noise, techniques called super-resolution can resolve objects more closely spaced than are resolvable by conventional techniques. The super-resolution techniques were initially developed and applied in nonradar fields. They were used in geology to aide in oil location. They have been applied to astronomy to more accurately investigate the varying rotation rate of the Earth on its axis. Conceptually, super-resolution techniques are applicable to radar in the domains of range or angle or frequency (Doppler). However, the implementations are expensive and the pay-off differs for the various domains. The techniques are more appropriate for the latter two domains.

10.1 Array Angle Resolution

The conventional array processor is the beam former depicted in Figure 10.1. It is termed a beam former because the weights form a beam directed to the steered spatial angle, θ_S. The voltage on the nth array element, y_n, of a uniformly spaced array due to a target of strength s at a spatial angle of θ_T is

$$y_n = n_n + s\exp[-j2\pi(n-1)(d/\lambda)\sin(\theta_T)] \tag{10.1}$$

where n_n denotes the thermal noise on the array element. For a thermal noise environment, the optimum weights to maximize the probability of detection for

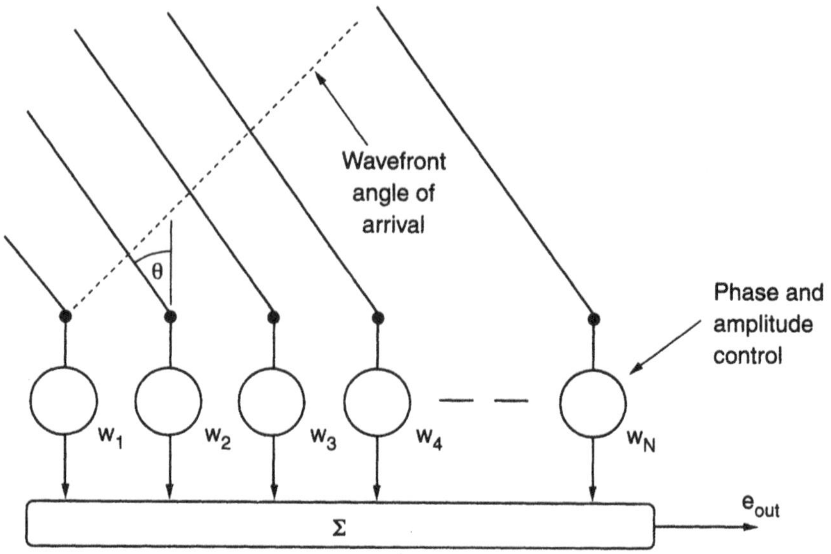

Figure 10.1 Line array processor.

a target at an assumed space angle of θ_S are

$$w_n = \exp[-j2\pi(n-1)(d/\lambda)\sin(\theta_S)] \tag{10.2}$$

The output voltage is

$$e_{OUT} = \sum_n w_n^* y_n = \sum_n w_n^* n_n + s\sum_n \exp[-j2\pi(n-1)(\phi_s - \phi_T)]$$
$$\phi_T = (d/\lambda)\sin(\theta_T) \qquad \phi_s = (d/\lambda)\sin(\theta_S) \tag{10.3}$$

The portion of the output due to target is a geometric series and sums to

$$e_{OUT} = s\{\sin[\pi N(\phi_s - \phi_T)]/[\pi N(\phi_s - \phi_T)]\}$$

The maximum response is at $\phi_s = \phi_T$. The first null of the response is at $\phi_s - \phi_T = 1/N$. The two-sided −3-dB mainlobe width is approximately equal to the distance between the peak angle and the first null, so the spatial angle width equals λ/Nd. This width is the conventional definition of the angle resolution. Equation (10.3) can be recognized as the digital Fourier transform (DFT) of the input. When each array element is processed by an A/D converter, the beam former can be implemented using the Fast Fourier Transform (FFT) algorithm, which is often available as a digital chip.

The super-resolution processor is based on the optimum array processor where the optimum weights are given by the previously derived equation

$$RW = S \tag{10.4}$$

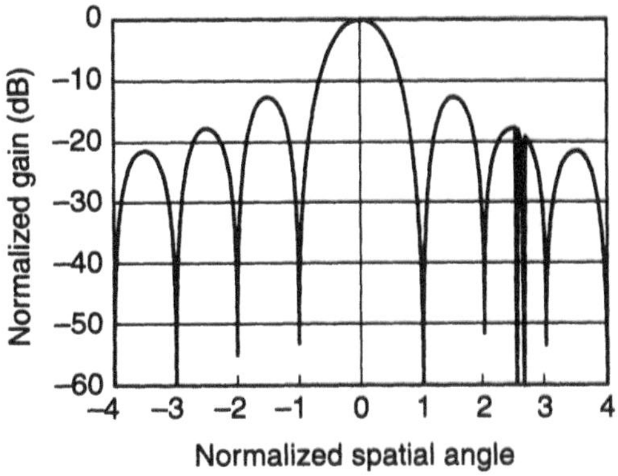

Figure 10.2 Two jammer optimum weights antenna pattern.

where R is the covariance matrix of the thermal noise plus interference and S is the steering vector whose components are identical to the optimum weight for the thermal noise-only environment and is given by (10.2). The optimum weights adapted array antenna pattern for a two-jammer environment is shown in Figure 10.2. Note that the antenna pattern has approximately a null response at the jammers' angular locations and that the jammers' angular spacing is much less than the nominal resolution beamwidth. The reciprocal of the antenna pattern is plotted as Figure 10.3. Note that this pattern shows large peaks at the jammer angles. This pattern can be used to determine the angular locations of the jammers. The jammers are resolved as two distinct sources even though their angular spacing is much less than the nominal resolution cell. The previous discussion of super-resolution in angle is via optimum weights. Since optimum weights are not realizable, the indicated resolution is an upper bound on performance.

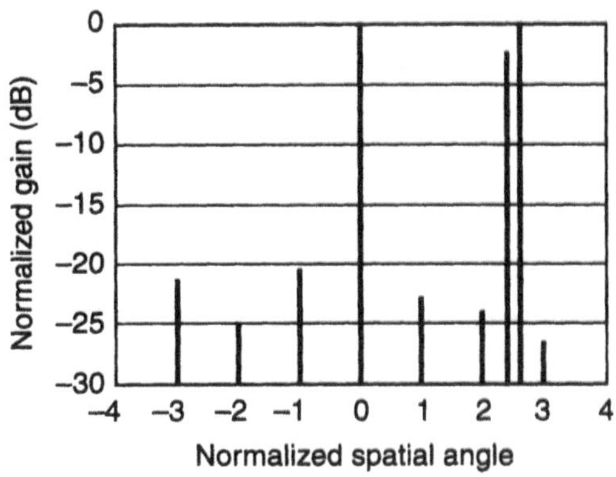

Figure 10.3 Resolved closely spaced jammers.

However, since efficient algorithms that can estimate the optimum weights with small error are known, resolution that is close to that predicted by optimum weight analysis is achievable.

10.2 Frequency Resolution

The classical processor for frequency estimation initially performs a Fourier transform of the data stream exactly like that indicated by (10.3) for the linear array processor. A major difference is that often the number of data points used is hundreds or thousands rather than the tens of points used for the linear array. After the initial Fourier transform, an estimate of the power spectrum is formed as the Fourier transform of the estimated autocorrelation function. Specifically, denote the observed data stream as $y(t)$. The Fourier transform of the data stream, $Y(f)$, is formed as

$$Y(f) = \int y(t) \exp(-j2\pi f t) dt$$

The estimate of the autocorrelation function, $R_{\text{est}}(\tau)$, is formed as

$$R_{\text{est}}(\tau) = w(\tau) \int |Y(f)|^2 \exp(j2\pi f \tau) df$$

where $w(\tau)$ is a weighting function that is used to reduce the estimation error. There are several weighting function choices with slightly different smoothing properties. The estimate of the power spectrum is formed as

$$P_{\text{est}}(f) = \int R_{\text{est}}(\tau) \exp(-j2\pi f \tau) d\tau$$

The resolution is again defined in terms of viewing two distinct peaks and a valley. The frequency resolution, δf, is approximately equal to ten times the reciprocal of the data window time duration.

Modern spectral estimation was conceived in 1969 by J. P. Burg[1] as the maximum entropy method (MEM). The technique is not based on a Fourier transform of the data but instead is based on maximizing a statistical measure of the data. After his conceptual breakthrough, many alternate spectrum estimation techniques were conceived. All are computationally complex. The achieved resolution depends upon the SNR but is often four to ten times better than the classical technique and is usually limited by implementation errors. For arrays, a

1. Burg, J. P., "Maximum Entropy Spectral Analysis," Ph.D Thesis, Stanford University, Stanford, CA, 1975.

fundamental limit on attainable super-resolution is the element mutual coupling and channel-to-channel mismatch. For Doppler frequency, the limitation is due to transmitter pulse-to-pulse amplitude and phase drifts.

Doppler frequency estimation can be used for adaptive Doppler processing or to obtain super-resolution for object identification. The spectral properties and the accuracy requirements are very different. The spectrum for object identification usually consists of a number of isolated lines. A high degree of precision is required to resolve the lines and estimate their frequencies. The spectrum of volume clutter is usually not several lines but looks like a Gaussian or a Rect function. The major spectral estimation requirement is to locate the center of the spectrum and the approximate bandwidth. This chapter only considers the second application. It is shown that the MEM algorithm successfully estimates the required Toeplitz covariance matrix.

10.3 Application to Adaptive Doppler Processing

Adaptive Doppler processing is needed primarily to detect targets in a volume clutter, rain, or chaff environment. The optimum weights, given by (10.4), are in terms of the unknown clutter covariance matrix. As discussed in Section 9.5, there are several algorithms that are candidates for approximating the optimum weights.

One is the LMS, or correlation feedback, algorithm. It is not a good choice for this application due to its slow convergence time. A slow convergence time algorithm is applicable only if the interference environment does not change with range. However, the characteristics of volume clutter will change substantially with range because the radar's antenna beam interrogates higher altitudes as the range increases. Since wind velocity varies with altitude, this means the spectrum of the volume clutter varies with range and so the LMS algorithm does not converge well.

The SMI algorithm is a better choice because its convergence time is shorter. For an N-pulse processor, the number of range cells of data required to achieve small loss as compared to the optimum weights processor is at least equal to $2N$. However, since there can be significant magnitude and spectral width changes, both bandwidth and center frequency, in $2N$ range cells, the attained performance often is significantly less than that of the optimum weights. Also, a significant defect of the SMI algorithm is that it is computationally complicated. It requires of the order of N^3 complex multiply operations per weight vector computation. Since the characteristics of the volume clutter environment change significantly with range, it is necessary to update the weights frequently, and this requires a relatively large special-purpose computer.

10.3.1 Modern Spectral Estimation and Adaptive Doppler Processing

As discussed in Section 9.5, the covariance matrix for an adaptive Doppler property has the special symmetry denoted as Hermitian-Toeplitz. The matrix is completely specified by the first column of the matrix, and this column is the sampled autocorrelation function. Since the autocorrelation function is the Fourier transform of the power density spectrum, spectral estimation techniques can be used to estimate a Hermitian-Toeplitz covariance matrix. Modern spectral estimation techniques, such as the Burg algorithm, are applicable. It can obtain a satisfactory estimate of the optimum weight vector using a single range cell of data. It is also computationally efficient relative to SMI because it requires only of the order of N^2 complex multiply operations per weight vector update rather than N^3.

The observed data stream, $y(n)$, has time spacing equal to the A/D sampling period. Modern spectral estimation techniques are based upon assuming that this data stream is generated by a digital filter. There are three possible types of generating filters. The first candidate is indicated as the upper network of Figure 10.4. The input stream, $z(n)$, is assumed to be white noise sampled at the A/D sampling frequency. The time spacing between taps of the modeling filter's delay line is equal to the A/D sampling period. The output is formed as the sum of weighted, delayed replicas of the output, and the filter output is assumed to be the observed data stream. This network is an all-pole network due to the feedback coefficients set $\{a_{\text{p}}\}$. The middle network of Figure 10.4 is an all-zero network. The network has Q zeros due to the feedforward coefficients set $\{b_{\text{q}}\}$. It is identical in form to a Doppler filter but is considerably different. The first difference is that the time delay between taps for the Doppler filter equals the PRI, whereas it equals the reciprocal (approximately) of the radar bandwidth for the filters of Figure 10.4. A second difference is that the Doppler filter is *implemented*, whereas the circuits of Figure 10.4 are not, but are merely models of how the observed time stream is created. Just as for the all-pole network, the spectral estimation procedure is to estimate the feedforward coefficients from the data and then compute the spectrum. The bottom network of the figure is a cascade of the all-pole and all-zero network. It has both poles and zeros. Again, the spectral estimation procedure is to estimate the feedforward and feedback coefficients from the data and then compute the spectrum. The all-pole network is most often used to model clutter spectra because its performance is satisfactory and its implementation equations are relatively simple.

The equation relating the input and output of an all-pole network is

$$y(n) = z(n) - \sum_{p=1}^{P} a_p y(n-p) \tag{10.5}$$

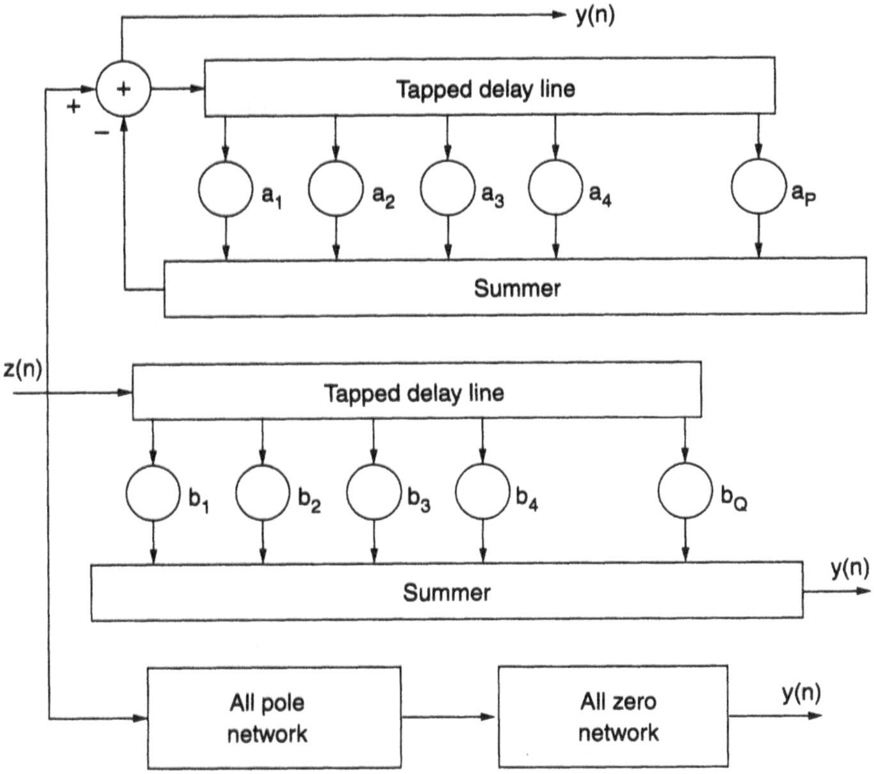

Figure 10.4 Assumed networks for modern spectral estimation.

The Z transform of this equation is

$$Y(z) = Z(z) - Y(z)\sum_{p=1}^{P} a_p z^{-p} \tag{10.6}$$

so the transfer function for the network, $H(z)$, is

$$H(z) = Y(z)/Z(z) = z^P \Big/ \sum a_p z^{P-p} \tag{10.7}$$

The denominator of $H(z)$ is a Pth degree polynomial with P roots. Thus, the Z transform of the filter's transfer function has P poles, due to the P feedback coefficients set $\{a_p\}$, and no zeros. The circuit is termed an all-pole network. The spectrum estimation procedure is to use the data to estimate the feedback coefficients and then to compute the spectrum using (10.7). One estimation procedure for an N-point data stream, with N greater than P, is to form the error

$$\varepsilon = \sum_{n=p+1}^{N} \left| y(n) - \sum_{p=1}^{P} a_p y(n-p) \right|^2 \tag{10.8}$$

and estimate $\{a_p\}$ as the values that minimize the error. From the estimated coefficient set, the estimated spectrum is computed using (10.7), and the estimated sampled autocorrelation function is the DFT of the estimated spectrum. This is then used as the estimate of the first column of the Hermitian-Toeplitz covariance matrix. A possible problem with this technique is that the derived network is not necessarily stable. The Burg algorithm is also based on an all-pole network assumption. However, this algorithm minimizes (10.8) subject to a constraint that guarantees the derived network is stable.

The estimation equations for either the all-zero network or the combined poles and zeros network are highly nonlinear and much more difficult to implement than those for the all-pole network. In addition, it can be shown that the inverse covariance matrix is easily computed from the feedback coefficients for the all-pole network but is computationally difficult to obtain for the other circuits. Therefore, the all-pole network is easier to apply for this application and only all-pole models are considered further.

10.3.2 Predicted Performance

A sketch of a typical clutter-plus-thermal noise spectrum and estimates based on the Burg algorithm are shown in Figure 10.5. The clutter spectrum is an approximation of that received from rain at a 75-nautical mile range by a search radar with a pencil beam. The clutter's nonzero center frequency and spread are due to wind velocity varying with altitude. The CNR equals 50 dB. A 16-pulse optimum weights filter would significantly cancel the clutter, and the optimum weights improvement factor equals 47.8 dB. The difference between the optimum improvement factor of 47.8 dB and the 50-dB CNR indicates that the residue power is approximately 2.2 dB above the thermal noise power level for the optimum filter.

The spectral estimates are based on the observed 16 voltages at the taps of the Doppler filter. Because the observed voltages are random, the estimated spectra are random. The estimates are shown for two sets of random data. Note that the estimate differs considerably from the true spectrum and is not too useful for object identification. However, the center frequency and bandwidth are fairly accurately identified, so the technique is useful in computing filter weights for clutter cancellation. The adaptive weights computed from the estimated spectrum are not equal to the optimum weights, so there is a loss due to spectral estimation errors.

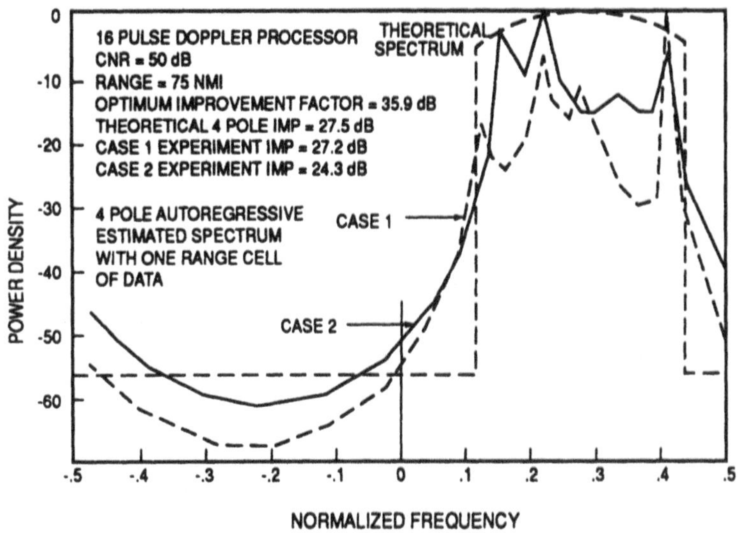

Figure 10.5 Estimated and true clutter spectrum.

There are several components to the error. One is that most spectra cannot be exactly produced by an all-pole filter. This is a modeling mismatch error and is not a function of random data. The estimated spectra of Figure 10.5 are computed using a four-pole model. The normalized improvement factor for a four-pole modeled spectrum and an infinite number of range cells of data equals 39.4 dB as compared to the optimum weights normalized improvement factor of 47.8 dB. The difference between 47.8 dB and 39.4 dB is caused by the mismatch error. The mismatch error increases as the CNR increases, increases as the clutter spectral width increases, and decreases as the number of poles of the modeling network increases. The mismatch error is graphed in Figure 10.6 for the clutter spectrum of Figure 10.5. It can be seen that for a CNR of 65 dB, the mismatch loss equals 20 dB for a four-pole model and 3.5 dB for an eight-pole model. Figure 10.7 shows the mismatch loss as a function of the normalized clutter bandwidth for a 50-dB CNR. Again improved performance is obtained as the number of poles of the modeling network increases. However, the number of complex multiples required to estimate a weight vector increases linearly with the number of poles, so there is a performance-implementation cost trade-off.

10.3.3 Mismatch Loss Computation Equations

The weights of the optimum Doppler processor are computed under the assumption that the spectrum of the clutter plus thermal noise is known exactly. Since the autocorrelation function is the Fourier transform of the spectrum, it is also known exactly. Since the Hermitian-Toeplitz covariance matrix is specified

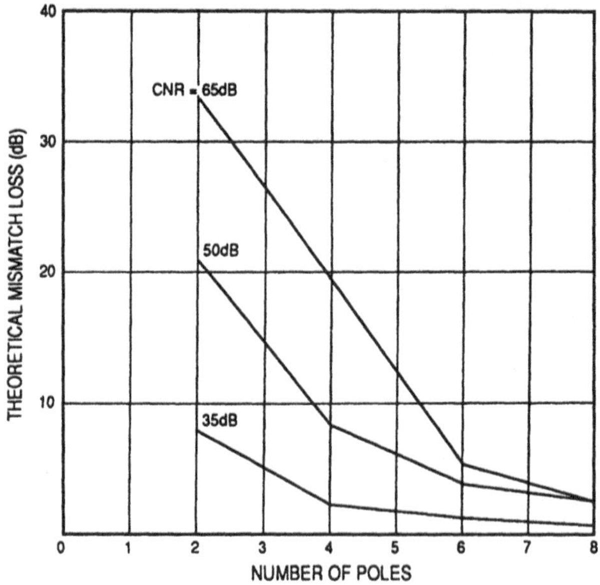

Figure 10.6 Mismatch loss vs. number of poles for various CNR values.

by the known values of the sampled autocorrelation function, the optimum weights are computed from (10.4).

The autocorrelation function for the P-pole all-pole network that approximates the true spectrum is computed using the Yule-Walker equations developed next. The optimum weights based on the all-pole spectral approximation are computed by again using (10.4) with the approximating covariance matrix. The mismatch losses of Figures 10.6 and 10.7 are the difference between the optimum output SINR and the all-pole output SINR.

The Yule-Walker equations relate the feedback coefficients of the unique P-pole all-pole network to a specified $P + 1$ length sampled autocorrelation function. These equations can be developed from (10.5). Thus, multiplying both sides of (10.5) by $y^*(n - k)$, for positive k, and taking expectations yield

$$R(k) + \sum a_p R(k - p) = 0 \tag{10.9}$$

Combining these equations as a matrix yields (10.10)

$$\begin{pmatrix} R(0) & R(-1) & R(-2) & \cdots & R(1-P) & a_1 \\ R(1) & R(0) & R(-1) & \cdots & R(2-P) & a_2 \\ \vdots & & & & & \\ R(P-1) & R(P-2) & R(P-3) & \cdots & R(0) & a_P \end{pmatrix} = \begin{pmatrix} R(1) \\ R(2) \\ \\ R(P) \end{pmatrix} \tag{10.10}$$

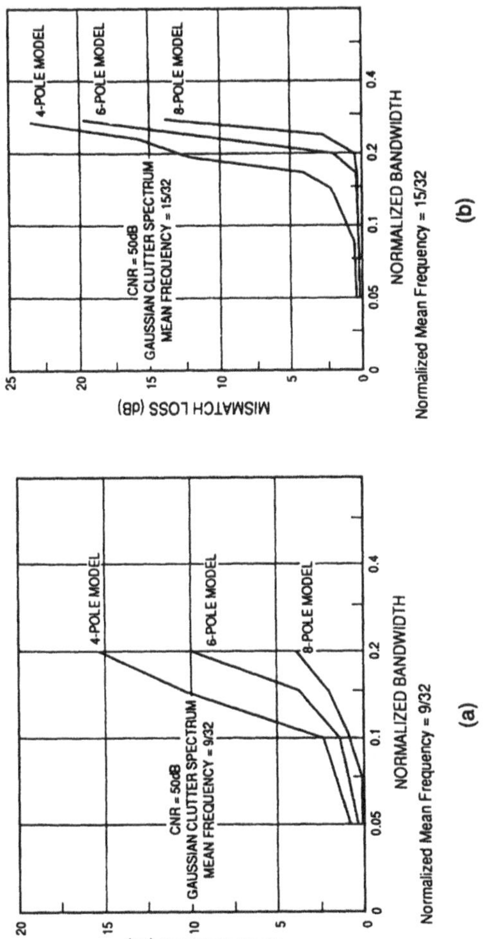

Figure 10.7 Mismatch loss vs. bandwidth for various number of poles.

Equation (10.10) can easily be solved for the set of feedback coefficients $\{a_p\}$. Then (10.9) can be used to compute the autocorrelation values for $R(P+1)$ through $R(N)$ in order to form the N by N Hermitian-Toeplitz matrix and solve for the approximating optimum weights.

10.3.4 Estimation Error Loss

Another error component is the number of range cells of data used to estimate the spectrum. The computational complexity of the algorithm increases linearly with the number of range cells of data. The estimated spectra of Figure 10.5 are based on a single range cell of data. The experimental normalized improvement factor for case 1 equals 39.1 dB, and for case 2 equals 36.2 dB. The difference between these values and the infinite number of range cells value of 39.4 dB is due to random data noise and is termed an estimation loss.

The data in Table 10.1 are a more extensive indication of the mismatch and estimation losses. It tabulates the losses for three different bandwidths and center frequencies corresponding to volume clutter at ranges of 75, 50, and 25 nautical miles, respectively, relative to optimum processing for a CNR equal to 50 dB. It is seen that the mismatch loss decreases monotonically with an increased number of poles for all three cases. However, there is a tendency for the estimation loss to increase with the number of poles used in the model. Thus, the combined loss sometimes increases with the number of poles and sometimes decreases with the number of poles.

Table 10.1
Mismatch and estimation loss

Bandwidth	Poles	Mismatch loss (dB)	Estimation loss (dB)	Combined loss (dB)
0.13	4	8.4	1.4	9.8
	6	4.1	4.4	8.5
	8	2.5	4.7	7.2
0.09	4	1.8	2.6	4.4
	6	1.0	2.0	3.9
	8	0.6	2.4	3.0
0.05	4	0.8	1.0	1.8
	6	0.3	2.0	2.3
	8	0.2	2.7	2.9

10.4 The Burg Algorithm

The set of feedback coefficients determines a power spectrum, $P(f)$, given by

$$P(f) = \left| 1 - \sum_{n=1}^{P} a_{nP} \exp(j2\pi nf) \right|^{-2} \tag{10.11}$$

The coefficients have double subscripts to accommodate the computational procedure. Denote the set of observed data as $y(1), y(2), \ldots, y(N)$. Define two sets, $\{p_1(n)\}$ and $\{q_1(n)\}$, as

$$p_1(n) = y(n) \qquad n = 1, \ldots, N$$
$$q_1(n) = y(n+1) \qquad n = 1, \ldots, N-1$$

Then compute for $M = 1$

$$a_{MM} = 2 \sum_{n=1}^{N-M} p_M^*(n) q_M(n) \Bigg/ \left[\sum_{n=1}^{N-M} |p_M(n)|^2 + |q_M(n)|^2 \right] \tag{10.12}$$

For M > 1, recursive relations to update the p and q sets are

$$p_M(n) = p_{M-1}(n) - a^*_{M-1,M-1}\, q_{M-1}(n)$$
$$q_M(n) = q_{M-1}(n+1) - a_{M-1,M-1}\, p_{M-1}(n+1)$$

and these relations are used in (10.12) to compute successive values of a_{MM} until $M = P$. After these values are computed the other coefficients are computed by

$$a_{M,k} = a_{M-1,k} - a_{MM} a_{M-1,M-k}$$

11

Polarization Rejection of Clutter

A propagating plane wave has an electric (E) field and an orthogonal magnetic (H) field in the plane perpendicular to the direction of propagation. The polarization of a propagating wave is defined in terms of the direction of the E-field. The top sketch of Figure 11.1 shows the sampled instantaneous voltage versus time at a specified test point for vertical polarization. The locus of the voltage vector, shown on the left in Figure 11.2, is a vertical line retraced at the carrier frequency. The bottom sketch of Figure 11.1 shows a conceptual antenna implementation for a dual polarization feed. The feed is composed of a horizontal dipole and a vertical dipole and the transmitter connects to both dipoles in parallel. The relative phase and amplitude of the horizontal and vertical feeds determine the transmit polarization. Figure 11.2 shows examples of the E-field locus for several polarizations. The left-hand example is a vertical polarization. It is generated when the transmitter is connected only to the vertical dipole. Horizontal polarization is generated when the transmitter is connected only to the horizontal dipole. The center sketch is for 45-degree linear polarization. It is generated when both dipoles are fed in phase and with equal magnitude. A negative 45-degree linear polarization is generated when the dipoles are fed with a relative phase shift of 180 degrees and equal magnitude. The right-hand sketch is for right-hand circular (RHC) polarization. It is generated when the dipoles are fed with a relative phase shift of 90 degrees and equal magnitude.

The locus point indicated as (1) corresponds to the time instant when the vertical E-field is a maximum and the horizontal E-field equals zero. The locus point indicated as (2) corresponds to the time instant when the vertical E-field equals zero and the horizontal E-field is a maximum. Thus, the locus is a clockwise circle. Left-hand circular (LHC) polarization is generated using

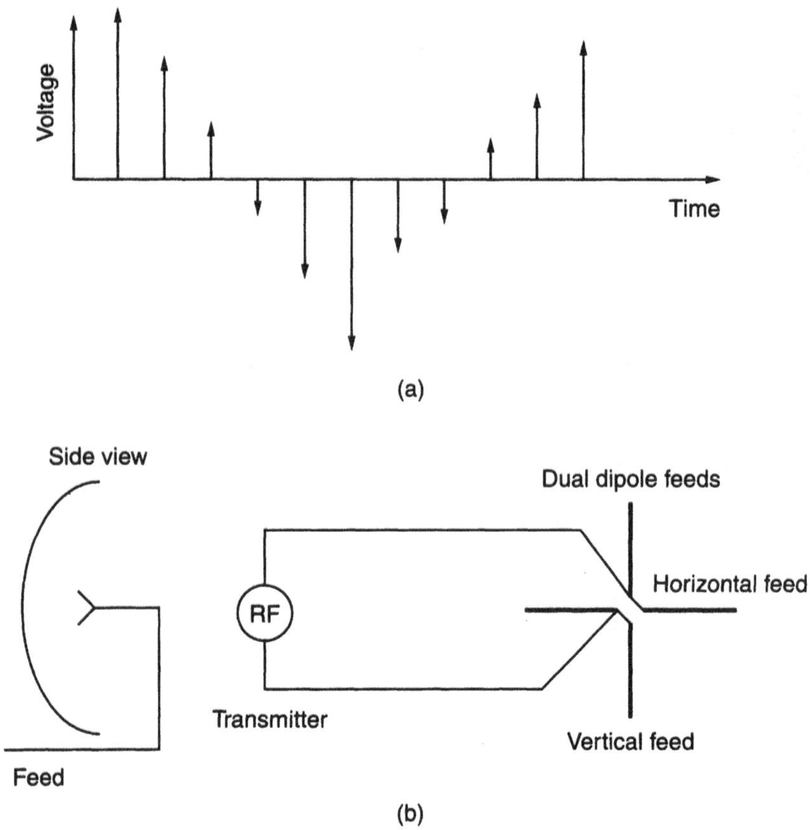

Figure 11.1 Conceptual dual polarization feed structure.

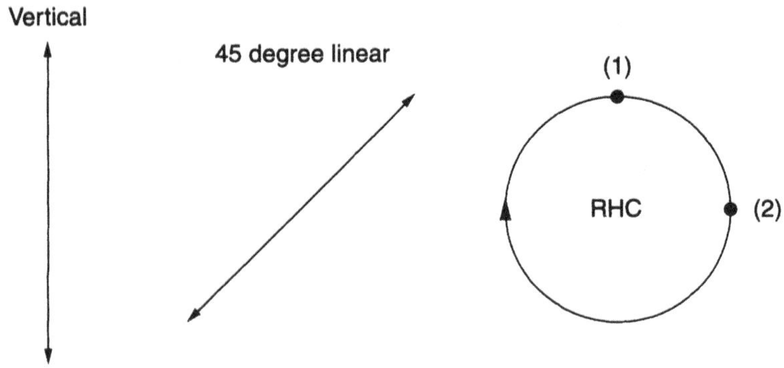

Figure 11.2 Polarization loci.

a relative phase shift of −90 degrees and equal magnitudes. The locus is a counterclockwise circle.

11.1 Reflected Wave Polarization

The polarization properties of a reflected electromagnetic wave are determined by the properties of the reflected E-field. As an example, the E-field at the surface of a perfect conductor equals zero. Thus, the reflected E-field equals the negative of the incident E-field. For vertical polarization, the direct wave has a polarization of V and the reflected wave has a polarization of $-V$. Similarly, for horizontal polarization the direct wave has a polarization of H and the reflected wave has a polarization of $-H$.

The situation is more complicated for circular polarization. The upper left-hand portion of Figure 11.3 indicates the direct vertical and horizontal E-fields and the resultant E-field at a particular time instant. The bottom left indicates the direct fields at a later time instant. The propagation is assumed to be downward and perpendicular to the plane of the figure, termed *into the paper*, and the rotation of the resultant field is clockwise. The upper right portion shows the reflected wave E-field components at the first time instant, and the lower right-hand portion shows the reflected fields at the latter time instant. The reflected wave propagates off the paper. The definition of the sense of rotation differs as defined by physicists and as defined by engineers. In engineering usage the direction of propagation is used in defining the polarization sense. Looking in the direction of propagation of the reflected wave, the rotation is counterclockwise. Thus, the direct wave is RHC and the reflected wave is LHC.

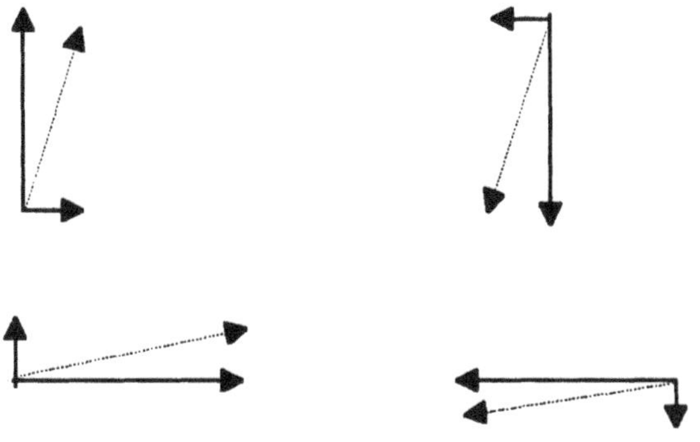

Figure 11.3 Direct and reflected polarization example.

11.1.1 Receive Antenna Polarization Response

Denote the transmitter generator waveform as $\exp(j\omega t)$. For a circular polarization transmission, the voltage inputs to the vertical and horizontal dipoles are

$$V_{\mathrm{T}} = \exp(j\omega t) \qquad H_{\mathrm{T}} = j\exp(j\omega t)$$

For reflection from a perfect conductor, the reflected voltages are proportional to

$$V_{\mathrm{R}} = -\exp(j\omega t) \qquad H_{\mathrm{R}} = -j\exp(j\omega t)$$

The antenna responses to the incident waveform components are

$$V_{\mathrm{A}} = V_{\mathrm{R}} = -\exp(j\omega t) \qquad H_{\mathrm{A}} = jH_{\mathrm{R}} = \exp(j\omega t)$$

where the 90-degree phase shift of H_{A} relative to H_{R} is identical to the 90-degree phase shift on transmission. Since the antenna output voltage is the sum of V_{A} and H_{A}, the output voltage equals zero.

11.1.2 Reflectivity Examples

As another example of the polarization reflection properties of objects, consider a thin wire. If the wire is vertical, it will have maximum reflectivity to vertical polarization and (because of the assumption of zero width) zero reflectivity to horizontal polarization. If the wire is horizontal, it will have maximum reflectivity to horizontal polarization and zero reflectivity to vertical polarization. If the wire is tilted at an angle of θ from the vertical and the incident polarization is vertical, the wire will have a current induced that is proportional to $\cos(\theta)$. The wire will reradiate an E-field that has both a vertical component and a horizontal component. The vertical component is proportional to $\cos^2(\theta)$ and the horizontal component is proportional to $\cos(\theta)\sin(\theta)$. If the incident polarization is horizontal, the wire will have a current induced that is proportional to $\sin(\theta)$. It will reradiate an E-field that has both a vertical component and a horizontal component. The horizontal component is proportional to $\sin^2(\theta)$ and the vertical component is proportional to $\cos(\theta)\sin(\theta)$.

A general transmit polarization has a representation of

$$e_{\mathrm{t}} = a_1 e_{\mathrm{tV}} + a_2 e_{\mathrm{tH}}$$

with a vertical and horizontal transmit polarization components of e_{tV} and e_{tH}, respectively. Pure vertical polarization occurs for $a_2 = 0$, and pure horizontal polarization occurs for $a_1 = 0$. RHC polarization is generated when $a_1 = (2^{1/2})/2$ and $a_2 = j(2^{1/2})/2$; LHC polarization is generated when $a_1 = (2^{1/2})/2$ and

$a_2 = -j(2^{1/2})/2$. For the general transmit polarization, the reflected polarization components are given by

$$e_{\rm rV} = a_1 \cos^2(\theta) + a_2 \cos(\theta)\sin(\theta)$$
$$e_{\rm rH} = a_1 \cos(\theta)\sin(\theta) + a_2 \sin^2(\theta)$$

In matrix notation this is

$$\begin{pmatrix} e_{\rm rV} \\ e_{\rm rH} \end{pmatrix} = \begin{pmatrix} \cos^2(\theta) & \cos(\theta)\sin(\theta) \\ \cos(\theta)\sin(\theta) & \sin^2(\theta) \end{pmatrix} \begin{pmatrix} a_1 \\ a_2 \end{pmatrix} \tag{11.1}$$

For a generalized reflector, the comparable matrix equation is

$$\begin{pmatrix} e_{\rm rV} \\ e_{\rm rH} \end{pmatrix} = \begin{pmatrix} a_{\rm VV} & a_{\rm HV} \\ a_{\rm VH} & a_{\rm HH} \end{pmatrix} \begin{pmatrix} a_1 \\ a_2 \end{pmatrix}$$

which may be rewritten as

$$E_{\rm r} = PA$$

$$P = \begin{pmatrix} a_{\rm VV} & a_{\rm HV} \\ a_{\rm VH} & a_{\rm HH} \end{pmatrix} \qquad A = \begin{pmatrix} a_1 \\ a_2 \end{pmatrix} \qquad E_{\rm r} = \begin{pmatrix} e_{\rm rV} \\ e_{\rm rH} \end{pmatrix}$$

The first subscript of the double subscripts denotes the transmit polarization and the second subscript denotes the receive polarization. Thus, as an example, $a_{\rm HV}$ denotes the vertical component reflection for a horizontal transmit polarization. When the transmit antenna is also used as the receive antenna, the antenna output voltage, v, equals

$$v = e_{\rm rV} a_1 + e_{\rm rH} a_2 = A^{\rm T} E_{\rm r} = A^{\rm T} PA \tag{11.2}$$

For an arbitrary linear transmit polarization, the vector form of the coefficients a_1 and a_2 are given by

$$A = \begin{pmatrix} \cos(\phi) \\ \sin(\phi) \end{pmatrix}$$

For vertical transmit polarization $\phi = 0$, for horizontal polarization $\phi = \pi/2$, for 45-degree linear polarization $\phi = \pi/4$, and for -45-degree polarization $\phi = -\pi/4$. For the wire tilted at an angle of θ from the vertical

$$P = \begin{pmatrix} \cos^2(\theta) & \cos(\theta)\sin(\theta) \\ \cos(\theta)\sin(\theta) & \sin^2(\theta) \end{pmatrix} \tag{11.3}$$

and from (11.2) the antenna output voltage is

$$\begin{aligned} v &= A^{\rm t} PA \\ &= \cos^2(\theta)\cos^2(\phi) + 2\cos(\theta)\sin(\theta)\cos(\phi)\sin(\phi) + \sin^2(\theta)\sin^2(\phi) \end{aligned} \tag{11.4}$$

Using trigonometric identities, a simpler form of (11.4) can be obtained as

$$v = \cos^2(\theta - \phi) \tag{11.4a}$$

As examples, for a vertical wire and horizontal transmit polarization $\phi = 0$ and $\theta = \pi/2$, so the output voltage evaluates to zero; for a horizontal wire and vertical transmit polarization $\phi = \pi/2$ and $\theta = 0$, so the output voltage evaluates to zero. When the angle of the linear transmit polarization equals the tilt angle of the wire, the output voltage evaluates to unity.

11.2 Circular Polarization Clutter Cancellation

For RHC transmission and reflection from a sphere

$$A = (2^{1/2})/2 \begin{pmatrix} 1 \\ j \end{pmatrix} \qquad P = \begin{pmatrix} 1 & 0 \\ 0 & 1 \end{pmatrix} \tag{11.5}$$

and (11.2) the output voltage evaluates to zero. Assume a target's reflection characteristic is like a thin wire. From (11.3) the output voltage for circular transmit and receive polarization equals

$$v = [\cos^2(\theta) + 2j\cos(\theta)\sin(\theta) - \sin^2(\theta)]$$

so the antenna output voltage does not equal zero. This example of total cancellation of the reflection from a spherical reflector when using circular polarization antennas but noncancellation of the reflection from a wire demonstrates the potential improvement obtainable when detecting targets in rain.

11.2.1 Error Effects on Cancellation

The preceding example of the complete cancellation of the reflection from rain and the partial cancellation of the reflection from the wire implies the potential of a very large improvement in the output CNR using polarization clutter cancellation. The example is overly optimistic for several reasons. First, as indicated in Section 3.3 and Figure 3.1, the transmitted polarization of a reflector antenna is not pure but contains significant cross-polarization components. An array antenna also does not transmit a pure polarization because the dipole feeds transmit a cross-polarized component. Additionally, a raindrop is not a perfect sphere, so it does not reflect LHC (RHC) polarization even if the incident polarization is pure RHC (LHC) polarization.

The quantitative effect of these errors can be approximated by assuming the average errored circular transmit/receive polarization is represented by

$$A = \begin{pmatrix} 1 \\ j(1-\varepsilon) \end{pmatrix} \Big/ [1 + (1-\varepsilon)^2]^{1/2}$$

and the reflection matrix for the raindrop is that of an ellipsoid with relative axes dimensions of 1 and $1 - \delta$. Thus,

$$P = \begin{pmatrix} 1 & 0 \\ 0 & 1-\delta \end{pmatrix}$$

so

$$v = A^{t} P A = [1 - (1-\varepsilon)^2(1-\delta)]/[1 + (1-\varepsilon)^2]$$

Note that $v = 0$ when both $\delta, \varepsilon = 0$. If both equal 0.01, v approximately equals 0.03 and so the clutter cancellation is approximately −30.5 dB.

The approximate raindrop properties are that for light rain because the drops are fairly spherical and good cancellation can be achieved. However, the radar cross section of light rain is relatively small, so good cancellation is not necessary. For heavy rain, the rain is approximately teardrop in shaped and the radar cross section is relatively large. Thus, good cancellation cannot be achieved but good cancellation is needed. Experimental results indicate that the cancellation achieved by polarization techniques is of the order of −15 dB to −20 dB.

A significant advantage of polarization clutter cancellation is that it is compatible with Doppler processing, so the techniques can be cascaded. A significant disadvantage of the technique is that it is relatively expensive to implement the antenna so that the technique's implementation is primarily limited by the economics of the equipment.

About the Author

Ramon Nitzberg has been active in the field of radar signal processing for almost 50 years. His work experience has primarily dealt with radar and electronic counter-countermeasures at the General Electric Company Military Electronics—Advance Development Section as a senior consulting engineer. He has also taught courses at General Electric and Syracuse University and short courses for research associates of Syracuse both in the United States and Europe. Dr. Nitzberg has been a frequent contributor to professional journals on signal processing and was elected Fellow of the IEEE for "contributions to radar signal processing using adaptive antennas and Doppler filtering." He earned the B.E.E degree from City College of New York in 1950 and a Ph.D. degree from Syracuse University in 1970.

Index

Acquisition radar, 2–3, 6–10, 59, 98
Acquisition sweep, 61, 63
Adaptive array, 207, 241, 291–92, 297
Adaptive combination, input, 216–17
Adaptive Doppler processing, 291–92, 297, 317
  estimation error loss, 324
  mismatch loss, 321–24
  predicted performance, 320–21
  spectral estimation, 318–20
Adaptive threshold, 248–53, 257–58, 260, 264–65
Adaptive weight, 155, 209, 212, 239, 284, 320–21
A/D converter. *See* Analog-to-digital converter
Airborne radar, 198–202
Aircraft detection, 34, 145, 256, 258
All-pole network, 318–22
All-zero network, 320
Ambiguity, 47–52
  Doppler, 52–53, 307
  range, 47–52
Amplifier mismatch, 216–19
Amplifier nonlinearity, 186–90
Amplitude drift effect, 183–84
Amplitude modulation, 4
Amplitude monopulse antenna, 65–66
Analog implementation
  Doppler processing, 43, 45
  least mean square, 288
  sidelobe weighting, 133
Analog-to-digital converter, 44, 183, 186, 227–32
  quantization effects, 194–98
  straddling loss, 190–92
  waveform sampling, 192–94
Angle ambiguity, 17, 47
Angle resolution cell, 46–47
Angle tracking, 60, 63–66
Anomalous propagation, 11
Aperiodic waveform, 178
Arbitrary function, 171
Arithmetic mean normalizer, 253
Array angle resolution, 313–16
Array antenna, 267
Array covariance matrix, 296–97
Array factor gain, 116
Associated eigenvalue, 123
Attenuation, clutter, 169–73, 187–88, 197–98
Autocorrelation function, 84, 87–88, 171–72, 181, 187, 215, 316, 318, 321–22
Autocovariance function, 298
Azimuth angle, 2, 15–16, 157

Azimuth interval bracketing, 243

Background energy, 8
Backscatter, 156–58, 162
Bandlimitation, 83, 193, 228
Bandspand waveform, 103–6
Bandwidth diversity, 98
Bandwidth effect
  multiple sidelobe canceler, 281–84
  radar system parameters, 215–16
Barker code, 101
Barrage jammer, 22–23
Beamformer, 313–14
Beam steering, 6, 14, 119–20
Bessel function, 146
Bifurcated track, 67, 73
Binomial expansion, 176–77
Biphase-coded waveform, 99–103, 171
Bivariate Gaussian interference, 146
Blind speed, 36–37, 177–79, 181
BT product. *See* Time-bandwidth product
Burg algorithm, 318, 320–21, 325

CACFAR. *See* Cell averaging constant false alarm rate
Canceler. *See* Delay line canceler; Sidelobe canceler
Cancellation, 28–29, 34, 43–45, 143
  A/D converter effect, 197–98
  amplifier mismatch effect, 216–19
  circular polarization effect, 332–33
  degradation effect, 212–19
  polarization, 163–64
  propagation delay effect, 213–16
Cancellation ratio, 213–17, 278–79
Candidate waveform, 54–56
Carrier frequency sensitivity, 157
Cascading, 173
Cell averaging constant false alarm rate, 29–31, 243–44, 264
  homogenous environment, 245–48
  nonhomogeneous environment, 248–54
Censoring technique, 259–60
Center frequency, 275–76, 304, 308
Center of gravity, 144
CFAR. *See* Constant false alarm rate
Channel-to-channel mismatch, 274–78, 281, 317
Channel-to-match, 278–81
Characteristic vector. *See* Eigenvector
Chirp, 91
Circular polarization, 5, 163, 327–30, 332–33
Clock frequency, 44
Clutter, 2, 5, 8–9, 31–33, 155–56
  attenuation, 169–73, 187–88, 197–98
  discriminants, 156–58
  Doppler filter, 38–42, 43–45, 43–50
  maximum detection range in, 20, 33–34
  noncoherent summing, 42–43
  point and distributed, 33, 143
  spectral density, 167–69
  See also Ground clutter; Rain clutter; Sea clutter
Clutter cross section, 33, 41, 298
  computation methods, 158–64
Clutter environment. *See* Nonhomogeneous environment
Clutter map constant false alarm rate, 260–64
Clutter reflector, 167
Clutter-to-noise ratio, 196–98, 303–8, 320–21
CNR. *See* Clutter-to-noise ratio
Code diversity, 98
Coherent integration, 26, 42
Coherent local oscillator, 7
COHO. *See* Coherent local oscillator
Comparator, 50–51, 238
Complex envelope, 79–80, 83, 86, 88–89, 91, 114–15, 119, 185, 197, 203, 209, 211, 221–22, 228
Composite constant false alarm rate technique, 257, 264–65
Conical scan angle tracking, 63–65
Conjugate reverse voltage order, 310
Constant false alarm rate, 29–31, 41, 203, 212, 236, 241–43, 303
  cell-averaging, 243–46
  clutter map, 260–24
  composite techniques, 264–65
  geometric mean, 254–56
  greatest-of, 257–58
  homogeneous environment, 247–48
  multiple target effect, 258–60

nonhomogeneous environment, 248–54
Constructive combination, 12
Continuous phase change, 55
Convergence time analysis, 223–26, 232–33, 235–36, 285–86, 288–89, 317
Coordinate transformation, 123
Correlation feedback loop, 220–21, 285
Correlation mixer, 220–22, 227
Correlation vector, 269–71, 275, 277
Cosine squared illumination function, 127–30
Covariance matrix, 270–71, 275, 277, 295–97, 301, 309–11
  array, 296–97
  Doppler, 298–99
Cross-polarization, 5, 114, 162, 332
Cross-range measurement, 144
Cross section. *See* Clutter cross section; Target cross section

Data processing, 2
Decibel loss, 247–48
Decision error, 147
Degradation effects, 212–19
Delay line canceler, 34–38, 143, 155, 165–67
  power gain, 175–76
  shortcomings, 177–78
  three-pulse, 173–76
  two-pulse, 170–73
  weights, 176–77
Delay line tap, 40, 50, 165, 297–98
Delay-versus-frequency characteristics, 96–98
Delta function, 84–86, 171, 277
Depression angle, 160, 162, 198
Destructive combination, 12
Detection. *See* Target detection
Deterministic weight, 155
DFT. *See* Digital Fourier transform
Digital Fourier transform, 314, 320
Digital implementation
  advantages, 190
  bandwidth and code diversity, 98
  Doppler processing, 43–45
  LFM waveform, 98–99
  sample matrix inversion, 288
  sidelobe weighting, 133
Digital least mean square algorithm, 227–32
Dipole, 18
Discrete phase code, 56
Discrete time sampling, 192
Discriminant, target, 32
Displaced phase center antenna, 199–202
Distributed clutter, 33, 143
Dolph-Chebyshev weighting function, 130–33
Doppler ambiguity, 52–53
Doppler covariance matrix, 298–99
Doppler filtering, 26, 38–42, 143, 178, 243
  airborne radar, 198–202
  implementation details, 43–45, 155
    optimum weighting, 297–303
  performance evaluation, 303–8
  processor response, 202–6
  statistical independence, 206
  theory, 164–65
  See also Adaptive Doppler processing
Doppler peak shift, 5, 32, 95–96
DPCA. *See* Displaced phase center antenna
Drift
  local oscillator frequency, 184–86
  pulse-to-pulse amplitude, 183–84
Dwell time, 20–21, 32, 41, 125–26, 225
Dynamic range, 196

Early-late gate tracking, 59–63
Earth effects, 135–40, 158–60
ECCM. *See* Electronic countercountermeasure
ECM. *See* Electronic countermeasure
Effective radiated peak power, 9
Eigenvalue, 123–24
Eigenvector, 122–25
Electronic countercountermeasure, 22, 58, 207, 210, 215–16, 237–38
Electronic countermeasure, 22, 207, 215
Element gain, 115–16
Element spacing, 117–19, 120, 272
Elevation angle, 2
Elevation beamwidth, 15–16
Envelope detector, 3–4, 48, 243–45

Envelope probability density function, 146–47
Estimation error loss, 212, 324–25
Estimator algorithm, 285
Exponential density function, 149–50

False alarm rate, 3–5
  actual versus design, 27–28, 249–52
  probability effects, 12–14, 24–25, 27–31, 147, 149
False target repeater-jammer, 237–39
False track initiation, 67, 70–71
Farfield region, 109–10, 112, 127
Far-range target, 10–11
Fast Fourier transform, 314
Feedback loop, 285–86
FFT. *See* Fast Fourier transform
Filter impulse response, 171–72
Filtering-mixing, 220–22
Fixed threshold, 265
Fixed weighting, 307
Fluctuating target, 24–26, 148–51, 244
Fluctuation loss, 24–25, 56–58, 149–50, 152
Fluctuation models, fast and slow, 26
Fourier transform, 80–87, 104–5, 123, 128, 171, 181–82, 192, 215, 218, 298, 316
Fourth-time-around target, 49
Fraunhofer region. *See* Farfield region
Free-space antenna pattern, 136–37, 139
Free-space equation, 10–11
Frequency agility waveform. *See* Frequency diversity waveform
Frequency diversity waveform, 56–58
Frequency resolution, 316–17
Fresnel integral, 92

Gated sinusoid waveform, 6
Gate size, 73
Gaussian distribution, 26, 146, 150–51, 168, 172, 187–88, 196, 226, 228, 234, 268, 299, 304
Geometric mean normalization, 253–56, 259, 265
Glint, 58
GOCFAR. *See* Greatest-of constant false alarm rate
Grating lobe, 117–20, 273–74
Greatest-of constant false alarm rate, 257–58
Ground clutter, 138–39, 143, 157–60, 162, 167–68, 172, 242, 260–64

Hamming weighting function, 129–31, 138–39, 140–41, 191–92
Hand-off, 59
Hard limiter, 225–27
Hermitian matrix, 124–25, 295, 308–10
Hermitian-Toeplitz matrix, 309–10, 318–24
Hilbert transform, 80, 83
Homogeneous environment, 245–48, 254, 256–57, 259–60, 262
Horizon effect, 10–11
Horizon null, 12, 136
Horizontal polarization, 114, 327, 329–31
Householders transformation, 288
Howells-Applebaum correlation feedback loop, 220

IAGC. *See* Instantaneous automatic gain control
I channel, 45, 189–90
I component, 228
Ideal antenna pattern, 17
Identity function, 85
IF. *See* Intermediate frequency
Illumination functions, 127–33, 207
Imaginary function, 82–82
Improvement factor
  implementation errors effect, 183–90
  single-pulse canceler, 169–73
  stagger-coded waveform, 181–83
  three-pulse canceler, 173–76, 199
Impulse response, 179, 192–93, 218
Incident polarization, 5
Independence of detection, filter, 206
Infrared, 2
Initiation, tracking, 66, 68–72
In-phase, 228
Input clutter power, 169–70
Instantaneous automatic gain control, 227
Instantaneous frequency, 55, 91, 95–96, 223, 248

Instrumented range. *See* Maximum range
Integral transformation, 123
Interference, statistical nature of, 23
  See also Jamming
Interference power level, 27–29, 146, 244
Interfering energy, 2
Intermediate frequency, 7, 43, 45, 86, 220–22, 225
Interrogation angle cell, 21
IR. *See* Infrared
Iterative algorithm, 289

Jamming, 3, 5, 9
  data processor overloading, 28–29
  maximum detection range in, 22–23
JNR, 210, 303

Klystron, 36, 53

Lagrange multiplier, 121, 294–95
Least mean square algorithm, 220–21, 317
  convergence time analysis, 223–25
  digital, 227–32
  with hard limiter, 225–26
  multiple sidelobe canceler, 285–86
  SLC weight estimation, 222–23
  weight noise, 233–35
Left-hand circular polarization, 327–30, 332
LFM. *See* Linear frequency modulation
L'Hopital's rule, 247
Limiter-configured algorithm, 225–27, 231, 234–36
Line antenna, 18–19, 109–11
Linear filter, 86, 120–21
Linear frequency modulation, 55–56, 91–93, 103, 170
  bandwidth effect, 215
  biphase-coded waveform, 99–103
  Doppler tolerance, 93–96
  generation techniques, 96–99
  sidelobe reduction, 140–41
  wideband, 103–6
Linear phased array, 114–17
  beam steering, 119–20
  eigenvector, 123–25
  grating lobe, 117–19
  matched filter, 120–23
Line of sight, 139-40, 198–99, 201
LMS algorithm. *See* Least mean square algorithm
Local oscillator frequency drift, 184–86
LOS. *See* Line of sight
Low-altitude target, 136, 139–40
Lowpass filter, 44, 62, 199, 223, 225
Low-speed target, 177–79

Magnetron, 36, 53
Mainlobe gain region, 17, 306–7
Maintenance, tracking, 66, 72–74
Marcum's Q function, 148
Matched filter, 4, 215
  comparison, 120–23
  processing, 86–88
  right triangle, 89–90
  sampled waveform, 193–94
  sidelobe reduction, 140–41
  simple pulse, 88–89, 93–95
  sin x/x, 90–91
Matrix transformation, 123–25
Maximum detection range, 5, 8–12, 19–20
  in clutter, 33–34
  in jamming noise, 22–23
  multipath effect, 10–12
  multiple beam radar, 125–27
  in thermal noise, 20–22
Maximum entropy method, 316–17
Maximum likelihood estimator, 287
Measurement accuracy, 45–53
Mechanical scanning, 125, 169
Median of reference cell technique, 260
MEM. *See* Maximum entropy method
Mismatched environment, 249–54
Mismatch error, 274–78, 281, 321–24
  channel-to-channel, 274–78, 317
Mixing, 7, 220–22, 227
Monopulse angle tracking, 65–66
Monostatic radar, 93
Monotonic function, 245
Moving target indicator, 156
  delay line canceler, 165–67
  implementation errors, 183–90
  stagger-coded, 178–83
Moving window indicator, 261
MSLC. *See* Multiple sidelobe canceler
MTI. *See* Moving target indicator

Multipath effect, 11–12, 135–139
Multiple beam radar, 125–27
Multiple-feed antenna, 65
Multiple jammer, 301–3
Multiple sidelobe canceler, 208–9, 212, 236–37, 267
  algorithms, 284–90
  correlation vector, 270–71
  optimum weight derivation, 268–70
  optimum weight performance, 271–84
Multiple subcarrier waveform, 56–58, 103–4
Multiple target effect, 258–60
Mutual coupling, 310, 317

Natural beamwidth, 17–18, 129–31, 138
Negative spike, 61
NMLS algorithm. *See* Normalized least mean square algorithm
Noise, 2, 5, 9, 46, 67
  See also Thermal noise; Weight noise
Noncoherent integration, 42–43
Nonfluctuating target, 24–25, 150–51, 205–6
Nonhomogeneous environment, 248–54, 256, 260
Nonlinearity, 170, 186–90
Nonzero bandwidth, 212–16, 301
Nonzero sidelobe gain, 17
Normalized correlation coefficient, 272
Normalized least mean square algorithm, 227, 232, 235–37, 285–86
Normalizer cell, 31, 245, 251–52
  See also Reference cell
Null, 117, 128, 138, 179, 209, 291–92, 303

Optimum weight derivation, 268–70, 286–87, 307
  Doppler processor, 297–303, 317
Optimum weight performance evaluation
  bandwidth effect, 281–84
  channel-to-channel match, 278–81
  channel-to-channel mismatch, 274–78
  multiple sidelobe canceler, 271–74
Order statistic of reference cell, 260
OSCFAR. *See* Order statistic of reference cell
Output clutter power, 86, 169–70, 172, 181–82, 197

Parabolic reflector antenna, 14–18, 113–14
Parsavel's theorem, 170, 218
Passive sensor, 2
Peak power, 8–9, 53
Performance degradation, 190
Performance evaluation. *See* Optimum weight performance evaluation
Periodic waveform, 178, 181
Perturbation beam, 267
Phase-amplitude monopulse, 66
Phase-coded waveform, 54–56, 98
Phased-array antenna, 6, 18–19
Phase modulation, 4
Phase monopulse antenna, 65–66, 157, 200
Planar array, 18–19, 111–14
Plan-position-indicator, 48–50, 52
Point clutter, 33, 143
Point target, 158
Polarization
  cancellation, 163–64
  circular, 5, 163, 327–30, 332–33
  reflected wave, 329–32
  rejection of clutter, 327–29
  as target discriminant, 5, 32–33
Polarization sensitivity, 156, 163
Positive spike, 61
Post-mixer filter, 45
PPI. *See* Plan-position-indicator
Pre-detector target return, 145–46
PRF. *See* Pulse repetition frequency
PRI. *See* Pulse repetition interval
Probability density function, 146–49
Propagation delay, 213–16
Pseudorandom code, 102
Pulse coding, 54
Pulse compression, 4, 47–48, 56
Pulse repetition frequency, 5, 37, 41–43, 52–53, 60, 62–63, 166, 170, 172, 182, 199, 206, 307
Pulse repetition interval, 5, 21, 62–63, 166, 203
Pulse-to-pulse phase coherence, 36
Pulse-to-pulse phase drift, 183–84

Q channel, 45, 189–90
Q component, 228
Quadratic phase, 92, 141, 228

Quality number, track, 73–74
Quantization, 194–98

Radar fundamentals, 1–5
Radian carrier frequency, 220–21
Radio frequency interference, 5, 207
Rain clutter, 5, 34, 157, 163–64, 332–33
Ramp environment, 250–54, 256
Range ambiguity, 5, 47–52, 181
Range-coupling, 95
Range-Doppler cross coupling, 95–96
Range interval bracketing, 242
Range location error, 46–47
Range resolution, 45–46
Range-time delay, 96
Range tracking, 59–63
Rayleigh density function, 147, 149–50
Real-valued waveform, 79, 81–83, 121
Receive antenna, 8, 330
Reciprocity theorem, 109
Rect function signal, 81, 85, 88–89, 93
Reference cell, 29, 31, 242–45, 248, 251–54, 256, 259
Reflected wave polarization, 329–32
Reflection. *See* Clutter; Reverberation
Repeater jammer, 98
Repetitive pulse train waveform, 49–52
Residue power, 209–14, 216–17, 224, 232–35, 241, 254, 267, 271–76, 278, 283–85, 289
Resolution, 313
Resolution cell, 3–4, 12–14, 45–53, 145
Reverberation, 2
RFI. *See* Radio frequency interference
Right-hand circular polarization, 327–30, 332

Sample matrix inversion, 285–90, 308, 317
Saturation, 195–96, 241–42
Scalar multiple, 122
Scaling factor, 86
Scanning, 6, 14, 20–21, 32, 126, 169

Scan-to-scan averaging, 261
Scattering, 144, 156–58, 163
    See *also* Backscatter
Schwarz inequality, 87, 295
SCR. *See* Signal-to-clutter ratio
Sea clutter, 157–58, 160–62, 261
Search radar. *See* Surveillance radar
Second-time-around effect, 5, 11, 49, 53, 181
Servo loop, 61
Sidelobe blanking, 237–39
Sidelobe canceler, 207–8
    amplifier mismatch effect, 216–19
    cancellation degradation, 212–13
    concept, 208–10
    optimum weight, 220–32
    propagation delay effect, 213–16
    residue power minimization, 210–12
    weight noise, 232–36
    See also Multiple sidelobe canceler
Sidelobe gain, 17
Sidelobe reduction weighting, 40, 56, 93, 101–2, 114, 116, 127–33, 155, 207–10
    range problems, 47–48
    steering weight errors, 133–35
    waveform, 140–41
Signal processing, 2, 7
Signal-to-clutter ratio, 33–34, 126, 130, 163, 170
Signal-to-interference plus noise ratio, 247, 252–53, 255–56, 258, 261, 267, 289, 291–92, 322
    array covariance matrix, 296–97
    optimum weighting, 297–303
    weight maximization derivation, 292–96
Signal-to-interference ratio, 30–31
Signal-to-noise ratio, 10, 24–27, 38, 40, 42, 46, 70, 86–87, 116, 136, 138, 145, 147, 149–50, 195–96, 316
    maximum output, 86–87, 121–23, 127–29, 202–3, 205
Simple pulse receiver, 88–89, 93–95
Single-pole filter, 225, 228–29
SINR. *See* Signal-to-interference plus noise ratio
Sin signal, 90–91, 93
Sinusoidal oscillator, 4, 6, 34, 43
Sin x/x function, 111, 127–28, 130
SIR. *See* Signal-to-interference ratio
SJNR, 210, 212

SLC. *See* Sidelobe canceler
SMI. *See* Sample matrix inversion
SNR. *See* Signal-to-noise ratio
Sonar, 2
Spectral density clutter, 167–69
Spectral estimation, 318–21
Specular reflection, 157
Spherical Earth, 136, 139–40
Split gate tracking, 59–60
Spot jammer, 22
Squared magnitude, 82
Square law envelope detector, 243–45
Squinted beam, 66
Stable local oscillator, 7
Stagger coding, 178–83
STALO. *See* Stable local oscillator
Standard atmosphere, 11, 140
Steady-state value, 224–26, 229–30, 234–35
Steering vector, 297–98, 301, 303
Steering weight error, 133–35
Step function environment, 249–50, 256–57, 263
Step scan, 20, 169
Stochastic differential equation, 222, 226, 231
Straddling loss, 190–92
Subtractor, 35, 60–62, 194, 212
Super-resolution, 313
  array angle, 313–16
  frequency, 316–17
Surface echo area, 159
Surveillance radar, 2–3, 6–10, 59, 98
Swerling 1 model, 24–27, 57, 149–51, 244–45, 303
Swerling 2, 3, 4 models, 151–53
Switching, 6
Synchronous demodulation, 35, 43–44, 227
System trade areas, 75–77

TACCAR. *See* Time-averaged clutter coherent radar
Tap voltage, 297–98
Target absent decision, 3–4, 12, 41, 145, 147, 203, 242, 245
Target cross section, 23–24, 26, 31, 148–51
Target detection, 1–5, 8
  in clutter, 31–45
  in noise, 143–53
  probability equations, 23–27
  theory, 145–48
  in thermal noise, 194–98
  tracking error, 20
Target present decision, 3–4, 12, 29, 41, 72, 145, 147, 203, 242, 244–45
Taylor weighting function, 40, 130–35, 182, 258, 278
Tchebychev weighting, 40, 305–6
Test cell, 242, 248, 254, 257
Thermal noise, 4, 8–10
  A/D converter effect, 194–98
  maximum detection range in, 20–22
  target detection in, 143–53
Third-time-around target, 49
Three-pulse canceler, 173–76, 178–79, 182, 187–88, 199
Threshold value, 12, 27, 29, 147, 242
Time-averaged clutter coherent radar, 199
Time-bandwidth product, 94–95
Time constant, 262–64
Time delay, 86, 96–98
Track association, 67, 73
Track dropping, 67–68, 73–74
Tracking radar, 2–3, 58–59, 114
  angle techniques, 63–66
  range, 59–63
Track-while-scan radar, 3
  data processing, 66–67
  multiple bandwidth, 98
  track initiation, 68–72
  track maintenance, 72–74
Transfer function, 86–87, 90, 92, 140–41, 172–73, 177, 179, 213, 215–16, 219, 275–76
Transmit antenna, 8, 14–18
Transmitter frequency drift, 184–86
Transmitter phase, 183–84
Transmit waveform, 53–54
  candidate waveform, 54–56
  multiple subcarrier, 56–58
  pulse coding, 54
Triangle signal, 89–90
Two-domain adaptivity, 293
Two-pulse canceler, 170–73, 175–76, 179, 182–83, 187–88, 197

TWS radar. *See* Track-while-scan radar

Unambiguous range, 49, 52, 306
Uniform illumination, 15, 17, 127–33
Uniformly illuminated line antenna, 110–11
Uniformly illuminated linear phased array, 116–17
Uniformly illuminated planar antenna, 112–14
Uniform sidelobe weighting, 138–39
Unity, 40, 120, 175–76, 179, 230, 234, 270, 275–77
Update period, 14, 20–21, 32, 126

Vertical polarization, 327, 329–31
Volume clutter, 143

Waveform sampling, 192–94
Waveform sidelobe reduction, 140–41
Weather clutter, 5, 34, 156–57, 162–64, 261, 298, 332–33
Weight derivation, multiple sidelobe canceler, 268–70
Weight estimation
  LMS algorithm, 222–32
  mixing and filtering, 220–22
Weighting function. *See* Sidelobe reduction weighting
Weight maximization, SINR, 292–96
Weight noise, 232–33
  complex-valued, 239–40
  LMS algorithm, 233–36
Weinstock model, 152
White noise, 86
Wideband linear frequency modulation, 103–6

Yule-Walker equation, 322, 324

Zero bandwidth residue power, 214
Zero bandwidth waveform, 212
Zero clutter, 36, 37
Zero–crossing detector, 60–61, 62
Zero–derivative condition, 211
Zero Doppler, 155, 167, 179
Zero Doppler frequency, 40, 42, 52, 198, 203
Zero jammer angle, 216

Recent Titles in the Artech House Radar Library

David K. Barton, Series Editor

Airborne Pulsed Doppler Radar, Second Edition, Guy V. Morris and Linda Harkness, editors

Digital Techniques for Wideband Receivers, James Tsui

Electronic Intelligence: The Analysis of Radar Signals, Second Edition, Richard G. Wiley

High-Resolution Radar, Second Edition, Donald R. Wehner

Introduction to Electronic Warfare, D. Curtis Schleher

Introduction to Multisensor Data Fusion: Multimedia Software and User's Guide, TECH REACH, Inc.

Millimeter-Wave and Infared Multisensor Design and Signal Processing, Lawrence A. Klein

Modern Radar System Analysis, David K. Barton

Modern Radar System Analysis Software and User's Manual, David K. Barton and William F. Barton

Principles of High-Resolution Radar, August W. Rihaczek

Radar Cross Section, Second Edition, Eugene F. Knott, *et al.*

Radar Evalution Handbook, David K. Barton, *et al.*

Radar Meteorolgy, Henri Sauvageot

Radar Signal Processing and Adaptive Systems, Ramon Nitzberg

Radar Technology Encyclopedia, David K. Barton and Sergey A. Leonov, editors

For further information on these and other Artech House titles, including previously considered out-of-print books now available through our In-Print-Forever® (IPF®) program, contact:

Artech House
685 Canton Street
Norwood, MA 02062
Phone: 781-769-9750
Fax: 781-769-6334
e-mail: artech@artechhouse.com

Artech House
46 Gillingham Street
London SW1V 1AH UK
Phone: +44 (0)171-973-8077
Fax: +44 (0)171-630-0166
e-mail: artech-uk@artechhouse.com

Find us on the World Wide Web at:
www.artechhouse.com

www.ingramcontent.com/pod-product-compliance
Lightning Source LLC
Chambersburg PA
CBHW021055300426
R18219400001B/R182194PG42580CBX00001B/1
* 9 7 8 1 5 8 0 5 3 0 3 4 7 *